NEUROBIOLOGY of HUNTINGTON'S DISEASE

APPLICATIONS TO DRUG DISCOVERY

FRONTIERS IN NEUROSCIENCE

Series Editors
Sidney A. Simon, Ph.D.
Miguel A.L. Nicolelis, M.D., Ph.D.

Published Titles

Apoptosis in Neurobiology
Yusuf A. Hannun, M.D., Professor of Biomedical Research and Chairman, Department
 of Biochemistry and Molecular Biology, Medical University of South Carolina,
 Charleston, South Carolina
Rose-Mary Boustany, M.D., tenured Associate Professor of Pediatrics and Neurobiology,
 Duke University Medical Center, Durham, North Carolina

Neural Prostheses for Restoration of Sensory and Motor Function
John K. Chapin, Ph.D., Professor of Physiology and Pharmacology, State University
 of New York Health Science Center, Brooklyn, New York
Karen A. Moxon, Ph.D., Assistant Professor, School of Biomedical Engineering, Science,
 and Health Systems, Drexel University, Philadelphia, Pennsylvania

Computational Neuroscience: Realistic Modeling for Experimentalists
Eric DeSchutter, M.D., Ph.D., Professor, Department of Medicine, University of Antwerp,
 Antwerp, Belgium

Methods in Pain Research
Lawrence Kruger, Ph.D., Professor of Neurobiology (Emeritus), UCLA School of Medicine
 and Brain Research Institute, Los Angeles, California

Motor Neurobiology of the Spinal Cord
Timothy C. Cope, Ph.D., Professor of Physiology, Wright State University, Dayton, Ohio

Nicotinic Receptors in the Nervous System
Edward D. Levin, Ph.D., Associate Professor, Department of Psychiatry and Pharmacology
 and Molecular Cancer Biology and Department of Psychiatry and Behavioral Sciences,
 Duke University School of Medicine, Durham, North Carolina

Methods in Genomic Neuroscience
Helmin R. Chin, Ph.D., Genetics Research Branch, NIMH, NIH, Bethesda, Maryland
Steven O. Moldin, Ph.D., University of Southern California, Washington, D.C.

Methods in Chemosensory Research
Sidney A. Simon, Ph.D., Professor of Neurobiology, Biomedical Engineering,
 and Anesthesiology, Duke University, Durham, North Carolina
Miguel A.L. Nicolelis, M.D., Ph.D., Professor of Neurobiology and Biomedical Engineering,
 Duke University, Durham, North Carolina

The Somatosensory System: Deciphering the Brain's Own Body Image
Randall J. Nelson, Ph.D., Professor of Anatomy and Neurobiology,
 University of Tennessee Health Sciences Center, Memphis, Tennessee

The Superior Colliculus: New Approaches for Studying Sensorimotor Integration
William C. Hall, Ph.D., Department of Neuroscience, Duke University,
   Durham, North Carolina
Adonis Moschovakis, Ph.D., Department of Basic Sciences, University of Crete,
   Heraklion, Greece

New Concepts in Cerebral Ischemia
Rick C.S. Lin, Ph.D., Professor of Anatomy, University of Mississippi Medical Center,
   Jackson, Mississippi

DNA Arrays: Technologies and Experimental Strategies
Elena Grigorenko, Ph.D., Technology Development Group, Millennium Pharmaceuticals,
   Cambridge, Massachusetts

Methods for Alcohol-Related Neuroscience Research
Yuan Liu, Ph.D., National Institute of Neurological Disorders and Stroke,
   National Institutes of Health, Bethesda, Maryland
David M. Lovinger, Ph.D., Laboratory of Integrative Neuroscience, NIAAA,
   Nashville, Tennessee

Primate Audition: Behavior and Neurobiology
Asif A. Ghazanfar, Ph.D., Princeton University, Princeton, New Jersey

Methods in Drug Abuse Research: Cellular and Circuit Level Analyses
Barry D. Waterhouse, Ph.D., MCP-Hahnemann University, Philadelphia, Pennsylvania

Functional and Neural Mechanisms of Interval Timing
Warren H. Meck, Ph.D., Professor of Psychology, Duke University, Durham, North Carolina

Biomedical Imaging in Experimental Neuroscience
Nick Van Bruggen, Ph.D., Department of Neuroscience Genentech, Inc.
Timothy P.L. Roberts, Ph.D., Associate Professor, University of Toronto, Canada

The Primate Visual System
John H. Kaas, Department of Psychology, Vanderbilt University
Christine Collins, Department of Psychology, Vanderbilt University, Nashville, Tennessee

Neurosteroid Effects in the Central Nervous System
Sheryl S. Smith, Ph.D., Department of Physiology, SUNY Health Science Center,
   Brooklyn, New York

Modern Neurosurgery: Clinical Translation of Neuroscience Advances
Dennis A. Turner, Department of Surgery, Division of Neurosurgery,
   Duke University Medical Center, Durham, North Carolina

Sleep: Circuits and Functions
Pierre-Hervé Luppi, Université Claude Bernard Lyon, France

Methods in Insect Sensory Neuroscience
Thomas A. Christensen, Arizona Research Laboratories, Division of Neurobiology,
   University of Arizona, Tuscon, Arizona

Motor Cortex in Voluntary Movements
Alexa Riehle, INCM-CNRS, Marseille, France
Eilon Vaadia, The Hebrew University, Jerusalem, Israel

Neural Plasticity in Adult Somatic Sensory-Motor Systems
Ford F. Ebner, Vanderbilt University, Nashville, Tennessee

Advances in Vagal Afferent Neurobiology
Bradley J. Undem, Johns Hopkins Asthma Center, Baltimore, Maryland
Daniel Weinreich, University of Maryland, Baltimore, Maryland

The Dynamic Synapse: Molecular Methods in Ionotropic Receptor Biology
Josef T. Kittler, University College, London, England
Stephen J. Moss, University College, London, England

Animal Models of Cognitive Impairment
Edward D. Levin, Duke University Medical Center, Durham, North Carolina
Jerry J. Buccafusco, Medical College of Georgia, Augusta, Georgia

The Role of the Nucleus of the Solitary Tract in Gustatory Processing
Robert M. Bradley, University of Michigan, Ann Arbor, Michigan

Brain Aging: Models, Methods, and Mechanisms
David R. Riddle, Wake Forest University, Winston-Salem, North Carolina

Neural Plasticity and Memory: From Genes to Brain Imaging
Frederico Bermudez-Rattoni, National University of Mexico, Mexico City, Mexico

Serotonin Receptors in Neurobiology
Amitabha Chattopadhyay, Center for Cellular and Molecular Biology, Hyderabad, India

TRP Ion Channel Function in Sensory Transduction and Cellular Signaling Cascades
Wolfgang B. Liedtke, M.D., Ph.D., Duke University Medical Center, Durham, North Carolina
Stefan Heller, Ph.D., Stanford University School of Medicine, Stanford, California

Methods for Neural Ensemble Recordings, Second Edition
Miguel A.L. Nicolelis, M.D., Ph.D., Professor of Neurobiology and Biomedical Engineering,
    Duke University Medical Center, Durham, North Carolina

Biology of the NMDA Receptor
Antonius M. VanDongen, Duke University Medical Center, Durham, North Carolina

Methods of Behavioral Analysis in Neuroscience
Jerry J. Buccafusco, Ph.D., Alzheimer's Research Center, Professor of Pharmacology
    and Toxicology, Professor of Psychiatry and Health Behavior,
    Medical College of Georgia, Augusta, Georgia

In Vivo Optical Imaging of Brain Function, Second Edition
Ron Frostig, Ph.D., Professor, Department of Neurobiology,
    University of California, Irvine, California

Fat Detection: Taste, Texture, and Post Ingestive Effects
Jean-Pierre Montmayeur, Ph.D., Centre National de la Recherche Scientifique, Dijon, France
Johannes le Coutre, Ph.D., Nestlé Research Center, Lausanne, Switzerland

The Neurobiology of Olfaction
Anna Menini, Ph.D., Neurobiology Sector International School for Advanced
Studies,(S.I.S.S.A.), Trieste, Italy

Neuroproteomics
Oscar Alzate, Ph.D., Department of Cell and Developmental Biology,
University of North Carolina, Chapel Hill, North Carolina

Translational Pain Research: From Mouse to Man
Lawrence Kruger, Ph.D., Department of Neurobiology, UCLA School of Medicine,
Los Angeles, California
Alan R. Light, Ph.D., Department of Anesthesiology, University of Utah,
Salt Lake City, Utah

Advances in the Neuroscience of Addiction
Cynthia M. Kuhn, Duke University Medical Center, Durham, North Carolina
George F. Koob, The Scripps Research Institute, La Jolla, California

Neurobiology of Huntington's Disease: Applications to Drug Discovery
Donald C. Lo, Duke University Medical Center, Durham, North Carolina
Robert E. Hughes, Buck Institute for Age Research, Novato, California

NEUROBIOLOGY of HUNTINGTON'S DISEASE

APPLICATIONS TO DRUG DISCOVERY

Edited by
Donald C. Lo
Duke University Medical Center
Durham, North Carolina

Robert E. Hughes
Buck Institute for Age Research
Novato, California

CRC Press
Taylor & Francis Group
Boca Raton London New York

CRC Press is an imprint of the
Taylor & Francis Group, an **informa** business

Cover image: a live corticostriatal brain slice transfected twice using a biolistic gene transfer device; once with cyan fluorescent protein to label medium spiny neurons in the striatum (with the cortex physically masked), and then again with yellow fluorescent protein to label cortical pyramidal neurons (with the striatum physically masked). **Photomicrograph credit:** Laura W. Shaughnessy, Peter H. Reinhart, and Donald C. Lo, Duke University Medical Center.

CRC Press
Taylor & Francis Group
6000 Broken Sound Parkway NW, Suite 300
Boca Raton, FL 33487-2742

First issued in paperback 2017

© 2011 by Taylor and Francis Group, LLC
CRC Press is an imprint of Taylor & Francis Group, an Informa business

No claim to original U.S. Government works

ISBN 13: 978-1-138-11662-7 (pbk)
ISBN 13: 978-0-8493-9000-5 (hbk)

Library of Congress Cataloging-in-Publication Data

Neurobiology of Huntington's disease : applications to drug discovery / edited by Donald C. Lo and Robert E. Hughes.
    p. ; cm. -- (Frontiers in neuroscience)
  Includes bibliographical references and index.
  ISBN 978-0-8493-9000-5 (hardcover : alk. paper)
  1. Huntington's chorea--Pathophysiology. 2. Huntington's chorea--Molelcular aspects.
3. Neuropharmacology. I. Lo, Donald C. II. Hughes, Robert E., Ph.D. III. Title. IV. Series:
Frontiers in neuroscience (Boca Raton, Fla.)
  [DNLM: 1. Huntington Disease--physiopathology. 2. Drug Discovery--methods. 3.
Huntington Disease--drug therapy. WL 390 N4948 2010]

RC394.H85N48 2010
616.8'51--dc22

2010000178

Visit the Taylor & Francis Web site at
http://www.taylorandfrancis.com

and the CRC Press Web site at
http://www.crcpress.com

To the HD family community

Contents

Series Preface

Our goal in creating the Frontiers in Neuroscience Series is to present the insights of experts on emerging fields and theoretical concepts that are, or will be, in the vanguard of neuroscience. Books in the series cover genetics, ion channels, apoptosis, electrodes, neural ensemble recordings in behaving animals, and even robotics. The series also covers new and exciting multidisciplinary areas of brain research, such as computational neuroscience and neuroengineering, and describes breakthroughs in classical fields like behavioral neuroscience. We hope every neuroscientist will use these books in order to get acquainted with new ideas and frontiers in brain research. These books can be given to graduate students and postdoctoral fellows when they are looking for guidance to start a new line of research.

Each book is edited by an expert and consists of chapters written by the leaders in a particular field. Books are richly illustrated and contain comprehensive bibliographies. Chapters provide substantial background material relevant to the particular subject. We hope that as the volumes become available, the effort put in by us, the publisher, the book editors, and individual authors will contribute to the further development of brain research. The extent to which we achieve this goal will be determined by the utility of these books.

Sidney A. Simon, Ph.D.
Miguel A.L. Nicolelis, M.D.,Ph.D.
Series Editors

Preface

At the time of this writing, 138 years have passed since the original published description of Huntington's disease (HD). This remarkable account of HD, written by George Huntington in 1872, remains a fascinating and accurate read today. One hundred twenty-one years later, in 1993, a landmark article was published by the Huntington's Disease Collaborative Research Group that reported the identification of the genetic mutation responsible for HD, a fundamental milestone in human genomics. The discovery of the primary cause of the disease, a CAG repeat expansion in a gene now known as *HTT*, catapulted the field of HD research into the modern era of genomics and, in so doing, created great expectations that the cutting-edge tools of molecular biology and chemistry would lead rapidly to the development of effective treatments for this devastating illness.

In the 16 years since the cloning of the gene, scientists in the HD field, as well as patients and their families, have witnessed an impressive expansion in our knowledge of the effects that this mutation in the *HTT* gene has on a wide range of neuronal functions and of its many deleterious effects outside of the nervous system. We have seen the field engage in extraordinary efforts to understand the nature of the HD mutation and how it can initiate and drive cellular dysfunction. From this work, the field has documented an unexpectedly large and diverse range of biochemical and genetic perturbations that seem to result directly from the expression of the mutant *HTT* gene. Indeed, at this point one is increasingly hard-pressed to name an aspect of cellular function that is *not* impacted by expression of mutant *HTT*.

Steady progress in elucidating the mechanistic consequences of the *HTT* mutation for cellular physiology has driven an explosive growth in translational research activities aimed at development of new drug therapies for HD. However, it is worth noting that effective treatments for HD remain elusive. The natural exuberance that followed the identification of the HD gene is now tempered by an appreciation of the complexities attending true understanding of *HTT* function and dysfunction. Distinguishing the biochemical perturbations merely associated with HD from those critical for disease progression is a major ongoing task.

Despite the challenges in developing drugs to treat HD, a strong basis for optimism stems from the monogenic nature of the disease. This aspect of HD offers significant advantages relative to other common neurodegenerative disorders, such as Alzheimer's disease (AD) and Parkinson's disease (PD). The sporadic and idiopathic natures of AD and PD make their underlying mechanisms fundamentally more uncertain. Despite all the complexities regarding the mechanistic consequences of mutant *HTT* expression, one thing is certain: HD is caused exclusively by a CAG expansion in the *HTT* gene. Furthermore, expression of the mutant *HTT* gene can consistently, indeed stereotypically, confer dysfunction and sickness when expressed in cultured cells, invertebrate animals, mice, and monkeys.

From the perspective of translational research, this confers enormous practical advantages in the face of the obstacles presented by HD. Ultimately, it has allowed the creation of robust experimental tools, models, and assays to facilitate discovery and validation of molecular targets and drug candidates for HD. Indeed, basic knowledge and assay development in the HD field have clearly matured to the point where we can begin to launch rational and effective drug discovery campaigns. This is, in part, evidenced by the proliferation of translational programs in HD, not just in academia but in biotechnology and pharmaceutical firms as well. The synergy of cell-based assays, mouse models, a highly organized consortium of clinical investigators, and increasing knowledge of the druggable target space around HD has brought us to the threshold of major breakthroughs in the quest for therapeutic agents.

As the editors of this volume, we have enlisted the help of key investigators in the field to give the reader an accurate and rigorous overview of an HD drug discovery pipeline that has been deliberately developed and coordinated in the century since George Huntington's seminal report. We express our sincere gratitude to our dedicated colleagues for contributing these chapters. Finally, we acknowledge the many researchers, consortia, organizations—and the HD community—who have worked so hard to make this field into what it has become. In our experience, the HD research field is notable for its sense of common purpose, its spirit of cooperation and sharing, and its steadfast commitment to scientific rigor and excellence. In the end, these human qualities will prove to be as critical as the impressive array of drug discovery resources described here to delivering the effective drug therapies long-awaited by HD patients and their families.

Donald C. Lo
Durham, North Carolina

Robert E. Hughes
Novato, California

Editors

Donald C. Lo is an associate professor in the Department of Neurobiology at Duke University Medical Center and has been engaged in drug discovery and development for neurodegenerative disorders and stroke for more than 12 years. In 1997, he cofounded and was chief scientific officer of the biotechnology company Cogent Neuroscience, which developed neurogenomic approaches to new drug and drug target discovery in stroke, Huntington's disease, and Alzheimer's disease, the latter two indications in joint ventures with Elan Pharmaceuticals. In 2002, Lo established the Center for Drug Discovery (CDD) at Duke, for which he currently serves as director. The Duke CDD pursues multiple neurological drug discovery and development programs in collaboration with major pharmaceutical firms, biotechnology companies, and nonprofit disease research foundations, with its major focus on Huntington's disease.

Robert E. Hughes is an assistant professor at the Buck Institute for Age Research in Novato, California. He has been actively engaged in Huntington's disease research for more than 10 years in both academia and the biotechnology industry. From 2002–2007, Hughes served on the scientific advisory board of the Hereditary Disease Foundation. Before joining the Buck Institute, he worked at Myriad Proteomics (later Prolexys Pharmaceuticals), where he directed a bioinformatics group and a Huntington's disease target discovery program. Currently, Dr. Hughes's laboratory has a major emphasis on the discovery and validation of novel targets in Huntington's disease and other age-related diseases.

Contributors

Roger L. Albin
Department of Neurology
University of Michigan Health System
and
Geriatrics Research, Education, and
    Clinical Center
Ann Arbor VAHS
Ann Arbor, Michigan

Neil Aronin
Department of Medicine
University of Massachusetts Medical
    School
Worcester, Massachusetts

Charles W. Bugg
Divisions of Chemistry and Biology
California Institute of Technology
Pasadena, California

E. Ray Dorsey
Department of Neurology
University of Rochester Medical Center
Rochester, New York

Michelle Gray
Center for Neurodegeneration and
    Experimental Therapeutics
Department of Neurology
University of Alabama at Birmingham
Birmingham, Alabama

Steven M. Hersch
Massachusetts General Hospital
Harvard Medical School
Boston, Massachusetts

David S. Howland
CHDI Foundation, Inc.
Princeton, New Jersey

Ali Khoshnan
Division of Biology
California Institute of Technology
Pasadena, California

Jan C. Ko
Division of Biology
California Institute of Technology
Pasadena, California

Seung P. Kwak
CHDI Foundation, Inc.
Princeton, New Jersey

Albert R. La Spada
Departments of Pediatrics and Cellular &
    Molecular Medicine
University of California, San Diego
La Jolla, California
and
Rady Children's Hospital
San Diego, California

Janet M. Leeds
CHDI Foundation, Inc.
Los Angeles, California

Douglas Macdonald
CHDI Foundation, Inc.
Los Angeles, California

John P. Miller
Buck Institute
Novato, California

Richard J. Morse
CHDI Foundation, Inc.
Los Angeles, California

Christian Neri
Center for Psychiatry and Neuroscience
  U894
National Institute of Health and
  Medical Research
Paris, France

Robert Pacifici
CHDI Foundation, Inc.
Los Angeles, California

Larry Park
CHDI Foundation, Inc.
Los Angeles, California

Paul H. Patterson
Division of Biology
California Institute of Technology
Pasadena, California

Henry L. Paulson
Department of Neurology
University of Michigan Health System
Ann Arbor, Michigan

Victor V. Pineda
Department of Pathology
University of Washington
Seattle, Washington

H. Diana Rosas
Massachusetts General Hospital
Harvard Medical School
Boston, Massachusetts

Meghan Sass
Department of Medicine
University of Massachusetts Medical
  School
Worcester, Massachusetts

Ira Shoulson
Department of Neurology
University of Rochester Medical Center
Rochester, New York

Amber L. Southwell
Division of Biology
California Institute of Technology
Pasadena, California

Brent R. Stockwell
Departments of Biological Sciences and
  Chemistry
Columbia University
New York, New York

Leticia Toledo-Sherman
CHDI Foundation, Inc.
Los Angeles, California

Hemant Varma
Department of Biological Sciences
Columbia University
New York, New York

James K. T. Wang
CHDI Foundation, Inc.
Boston, Massachusetts

Patrick Weydt
Department of Neurology
University of Ulm
Ulm, Germany

X. William Yang
David Geffen School of Medicine
University of California at Los Angeles
Los Angeles, California

1 Huntington's Disease
Clinical Features and Routes to Therapy

Henry L. Paulson and Roger L. Albin

CONTENTS

OVERVIEW

In 1872, physician George Huntington reported a familial form of chorea noted previously on Long Island by his father and grandfather, also physicians. More than a century later his comments about the disease now carrying his name, Huntington's disease (HD), remain a clear description of its major clinical features (Huntington, 1872; reprinted in Huntington, 2003). Huntington described chorea in general as the "dancing propensities of those ... affected," in whom there "seems to exist some hidden power, something that is playing tricks, as it were, upon the

1

will." The familial form started "as an ordinary chorea might begin, by the irregular and spasmodic action of certain muscles as of the face, arms, etc. These movements gradually increase when muscles hitherto unaffected take on the spasmodic action...."

The disease, he further noted, "seems to obey certain fixed laws." It is "confined to ... a few families, is attended generally by all the symptoms of common chorea, hardly ever manifesting itself until adult or middle life, and then coming on gradually but surely, increasing by degrees, and often occupying years in its development, until the hapless sufferer is but a quivering wreck of his former self." The "tendency to insanity," Huntington observed, progresses so that the "mind becomes more or less impaired, in many amounting to insanity while in others, mind and body both gradually fail until death relieves them of their sufferings."

The inheritance pattern was also clear to Huntington: "When either or both the parents have shown manifestations ... one or more of the offspring almost invariably suffer from the disease ... but if by any chance these children go through a life without it, the thread is broken...." Finally, he noted the relentless, fatal course: "I have never known a recovery ... it seems at least to be one of the incurables." Commenting later on Huntington's description, Sir William Osler noted that "there are few instances in which a disease has been more accurately, more graphically, and more briefly described" (Stevenson, 1934).

The cardinal features of HD so aptly described by Huntington are even clearer to us now, nearly 140 years later. It is a dominantly inherited, neuropsychiatric disorder that affects successive generations of afflicted families. It progresses slowly over years with symptoms typically, but not always, beginning in adulthood. Although HD usually involves chorea and other abnormal movements, the progressive cognitive impairment and behavioral problems are perhaps even more disabling. The remarkable clinical variability and wide range in age of onset, even among affected members in the same family, are now recognized to stem directly from the type of mutation in HD, a dynamic repeat expansion in a polyglutamine-encoding CAG tract. Likewise, the clinical phenomenon of anticipation—the tendency for disease to begin earlier in successive generations—is the result of the propensity for further expansion of the disease-causing repeat on transmission to offspring. We now also recognize that not *all* HD is familial, as a small percentage of cases arises sporadically from newly expanded alleles. Naturally, such sporadic HD cases then constitute the first generation of new HD families.

The HD mutation, a CAG repeat expansion in the *HTT* gene (Huntington's Disease Collaborative Research Group, 1993), is present in all affected persons. The high sensitivity and specificity of genetic testing for the mutation ensure that molecular confirmation of HD is straightforward in the clinic (Davis et al., 1994; Kremer et al., 1994). For persons in whom HD is clinically suspected, a gene test can quickly prove or disprove the diagnosis. For those who are known to be at risk for disease, presymptomatic testing is available after appropriate counseling (WFN Research Group on Huntington's Chorea, 1994). Currently, in the absence of a preventive medication or disease-modifying therapy, only a small minority of at-risk persons choose to undergo presymptomatic testing (Quaid and Morris, 1993). But the ability to track presymptomatic HD gene carriers in long-term observational studies has greatly improved researchers' and clinicians' ability to chart the course of disease symptoms and signs, both before and after diagnosis. This knowledge will be crucial to the success of future preventive and symptomatic trials.

The discovery of the HD mutation has also shed light on possible disease mechanisms (Gatchel and Zoghbi, 2005; Nakamura and Aminoff, 2007; Todi et al., 2007). The CAG repeat expansion encodes a polymorphic stretch of the amino acid glutamine in the disease protein, a large polypeptide known as huntingtin (htt). This expanded polyglutamine tract is widely believed to be the key molecular culprit in HD (Mangiarini et al., 1996). Although the mechanism by which expanded polyglutamine in huntingtin causes brain dysfunction and neuronal cell death remains uncertain, growing knowledge about both the aberrant properties of the disease protein and the havoc it wreaks on cells has provided new insight into potential rational therapies (Di Prospero and Fischbeck, 2005), as discussed elsewhere in this book.

Although HD is a clinically distinctive neurodegenerative disease, polyglutamine expansions are not unique to HD. At least eight other neurodegenerative disorders are caused by polyglutamine expansions (Table 1.1) (Gatchel and Zoghbi, 2005; Riley and Orr, 2006; Todi et al., 2007). In many important ways, HD can be considered a flagship for other polyglutamine diseases. It is the most common among them: roughly 30,000 people are diagnosed with HD in the United States and Canada, with another ~150,000 at risk for disease. More than any other polyglutamine diseases, HD is studied by a vast array of clinicians and scientists who explore disease mechanisms, the natural course of disease, and potential therapies. In particular, the highly

---

TABLE 1.1

List of Known Polyglutamine Diseases, Their Disease Protein, Threshold Disease Repeat, and Major Clinical Features

Disease	Disease Protein	(CAG)N Disease Threshold	Major Clinical Features
HD	Huntingtin	>35	Motor, cognitive, psychiatric triad
SCA1	Ataxin-1	>38	Ataxia, brainstem dysfunction
SCA2	Ataxin-2	>31	Ataxia, brainstem dysfunction
SCA3	Ataxin-3	>50	Ataxia, brainstem dysfunction, dystonia, parkinsonism, neuropathology
SCA6	Ca++ channel subunit	>20	Ataxia
SCA7	Ataxin-7	>36	Ataxia, brainstem dysfunction, blindness
SCA17	TATA binding protein	>42	Ataxia, chorea, brainstem dysfunction, dystonia, dementia
DRPLA	Atrophin-1	>47	Ataxia, chorea, seizures, dementia
SBMA	Androgen receptor	>39	Motor neuron loss

Note that the threshold for a repeat to cause disease varies slightly among the diseases. This likely reflects the specific gene and protein context in which each repeat occurs.

organized HD clinical research community is driving the push toward biomarkers and quantifiable clinical measures that will serve as the basis for successful human trials. Importantly, because all polyglutamine diseases may share common elements of pathogenesis, insights from the HD field could prove valuable for the thousands of people affected by other polyglutamine diseases.

Although HD and other polyglutamine disorders are relatively uncommon, they have an impact disproportionate to their prevalence. In populations of European descent, the estimated prevalence of HD is approximately 4–9/100,000. Prevalence rates are much lower in populations of non-European ancestry. The aggregate prevalence of all other polyglutamine diseases, primarily spinocerebellar ataxias, may approach the lower bound of prevalence estimates for HD. Because HD and other polyglutamine diseases tend to strike in middle age and are slowly progressive, the economic impacts as a result of lost earnings and costs of care are enormous. The personal suffering and family disruption imposed by HD are devastating. The combination of usual midlife onset and dominant inheritance tends to pull HD families down the social scale, compromising the lifetime opportunities of both mutant allele carriers and their unaffected relatives.

Despite the many advances in our understanding of HD since discovery of the disease gene, HD remains, in Huntington's words, "one of the incurables." As the existence of this book attests, however, the knowledge gained since the gene discovery puts us in a position where this incurable status may soon change. Here we review HD, focusing on its major clinical features, neuropathological hallmarks, and current and possible future therapies. We emphasize literature from the past 15 years. Earlier studies are reviewed thoroughly in other books and chapters (Harper, 1991; Marshall, 2004). In keeping with this book's purpose, we stress clinical and neuropathological aspects of HD that should be kept in mind as the field moves toward preventive therapy.

CLINICAL FEATURES

HD is, in every sense of the word, a "neuropsychiatric" disorder. No other disease so firmly bridges the disciplines of psychiatry and neurology. The classic clinical triad in HD is (1) progressive movement disorder, most commonly chorea; (2) progressive cognitive disturbance culminating in dementia; and (3) various behavioral disturbances that often precede diagnosis and can vary depending on the state of disease.

LEARNING FROM WOODY

Physicians who run HD clinics are continually astonished by the diverse array of symptoms and signs that affected persons can develop. In one respect or another, each person with HD is clinically unique. But if there is a "typical" case of HD, it would be adult-onset disease manifested by chorea and behavioral changes that, over a 15–20-year period, slowly rob the affected person of his or her professional skills and social graces. Woody Guthrie, the famous activist and folksinger who died from HD in 1967, was by no means a typical person. Yet the course of his HD fits the description above (Arevalo et al., 2001). Reviewing his life story may help readers gain a personal perspective on this devastating disease.

Born in 1912, Guthrie proved to be a precocious boy and later a creative and prolific folksinger (among his thousands of songs are "This Land Is Your Land" and "So Long, It's Been Good to Know You"). By the time Guthrie was a young adult, his mother had passed away from progressive dementia accompanied by disruptive behavior that led to her institutionalization. When Guthrie was 39, he was admitted to a detoxification center after an angry outburst toward his wife. This was the first of many institutions he would enter in the final 15 years of his life. At that time, Guthrie's own notes described his various symptoms and his psychological state very well (Arevalo et al., 2001). He felt "terribly restless always. I get here and I want to be yonder. I get over yonder and I want to be back over here ... I don't trust anybody I see." Later the same year he wrote:

> Here's my funny feeling over me again. That lost feeling. That gone feeling. That old empty whipped feeling. Shaky. Bad control. Out of control. Jumpy. Jerky. High tension. Least little thing knocks my ego down below zero mark. Everything cuts into me and hurts me several times more than it should ... no bodily (physical) pains; just like my arms and legs and hands and feet and my whole body belongs to somebody else and not to me; so ashamed of myself ... I spend every drop of my bodily strength trying to hide my trouble away so you can't see it; trying to keep you from reading it in my face, or my eyes, or in any words I'd say ... or in that stumbly way I walk around.

A short time later Guthrie wrote, "My chorea sure isn't kidding these days. I feel it as a nervous fluttery heart condition along with a slight lack of control over my body at times." As his disease progressed, the diagnosis was soon established as HD based on his progressive clinical signs (including chorea) and his mother's illness. Eleven years later, Guthrie's wife Marjorie described his now advanced symptoms at age 51:

> Woody's muscular condition continues to deteriorate. I find it more and more difficult to understand him when he speaks and this is when his fiery temper jumps out. His memory remains uncanny, but he just hasn't got the muscular control to mouth the words. His balance is very poor ... forever lighting up a new cigarette because the old one fell out of his mouth.

By age 53, Guthrie had stopped speaking entirely and could only communicate by pointing a wildly flailing arm at "yes" and "no" cards that Marjorie made for him. Eight months before he died in 1967 at 56 years of age, the folksinger Pete Seeger visited him. Woody was, Seeger noted,

> in a wheelchair. He couldn't walk anymore so the hospital attendant wheeled him out under a porch where it was warm.... You could see how much he wanted to be part of our little [group] ... he tried to get his arms going but they were just flailing around like a windmill. He got to the point where it looked as though he might hurt himself so the attendant said, "You'd better quit playing." (Arevalo et al., 2001)

AGE OF ONSET, NATURAL HISTORY, AND DIAGNOSIS

Onset of manifest HD is defined conventionally by the presence of a motor disorder, usually the involuntary movements known as chorea (see "Movement Disorder" below). In some patients, other motor abnormalities may lead to a diagnosis of HD (Louis et al., 2000). Because the motor features of HD are often preceded by cognitive or behavioral features (see "Movement Disorder" below), the diagnostic criterion of a discernable motor disorder is a fairly conservative standard. In the earlier stages of HD, motor abnormalities may be subtle and may fluctuate depending on the state of arousal of the patient. Median age of diagnosis is approximately 40 years of age with a wide range in ages of onset. Onset before age 20 or after age 65 is relatively rare, but both very young and elderly new-onset HD patients are seen in tertiary referral clinics. Death generally occurs 15–20 years after diagnosis.

When the characteristic motor disorder is present in an individual from a well-characterized HD pedigree, diagnosis of manifest HD is straightforward (Kremer et al., 1994). Circumstances may arise, however, where information about the family history is limited or absent. In some individuals without an apparent family history of HD, covert adoption or nonpaternity, as well as a new mutation, may have occurred. Molecular testing is invaluable in these situations.

MOVEMENT DISORDER

Although the range of movement disorders in HD extends beyond chorea, it remains the classic motor sign of HD. Derived from the Greek word for "dance," chorea often begins as fleeting, suppressible movements that may appear as fidgetiness. Over the course of disease, chorea becomes more overt as it involves larger muscle groups. Patients often will incorporate chorea into purposeful movements, something known as parakinesia. Motor impersistence, the inability to sustain a voluntary muscular effort, is common in HD and often goes hand in hand with chorea. The "fly-catcher's tongue" describes the difficulty many moderately advanced HD patients have in keeping their tongue from protruding beyond the lips.

By itself, chorea is usually not greatly disabling. For many HD clinicians chorea is not something that becomes a major target of treatment. However, flailing and continual chorea can be disabling or physically harmful, requiring treatment with dopamine-depleting agents such as tetrabenazine. In past decades, potent dopamine-blocking neuroleptic medications (e.g., haloperidol) were routinely used to control chorea in HD patients. More recently, however, clinicians have tended first to use newer atypical antipsychotic drugs in persons who experience severe chorea, especially when it is accompanied by psychiatric symptoms warranting antipsychotic use, such as delusions.

Dystonia, the involuntary contraction of muscles, is seen in most persons with HD but is especially prominent in younger-onset individuals. As "typical" adult-onset HD progresses, straightforward chorea can evolve into a more complicated constellation of movement disorders that increasingly include dystonia (Feigin et al., 1995). In addition, clumsiness and slowness of movements (bradykinesia) are common in those with HD, the latter especially in earlier-onset disease. Indeed, the juvenile-onset form of disease that manifests initially with bradykinesia, rigidity, and little or no chorea has been given its own name, the "Westphal variant." Very early-onset cases also have a high incidence of epilepsy. The distinctive phenotype of very early-onset HD probably reflects the effects of the mutant allele on developing brains. Rather than single out one clinical subset of HD, it is perhaps better to think of motor signs in HD as spanning a spectrum—earlier-onset cases tending to have more bradykinesia and dystonia and later-onset cases tending to have more chorea (Louis et al., 2000). In many adult-onset HD patients, chorea will gradually progress and then begin to subside as dystonia and bradykinesia become more prominent motor features. The combination of chorea, bradykinesia, and dystonia sometimes leads to very unusual gait patterns that leave patients highly prone to falls. Ataxia and marked postural instability are common in those with more advanced HD.

Eye movement abnormalities occur early and persist throughout the course of disease. Persons with HD have difficulty maintaining fixation and develop slowed initiation and velocity of the rapid eye movements known as saccades (Lasker et al., 1987). As this worsens, they may need to thrust their head or blink to break fixation and move their eyes to gaze at a new target.

In summary, HD manifests with a remarkably wide range of movement disorders, partly because of the phenotypic heterogeneity stemming from differences in expanded repeat lengths and partly reflecting the fact that disease symptoms and signs change over time. The progressive movement disorder and increasingly widespread failure of the motor system contribute greatly to the physical disability and decreased life expectancy of individuals with HD. In particular, difficulties with swallowing (dysphagia), speech (dysarthria), and balance (frequent falls) become very debilitating. A common cause of death in those with advanced HD is aspiration pneumonia, reflecting severe dysphagia and general immobility (Sorensen and Fenger, 1992).

Cognitive Disorder

It was once suspected that cognitive decline only began when, or soon before, manifest HD was apparent. Recently, however, large-scale observational studies of HD gene carriers before diagnosis make it clear that subtle cognitive impairment is among the earlier manifestations in the disease process and is associated with progressive caudate atrophy (Aylward, 2007; Aylward et al., 2000). By the time of diagnosis, most subjects with HD have significant cognitive impairment readily measurable by neuropsychological testing. These early signs do not impede most activities of daily living, but individuals with demanding jobs requiring sustained concentration often find work increasingly stressful and difficult. The cognitive impairment progresses slowly over many years to frank dementia in most persons with HD.

In contrast to the dementia of Alzheimer's disease (AD), HD results in a largely "subcortical" dementia characterized by slowness in initiating thought processes, difficulties with executive function, and problems with attentional and sequencing tasks (Paulsen et al., 1995; Rohrer et al., 1999). Although episodic memory is impaired, memory in general is relatively well preserved compared with, for example, AD.

Because a battery of cognitive tests can detect abnormalities before diagnosis, elements of cognitive impairment likely will serve as important quantifiable measures in future clinical trials of candidate preventive agents in prediagnosis individuals.

One intriguing feature of the cognitive impairment in HD is the lack of insight subjects may have regarding their own symptoms (Hoth et al., 2007). Some individuals with overt signs of HD, evident to other family members, deny that they experience any motor or cognitive difficulties. Lack of insight is a typical sign of impaired frontal lobe function and may reflect early dysfunction of striatal neurons receiving prominent frontal lobe inputs.

Behavioral Disorder

The behavioral problems arising in HD can be the most vexing to the patient, family, and physician. These range from affective illness, most notably depression and apathy, to delusional behavior that can include, rarely, hallucinations (Caine and Shoulson, 1983; Paulsen et al., 2001). As is true of other progressive neurodegenerative diseases, the behavioral disorders of HD evolve during the course of illness. Most HD gene carriers will experience some behavioral symptoms before establishing the diagnosis (Close Kirkwood et al., 2002; Duff et al., 2007; Kirkwood et al., 2002). These may include depression, obsessive–compulsive behaviors (OCBs), irritability, and behavioral outbursts (Duff et al., 2007; Kirkwood et al., 2002). Family members often comment that patients' personalities change in the years leading up to diagnosis, although this may be apparent only in retrospect.

Depression is particularly common in those with HD: between 30% and 50% of patients will develop depressive symptoms in the course of disease. The depression often responds very well to treatment, with selective serotonin reuptake inhibitor antidepressants commonly being the first agents tried. Often, however, there is superimposed apathy that is difficult to treat. Personality changes can occur that impair work performance and social interactions well before the diagnosis is established. This can lead to behavioral agitation, anxiety, alcohol abuse, marital problems, and antisocial behavior. Also common in HD patients are OCBs. These may overlap with features of rigidity and perseveration that also can be seen in individuals with frontal lobe dysfunction. OCBs, rigidity, and perseverative features probably all reflect striatal dysfunction.

The severity of psychiatric symptoms varies greatly in those with HD and does not correlate with dementia and chorea (Paulsen et al., 2001). Recent findings suggest that psychiatric problems are especially problematic in juvenile-onset cases (Ribai et al., 2007).

Some psychiatric symptoms tend to run in specific HD families. For example, overt psychotic behavior more likely will occur in someone from a family where one or more affected persons also suffer from psychosis (Tsuang et al., 2000).

An important cautionary note is the high rate of suicide in patients with HD (Almqvist et al., 1999; Lipe et al., 1993). The risk is significantly higher for those with HD gene-positive status, including both those with diagnosed HD and those who are prediagnosis. The risk may also be higher for family members who are *not* carriers of the gene, underscoring the tremendous stress and burden placed on families with HD

(Lipe et al., 1993). Factors that increase the risk of suicide in gene-positive individuals include being single, lacking children, living alone, having depression, and already having manifest signs of HD. Degeneration of striatal neurons involved in limbic circuits is a possible substrate of depression in those with HD (Tippett et al., 2007).

Clinical Heterogeneity

In patients with HD, the remarkable range in symptoms and age of onset largely reflects differences in the length of the repeat expansion. As shown in Figure 1.1 (Langbehn et al., 2004) and documented in many publications, repeat length inversely

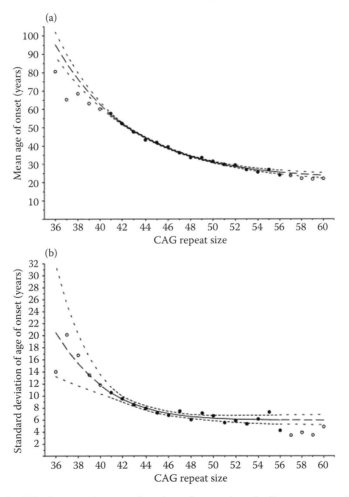

FIGURE 1.1 HD disease onset as a function of repeat length. Shown are population estimates based on an analysis of nearly 3,000 individuals seen at 40 HD centers. (a) Inverse correlation between mean age of onset and CAG repeat length. (b) Standard deviation of age of onset versus repeat length. Dashed lines indicate 95% confidence intervals. Data from 41–56 repeats were used to fit exponential curves, which were then extrapolated for repeats of 40 or less. (From Langbehn, D. R., Brinkman, R. R., Falush, D., Paulsen, J. S., and Hayden, M. R., *Clin Genet* 65, 267, 2004. With permission.)

correlates with age of onset (Andrew et al., 1993; Ashizawa et al., 1994; Duyao et al., 1993; Kieburtz et al., 1994; Stine et al., 1993). Because earlier- versus later-onset forms of disease typically manifest different motor symptoms, the differences in repeat length also explain, in part, the heterogeneous phenotype.

Expanded repeat length accounts for approximately 50% of variability in age of onset. However, also note in Figure 1.1b that the *variance* in age of onset for a given repeat length is much greater with small expansions (e.g., 39) than with the rarer, large expansions (e.g., ~56). In other words, the range in age of disease onset for a repeat of 39 can span four decades or more, whereas a repeat of 56 has a narrow range. Clearly other genetic and environmental factors contribute to age of disease onset, particularly for the more common, smaller expansions (<45). Indeed, genetic modifiers of disease onset have already been identified. For example, Meyers and colleagues (Li et al., 2006) and Rubinsztein and colleagues (Rubinsztein et al., 1997; Zeng et al., 2006) separately have identified loci that influence the rate of onset. As mentioned earlier, familial aggregation of certain symptoms (e.g., psychosis) occurs in HD, and this too likely reflects genetic modifiers. It is hoped that further success in defining genetic risk factors/modifiers will shed light on biological pathways that contribute to disease.

Given the relationship of repeat length to age of onset, one can generate "probability curves" that, for a given repeat length, plot the probability of developing manifest disease by a particular age (Langbehn et al., 2004). A series of such probability curves is shown in Figure 1.2. For larger expansions, the narrower range of disease onset is again clear (i.e., steeper slope). An important point illustrated by this figure is that rather modest expansions (≤40) may *not* lead to manifest disease until very late in life, and in some cases, HD symptoms may not surface before one dies from other age-related causes. In other words, shorter HD expansions are not fully penetrant alleles (Quarrell et al., 2007). There is also an underappreciation of the fact that HD manifesting after age 50 years is not as rare as once thought. Because

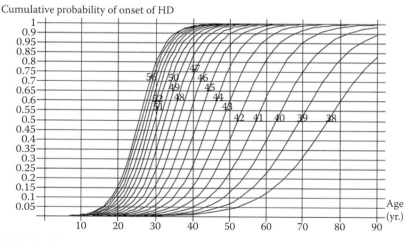

FIGURE 1.2 Cumulative probability curves of HD onset for various CAG repeat lengths. Numbers shown indicate repeat length for a given curve. (From Langbehn, D. R., Brinkman, R. R., Falush, D., Paulsen, J. S., and Hayden, M. R., *Clin Genet* 65, 267, 2004. With permission.)

a higher percentage of late-onset cases will be sporadic cases from newly expanded alleles, clinicians need to keep in mind that late-onset HD often does *not* have the typical positive family history (Falush et al., 2001).

Unlike age of *onset,* which strongly correlates with repeat length, rate of disease *progression* is not convincingly linked to repeat length. Some studies have found no correlation (Kieburtz et al., 1994), whereas others have, especially with late-stage outcomes (Marder et al., 2002). Ongoing, large observational studies of gene carriers may provide new data that resolve this uncertainty.

HUNTINGTON'S DISEASE MIMICS

HD is a distinctive phenotype, but mimics of HD are encountered occasionally in clinical practice (Rosenblatt et al., 1998; Schneider et al., 2007). The list of chore-iform disorders in the differential diagnosis is extensive. Many can be excluded based on the history or features of the physical examination. In doubtful cases, molecular testing is invaluable for identifying or excluding HD. More difficult to evaluate are a few inherited neurodegenerations that can mirror HD. These include other domi-nant disorders such as the polyglutamine diseases dentatorubral-pallidoluysian atro-phy (DRPLA) and SCA17; Huntington's disease-like 2, which is another expanded repeat disorder; and members of the neurodegeneration with brain iron accumulation spectrum such as neuroferritinopathy. Recessive and X-linked disorders in the neu-roacanthocytosis family of diseases can also mimic HD. Molecular diagnoses are available for all of these diseases. Because among these disorders HD is by far the most common, a reasonable strategy is first to test for the presence or absence of the HD mutation. If absent, more extensive testing will be needed to establish a diagno-sis. Establishing an accurate diagnosis is crucial for the family and the patient. The clinical implications are vastly different for a dominant versus a recessive disorder, and appropriate presymptomatic testing in other family members can be pursued only with an accurate diagnosis.

NEUROPATHOLOGICAL FEATURES

NON–CENTRAL NERVOUS SYSTEM CHANGES

As with the other polyglutamine disorders, HD is primarily a disorder of the central nervous system (CNS). Huntingtin is expressed in many tissues, but the clinical fea-tures of HD reflect CNS dysfunction, and almost all histopathologic abnormalities are restricted to the brain. It is important to note that although clinical features and traditional histopathologic analyses point to HD as a CNS disease, investigations with more sensitive measures point to widespread, albeit subclinical, effects of expanded htt. Studies of peripheral blood and lymphoblasts derived from HD subjects have revealed aberrant gene regulation and mitochondrial abnormalities (Borovecki et al., 2005; Sawa et al., 1999). A recent study of urea cycle metabolites suggests a subclini-cal but detectable effect of expanded htt on hepatic function (Chiang et al., 2007). Several studies suggest abnormalities of another long-lived postmitotic tissue, muscle. Gene regulation studies, magnetic resonance spectroscopic studies, and biochemical studies suggest muscle mitochondrial abnormalities in HD, and muscle biopsy histology indicates the presence of significant but nonspecific abnormalities (Arenas et al., 1998; Lodi et al., 2000; Saft et al., 2005; Strand et al., 2005). An interesting recent observation is the description of testis abnormalities in HD. Testes are characterized by high-level expression of *HTT* mRNA. Van Raamsdonk et al. (2007) describe reduced numbers of germ cells and abnormal seminiferous tubule morphology.

GROSS CNS PATHOLOGY

CNS changes, however, drive the mortality and morbidity associated with HD. A plausible explanation for the CNS selectivity of HD is that cell populations with limited lifespans do not live long enough to experience the damage caused by expanded polyglutamine htt. In this model, neurons are preferentially susceptible to the toxicity of expanded htt because of their long lifespan. Although this could explain the selective CNS pathology of HD, it explains neither the regional selectivity within the CNS nor the midlife onset of HD. Some unexplained interaction between expanded polyglutamine htt effects and normal aging phenomena must be invoked to explain the relatively late onset of HD.

Within the CNS, htt is expressed by neurons throughout the neuraxis without dramatic regional differences. Despite this, there is a definite regional pattern to HD pathology (Vonsattel and DiFiglia, 1998). Classic descriptions emphasize that HD is a striatal degeneration. This is true to a first approximation, but this traditional teaching point should not be exaggerated. Gross striatal atrophy is a prominent feature of HD. Careful pathologic analyses, performed usually on specimens of advanced HD, reveal a more complex picture (Figure 1.3). Thinning of the cortical mantle and decreased brain weights and volumes are common as well. Consistent with diffuse loss of neurons is the diffuse loss of cerebral white matter. Careful studies reveal neuronal loss in many regions, including the neocortex, cerebellum, hippocampal formation, substantia nigra, and brainstem nuclei. These findings reflect the widespread expression of htt and correlate well with the profound and diffuse clinical deficits found in patients with advanced HD, including pyramidal signs, ataxia, severe bulbar dysfunction, marked incoordination, and profound dementia.

An important question to consider is what structures are affected earlier in the course of HD. Identifying neuronal populations that are relatively susceptible to expanded polyglutamine htt effects might provide important clues to pathogenic mechanisms. The prevailing impression is that the striatal complex is affected early in HD, although there is also increasing evidence of early neocortical degeneration (Vonsattel and DiFiglia, 1998; see discussion of imaging results under "Striatal Pathology" below). An important development in the study of postmortem changes in HD was the introduction by Vonsattel et al. (1985) of a pathologic grading system based on the degree of striatal pathology. In earlier (lower) Vonsattel grades of pathology, the striatum does seem to be more affected than other brain regions. Vonsattel and colleagues also documented subregional differences in HD pathology in the striatal complex. In the extensive series of specimens examined by Vonsattel and colleagues, neurodegeneration within the striatal complex progressed in caudal-to-rostral and dorsal-to-ventral gradients.

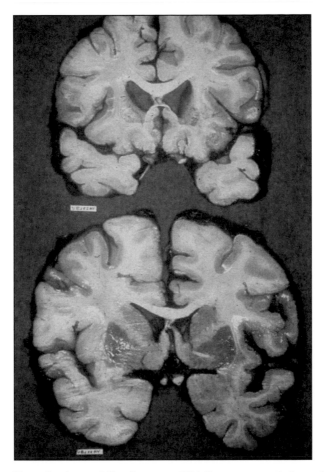

FIGURE 1.3 **(See color insert following page 172.)** Gross neuropathology of HD. The top section is from an HD subject, whereas the lower section is from a control without neurologic disease. There is marked striatal atrophy with corresponding ventriculomegaly. There is also diffuse thinning of the cortical mantle. Although sections are not at precisely the same level, diffuse loss of cerebral volume, including loss of white matter, is evident in the HD specimen. (Courtesy of Dr. Andrew Lieberman, Department of Pathology, University of Michigan, Ann Arbor, MI.)

STRIATAL PATHOLOGY

Initial explorations of HD striatal pathology beyond classic histopathology involved biochemical assessments of neurotransmitter systems. These initial studies revealed loss of γ-aminobutyric acid (GABA)ergic and cholinergic markers with relative sparing of dopaminergic markers. These results suggested loss of intrinsic GABAergic and cholinergic neurons with relative sparing of extrinsic dopaminergic terminals. The similarity of these changes to the results of experimental excitotoxic lesions gave rise to the still popular excitotoxic hypothesis of neuronal death in HD.

The identification of striatal neuron subpopulations made it possible to pursue a fine-grained analysis of striatal changes in HD. Striatal neuron populations are divided into two major groups: (1) aspiny interneurons whose projection arbors are restricted to the striatum; and (2) medium spiny projection neurons whose primary axons synapse in the downstream targets of the striatal complex. Striatal interneurons are further subdivided based on the morphology and neurochemical phenotype. Well-studied populations of striatal interneurons include the large aspiny cholinergic interneurons, which are probably synonymous with the tonically active neurons described in extracellular recording studies. These neurons are virtually spared in HD striatum (Ferrante et al., 1987). Interestingly, although there is solid evidence that cholinergic neuron perikarya persist even in those with advanced HD, striatal choline acetyltransferase (ChAT) levels decrease markedly. The discrepancy between decreasing ChAT activity and preserved perikarya suggests significant striatal cholinergic interneuron dysfunction, as opposed to neurodegeneration, in HD. Another spared subpopulation of striatal interneurons are those containing somatostatin, neuropeptide Y, and nitric oxide synthase, which persist even in advanced HD striatum (Ferrante et al., 1985).

Subpopulations of striatal projection neurons, all of which are GABAergic, are defined by their primary projection targets, coexpressed neuropeptides, and neurotransmitter receptors. Striatal neurons projecting to the different segments of the globus pallidus (external [GPe] and internal [GPi]) and the different components of the substantia nigra (dopaminergic pars compacta [SNc] and GABAergic pars reticulata [SNr]) form relatively segregated pools of neurons. Striato-GPe neurons are distinguished by the expression of enkephalins, dopamine D2 receptors, and adenosine A2a receptors, whereas the other striatal projection neuron pools tend to express tachykinins and dopamine D1 receptors. Some recent studies in nonhuman primates using sensitive tract tracing methods have cast doubt on the idea of segregated striatal projection neuron pools (Levesque and Parent, 2005). However, recent data from mice engineered to express green fluorescent protein under the control of the D1 or D2 promoters strongly support the existence of distinct pools of striatal projection neurons (Lobo et al., 2006). The existence of different pools of striatal projection neurons overlaps to some degree with another important aspect of striatal organization: the patch (striosome)–matrix differentiation. A variety of neurochemical markers differentiate the striatum into two compartments: (1) patches (striosomes), a geometrically complex interconnected tubular compartment; and (2) the surrounding matrix compartment. Striato-SNc neurons are disproportionately represented within the patch compartment.

Examination of postmortem HD material across a full spectrum of pathologic grades suggests a specific temporal order in degeneration of striatal projection neuron subpopulations. The early changes appear to be loss of striato-GPe neurons and perhaps striato-SNr neurons (Albin et al., 1992; Deng et al., 2004; Reiner et al., 1988). In contrast, striato-GPi neurons are relatively spared until later in the course of HD. This sequential pattern of neuronal loss correlates broadly with certain features of the natural history of HD. Oculomotor abnormalities, particularly abnormalities of saccadic eye movements, are early features of HD. As basal ganglia inputs to the superior colliculus come from SNr, the early loss of striato-SNr projection neurons correlates nicely with early saccadic abnormalities. The evolution of involuntary movements in HD correlates also with the evolution of striatal projection neuron changes. One predicted downstream consequence of initial degeneration of striato-GPe neurons is inhibition of the subthalamic nucleus. Decreased subthalamic activity is associated with choreoathetoid movements. In many patients, disease progression is associated with gradual worsening of choreoathetoid movements, which then peak in intensity and gradually decrease. The decrease in choreoathetosis is accompanied by worsening dystonia and bradykinesia. The apparent correlate in striatal pathology is generalized loss of striatal projection neurons and probably neurons within other nuclei of the basal ganglia (Albin et al., 1990). In a recent, interesting study, Tippett et al. (2006) looked specifically at a marker of striosomal striatal projection neurons in a broad spectrum of HD postmortem specimens. This work suggests a correlation between mood disorders in HD and striosomal pathology.

Neocortex is also affected significantly in HD. Recent neuroimaging data (see "Neuroimaging" below) suggest that this may be an early event in HD. Some data suggest that neocortical atrophy proceeds in tandem with striatal atrophy (Halliday et al., 1998; Macdonald and Halliday, 2002). As in the striatum, there is evidence for differential involvement of different cortical neuron populations in HD. Careful studies looking at changes in cortical laminae reveal differential loss of neurons in different layers. Pyramidal neurons of deeper cortical layers appear to be more affected in HD, although this may vary by cortical field (Hedreen et al., 1991; Sieradzan and Mann, 2001; Sotrel et al., 1991, 1993).

Unlike the neurofibrillary tangles of AD or the Lewy bodies of Parkinson's disease, HD was believed historically not to exhibit characteristic abnormalities at the cellular level. This perception changed radically with the discovery in the R6/2 transgenic mouse model of neuronal intranuclear and cytoplasmic aggregates containing expanded polyglutamine htt, ubiquitin, and other proteins. These types of inclusions were also found in HD brain and in most other polyglutamine disorders. Although inclusions are found in the HD brain, their frequency is less than that seen in many of the murine genetic models (DiFiglia et al., 1997; Gutekunst et al., 1999). Perhaps because of the relatively short lifespan of mice, neuronal loss in murine genetic models is less severe than that seen in end-stage HD. It is possible that neuronal loss in HD is accompanied by a corresponding loss of neuronal inclusions.

Debate has arisen about the role, if any, of neuronal inclusions in disease pathogenesis. Suggestions for the role of inclusions include that they are (1) directly pathogenic structures; (2) markers of failed clearance of abnormal disease protein; (3) protective "sinks" for toxic polyglutamine proteins and associated factors; and (4) epiphenomena of the disease process (Ross and Poirier, 2005). A relevant fact from postmortem analysis is that neuronal inclusions are more readily found in some of the spared striatal interneuron populations than in the vulnerable striatal interneurons (Kuemmerle et al., 1999).

Neuroimaging

Whereas imaging methods were pursued originally to improve diagnosis of HD, recent studies have pursued imaging methods as part of efforts to develop biomarkers for HD. The hope is that novel imaging methods will provide sufficiently precise measures of disease state and progression to serve as state markers, as response markers, or even as surrogate endpoints in trials.

Computed tomography, magnetic resonance imaging (MRI), and positron emission tomography (PET) imaging methods have all been applied to HD. Most recent studies have concentrated on MRI morphometric analyses of regional brain volumes in the hope of establishing changes over time that will be useful as a marker of disease progression in symptomatic or presymptomatic HD subjects. Using observer-identified volumes of interest, Aylward and colleagues presented data suggesting progressive decrease in striatal volume well before the onset of manifest HD (Aylward, 2007). In their dataset, approximately 50% of the striatum had degenerated by onset of manifest HD. This approach seems promising for the development of a surrogate marker of disease progression that could be applied in both prediagnosis and manifest HD populations. On the other hand, the relatively small volume of the striatum and the apparent fact that considerable striatal atrophy occurs early in the course of HD may limit the dynamic range of striatal volume as a biomarker in studies of manifest HD. For presymptomatic subjects, the imaging arm of the PREDICT-HD observational study (Paulsen et al., 2008) will provide definitive data about the potential use of this approach as a biomarker.

Improved MRI morphometry analysis methods may provide additional or even superior biomarkers. Cortical mantle atrophy is a significant feature of HD pathology. The large extent of the cortex and the possibility of assessing many cortical regions offer the theoretical hope of a morphometric marker with very wide dynamic range. Improved morphometry methods make it possible to measure cortical thickness of many regions simultaneously. Rosas and colleagues pursued cortical morphometry in both symptomatic and presymptomatic HD subjects (Rosas et al., 2002, 2005). Their work suggests that cortical atrophy is an early event in HD, that it occurs in presymptomatic subjects, and that the magnitude of changes over relatively short intervals (e.g., 2–3 years) is sufficiently great to provide a biomarker of disease progression. In addition, this group's work suggests that cortical degeneration is not uniform throughout the cortical mantle but occurs in a stereotyped sequence, with some cortical fields affected earlier than others (HD Rosas, personal communication). This exciting work awaits validation in suitable prospectively collected large datasets.

There has been less activity with PET imaging in HD research, but this remains an interesting potential imaging modality for HD. PET availability was limited historically to research centers with cyclotrons, but PET imaging has become a standard modality for many medical applications, and both scanner availability and the number of easily used tracers is expanding rapidly. The theoretical advantage of PET is its ability to image specific molecular targets. In HD, several PET studies have demonstrated decreases in striatal neurotransmitter markers, particularly striatal dopamine receptors. Changes in these markers appear to be an early feature of HD, consistent with the MRI morphometry data of Aylward and colleagues (Ginovart et al., 1997; van Oostrom et al., 2005; Weeks et al., 1996). Counterintuitively, some PET studies also demonstrate loss of striatal dopamine terminal markers in HD (Backman et al., 1997; Bohnen et al., 2000). This loss seems to be associated with more severe disease. It may reflect the loss of nigrostriatal dopaminergic neurons as a result of direct effects of perikaryal nigral polyglutamine toxicity or the loss of trophic support from striatal neurons.

Ongoing studies with MRI morphometry are likely to provide the critical data within the next 4–5 years. There is a very good chance that MRI morphometry will provide adequate biomarkers for evaluating progression and treatment response in HD.

THERAPY: PRESENT AND FUTURE

SYMPTOMATIC THERAPY

Although HD may not yet be curable, some features are amenable to therapy (Bonelli and Hofmann, 2007). These include depression, which responds often to antidepressants, most commonly selective serotonin reuptake inhibitors or other, newer antidepressants. OCBs may also respond to antidepressants, as may agitation and irritability. Irritability and impulsive behavior are sometimes treated with anticonvulsants, such as valproic acid or carbamazepine. Anxiety can benefit from anxiolytic medications. When HD is accompanied by delusions or paranoid thoughts, atypical antipsychotic drugs are frequently used (less so the classic neuroleptic agents), and appropriate antipsychotic medications can improve behavioral outbursts. Thus far, limited trials of cognitive-enhancing agents used primarily in patients with AD, such as memantine (Beister et al., 2004), rivastigmine (de Tommaso et al., 2007), and donepezil, have shown only modest benefit. Bradykinesia and rigidity in younger-onset individuals can respond to dopaminergic agents used in parkinsonism (Jongen et al., 1980).

Chorea will respond to dopamine-blocking agents; for example, a recent well-controlled clinical trial proved that tetrabenazine reduces chorea in a dose-dependent manner (Huntington Study Group, 2006). Many other medications have shown modest benefit for chorea (Bonelli et al., 2002; de Tommaso et al., 2005; Huntington Study Group, 2003; Kremer et al., 1999; O'Suilleabhain and Dewey, 2003; Verhagen Metman et al., 2002), but not all have been tested in well-controlled trials. Suppressing chorea per se is not an endpoint sought by many physicians who treat HD unless the chorea is disabling or so severe as to increase risk of injury. Although traditional neuroleptic agents or dopamine blockers such as tetrabenazine can be effective in reducing chorea, these medications are not without their side effects, including parkinsonism, gait disturbance, somnolence, and depression. Myoclonus, which is rare in HD and sometimes mistaken for chorea, can respond to valproic acid (Saft et al., 2006).

Assistive therapies such as physical, occupational, and speech therapy play an important role in the care of HD patients. In many cases, these are the primary therapeutic interventions and can be very useful for some patients. The efficacy of assistive interventions is often limited, however, by the cognitive and behavioral impairments of HD patients. As with many aspects of HD, the burden of maximizing benefits from assistive interventions falls on caregivers.

Although the focus of this book and the HD research community is on identifying a preventive medicine for HD, we must also continue efforts to seek out symptomatic medications and enhance practical measures (Nance, 2007) that can improve the lives of those currently affected by HD.

TOWARD PREVENTIVE THERAPY

As our understanding of the molecular basis of HD becomes clearer, routes to preventive therapy are being considered (Di Prospero and Fischbeck, 2005; Feigin and Zgaljardic, 2002; Nakamura and Aminoff, 2007). Potential contributing factors to HD include perturbations in protein homeostasis, mitochondrial dysfunction, excessive or aberrant corticostriatal input, transcriptional dysregulation, loss of neurotrophic support, and impairments in axonal trafficking. Correspondingly, potential compounds or strategies considered for preventive treatment include histone deacetylase inhibitors (e.g., phenylbutyrate [Gardian et al., 2005]); antioxidants, mitochondrial enhancers, and energy substrates (e.g., coenzyme Q_{10} [Huntington Study Group, 2001] and creatine [Hersch et al., 2006]); neuroprotective compounds (e.g., lithium [Wei et al., 2001]); antiapoptotic compounds (e.g., minocycline [Huntington Study Group, 2004]); transglutaminase inhibitors (Dubinsky and Gray, 2006); reagents that block protein aggregation or assist protein folding; compounds to enhance autophagic clearance of mutant protein (Sarkar et al., 2007); inhibitors of the kynurenine 3-monooxygenase pathway (Giorgini et al., 2005); cell or gene replacement therapy (Bachoud-Levi et al., 2000; Bloch et al., 2004; Keene et al., 2007); and RNA interference reagents that "knock down" disease gene expression (Harper et al., 2005). The availability of numerous mouse models of HD in which disease progression can reliably be measured has allowed researchers to identify more than 20 compounds or reagents that slow disease in these models and therefore may warrant human trials (reviewed in Beal and Ferrante, 2004; Hersch and Ferrante, 2004). Two challenges at hand are (1) determining which molecules and compounds should be tried in human trials and (2) refining quantitative measures of HD in humans so that clinical trials can be performed faster and in fewer subjects without sacrificing statistical power.

To date, human clinical trials testing candidate preventive compounds have been discouraging. In the largest published study to date, the CARE-HD study, in which both coenzyme Q_{10} and remacemide were tested, coenzyme Q_{10} (600 mg) showed a statistically insignificant trend toward slowing disease (Huntington Study Group, 2001). A similar study of a higher dose of coenzyme Q_{10}, the 2CARE trial (also sponsored by the Huntington Study Group), recently began. This study tests the effects of 2,400 mg/day on the progression of functional decline in HD subjects.

The field is now witnessing human trials that include potential biomarkers of disease as outcome measures. For example, a short-term study of creatine in a small number of HD subjects led to a reduction in a marker of DNA damage, 8-OH-deoxyguanosine (Hersch et al., 2006). We caution the reader that even though reducing levels of a peripheral marker of DNA damage sounds promising, we do not know that doing so would alter the disease process or disease progression in any way. As more potential biomarkers of disease surface, it will be vitally important to correlate changes in them with disease progression and with the clinical response to any preventive medication that does surface.

Search for Quantifiable Measures

This latter study points to the growing interest in identifying biomarkers for HD—for example, changes in brain imaging (reviewed earlier), metabolic/proteomic profiles (Dalrymple et al., 2007; Gomez-Anson et al., 2007), or gene expression (Borovecki et al., 2005) that might one day serve as reliable proxies for clinical benefit. To get there, however, we first need to know more about disease progression.

In the HD research community, efforts to define disease progression and response to medications have been aided greatly by a widely used standardized clinical rating scale, the Unified Huntington's Disease Ratings Scale (UHDRS) (Huntington Study Group, 1996; Siesling et al., 1998). The UHDRS has four major components: motor function and cognitive, behavioral, and functional ability. It has been used by the large consortium of HD investigators known as the Huntington Study Group for 15 years and serves as the data collection base for two ongoing, large-scale observational trials: PREDICT-HD (Paulsen et al., 2006) and PHAROS-HD (Huntington Study Group PHAROS, 2006). The movement disorder of HD has classically been the barometer by which neurologists establish the diagnosis. Reliance on movement disorder as the *sine qua non* of HD diagnosis may change as these studies identify a nonredundant set of cognitive, psychiatric, and functional measures that are associated with motor scores.

With respect to developing therapeutics, it is important to recognize that clinical progression in HD and accompanying brain changes detectable by imaging can be quantified both before and after diagnosis. Studies have shown measurable decreases in total functional capacity, particularly in the early to middle years after diagnosis (Marder et al., 2000). Because the limited dynamic range of these measures in advanced disease makes it much more difficult to assess progression late in disease, effective preventive trials will almost certainly be carried out in early HD or prediagnosis subjects who are relatively close to the predicted age of onset. A subset of measures in the full UHDRS, perhaps supplemented with additional cognitive tests, likely will constitute a reliable, straightforward battery of tests with which clinicians in future trials can measure changes in rate of progression and more accurately pinpoint phenoconversion to manifest disease. As shown in Figure 1.4, an effective preventive medication would slow the rate to "phenoconversion."

Ongoing observational studies likely will generate the longitudinal data that allow the HD research community to establish associations between specific biomarkers and either conversion to manifest disease or progression of disease. Biochemical markers like 8-OH-deoxyguanosine or other metabolomic or genomic markers are being sought as state markers, progression markers, and response markers. Imaging markers look directly at brain changes and may prove to be particularly useful as progression/response markers and even prove to be suitable as surrogate endpoints. All putative biomarker methods will need to be validated against the well-developed clinical dataset and carefully evaluated to ensure that they have suitable statistical properties. Certain such markers may then become true surrogates for a clinical endpoint and thus permit trials of shorter duration in smaller numbers of subjects to identify preventive medications or experimental treatments. In short, "one of the incurables" might just become treatable after all.

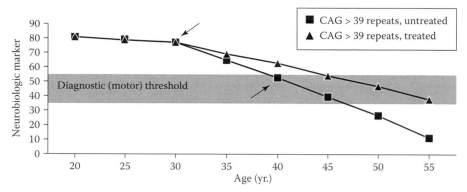

FIGURE 1.4 Intervention model for adult-onset HD. In this illustrated scenario, the downward-pointing arrow indicates administration of an effective neuroprotective agent in a treated individual, whereas the upward-pointing arrow indicates the point of clinical diagnosis in an untreated individual. In this scenario, treatment delays progression to manifest HD. (From Paulsen, J. S., et al., *Arch Neurol* 63, 883, 2006. With permission.)

SUMMARY

HD is a complex and clinically demanding disorder. Work over the past two decades has defined the clinical phenotype and neuropathology with increasing precision. This essential descriptive work provides the platform for an increasingly sophisticated body of clinical research aimed at developing tools for characterizing clinical, imaging, and other features of HD. These new tools will facilitate researchers' ability to translate insights from the laboratory into rigorous clinical research and potential therapies for HD.

ACKNOWLEDGMENTS

HD and polyglutamine disease research in the Paulson and Albin laboratories is funded by RO1-NS38712 (HLP) and a VA Merit Review grant, the High Q Foundation, and RO3-NS054810 (RLA). We also thank Drs. Jane Paulsen, Doug Langbehn, and Andrew Lieberman for providing figures.

REFERENCES

Albin, R. L., A. Reiner, K. D. Anderson, L. S. Dure IV, B. Handelin, R. Balfour, W. O. Whetsell, Jr., J. B. Penney, and A. B. Young. Preferential Loss of Striato-External Pallidal Projection Neurons in Presymptomatic Huntington's Disease. *Ann Neurol* 31, no. 4 (1992): 425–30.

Albin, R. L., A. Reiner, K. D. Anderson, J. B. Penney, and A. B. Young. Striatal and Nigral Neuron Subpopulations in Rigid Huntington's Disease: Implications for the Functional Anatomy of Chorea and Rigidity-Akinesia. *Ann Neurol* 27, no. 4 (1990): 357–65.

Almqvist, E. W., M. Bloch, R. Brinkman, D. Craufurd, and M. R. Hayden. A Worldwide Assessment of the Frequency of Suicide, Suicide Attempts, or Psychiatric Hospitalization after Predictive Testing for Huntington Disease. *Am J Hum Genet* 64, no. 5 (1999): 1293–304.

Andrew, S. E., Y. P. Goldberg, B. Kremer, H. Telenius, J. Theilmann, S. Adam, E. Starr, F. Squitieri, B. Lin, M. A. Kalchman, et al. The Relationship between Trinucleotide (Cag) Repeat Length and Clinical Features of Huntington's Disease. *Nat Genet* 4, no. 4 (1993): 398–403.

Arenas, J., Y. Campos, R. Ribacoba, M. A. Martin, J. C. Rubio, P. Ablanedo, and A. Cabello. Complex I Defect in Muscle from Patients with Huntington's Disease. *Ann Neurol* 43, no. 3 (1998): 397–400.

Arevalo, J., J. Wojcieszek, and P. M. Conneally. Tracing Woody Guthrie and Huntington's Disease. *Semin Neurol* 21, no. 2 (2001): 209–23.

Ashizawa, T., L. J. Wong, C. S. Richards, C. T. Caskey, and J. Jankovic. Cag Repeat Size and Clinical Presentation in Huntington's Disease. *Neurology* 44, no. 6 (1994): 1137–43.

Aylward, E. H. Change in MRI Striatal Volumes as a Biomarker in Preclinical Huntington's Disease. *Brain Res Bull* 72, no. 2–3 (2007): 152–58.

Aylward, E. H., A. M. Codori, A. Rosenblatt, M. Sherr, J. Brandt, O. C. Stine, P. E. Barta, G. D. Pearlson, and C. A. Ross. Rate of Caudate Atrophy in Presymptomatic and Symptomatic Stages of Huntington's Disease. *Mov Disord* 15, no. 3 (2000): 552–60.

Bachoud-Levi, A. C., P. Remy, J. P. Nguyen, P. Brugieres, J. P. Lefaucheur, C. Bourdet, S. Baudic, V. Gaura, P. Maison, B. Haddad, et al. Motor and Cognitive Improvements in Patients with Huntington's Disease after Neural Transplantation. *Lancet* 356, no. 9246 (2000): 1975–79.

Backman, L., T. B. Robins-Wahlin, A. Lundin, N. Ginovart, and L. Farde. Cognitive Deficits in Huntington's Disease Are Predicted by Dopaminergic Pet Markers and Brain Volumes. *Brain* 120 (Pt 12) (1997): 2207–17.

Beal, M. F., and R. J. Ferrante. Experimental Therapeutics in Transgenic Mouse Models of Huntington's Disease. *Nat Rev Neurosci* 5, no. 5 (2004): 373–84.

Beister, A., P. Kraus, W. Kuhn, M. Dose, A. Weindl, and M. Gerlach. The N-Methyl-D-Aspartate Antagonist Memantine Retards Progression of Huntington's Disease. *J Neural Transm Suppl*, no. 68 (2004): 117–22.

Bloch, J., A. C. Bachoud-Levi, N. Deglon, J. P. Lefaucheur, L. Winkel, S. Palfi, J. P. Nguyen, C. Bourdet, V. Gaura, P. Remy, et al. Neuroprotective Gene Therapy for Huntington's Disease, Using Polymer-Encapsulated Cells Engineered to Secrete Human Ciliary Neurotrophic Factor: Results of a Phase I Study. *Hum Gene Ther* 15, no. 10 (2004): 968–75.

Bohnen, N. I., R. A. Koeppe, P. Meyer, E. Ficaro, K. Wernette, M. R. Kilbourn, D. E. Kuhl, K. A. Frey, and R. L. Albin. Decreased Striatal Monoaminergic Terminals in Huntington Disease. *Neurology* 54, no. 9 (2000): 1753–59.

Bonelli, R. M., and P. Hofmann. A Systematic Review of the Treatment Studies in Huntington's Disease since 1990. *Expert Opin Pharmacother* 8, no. 2 (2007): 141–53.

Bonelli, R. M., F. A. Mahnert, and G. Niederwieser. Olanzapine for Huntington's Disease: An Open Label Study. *Clin Neuropharmacol* 25, no. 5 (2002): 263–65.

Borovecki, F., L. Lovrecic, J. Zhou, H. Jeong, F. Then, H. D. Rosas, S. M. Hersch, P. Hogarth, B. Bouzou, R. V. Jensen, et al. Genome-Wide Expression Profiling of Human Blood Reveals Biomarkers for Huntington's Disease. *Proc Natl Acad Sci U S A* 102, no. 31 (2005): 11023–28.

Caine, E. D., and I. Shoulson. Psychiatric Syndromes in Huntington's Disease. *Am J Psychiatry* 140, no. 6 (1983): 728–33.

Chiang, M. C., H. M. Chen, Y. H. Lee, H. H. Chang, Y. C. Wu, B. W. Soong, C. M. Chen, Y. R. Wu, C. S. Liu, D. M. Niu, et al. Dysregulation of C/Ebpalpha by Mutant Huntingtin Causes the Urea Cycle Deficiency in Huntington's Disease. *Hum Mol Genet* 16, no. 5 (2007): 483–98.

Close Kirkwood, S., E. Siemers, R. J. Viken, M. E. Hodes, P. M. Conneally, J. C. Christian, and T. Foroud. Evaluation of Psychological Symptoms among Presymptomatic HD Gene Carriers as Measured by Selected Mmpi Scales. *J Psychiatr Res* 36, no. 6 (2002): 377–82.

Dalrymple, A., E. J. Wild, R. Joubert, K. Sathasivam, M. Bjorkqvist, A. Petersen, G. S. Jackson, J. D. Isaacs, M. Kristiansen, G. P. Bates, et al. Proteomic Profiling of Plasma in Huntington's Disease Reveals Neuroinflammatory Activation and Biomarker Candidates. *J Proteome Res* 6, no. 7 (2007): 2833–40.

Davis, M. B., D. Bateman, N. P. Quinn, C. D. Marsden, and A. E. Harding. Mutation Analysis in Patients with Possible but Apparently Sporadic Huntington's Disease. *Lancet* 344, no. 8924 (1994): 714–17.

de Tommaso, M., O. Di Fruscolo, V. Sciruicchio, N. Specchio, C. Cormio, M. F. De Caro, and P. Livrea. Efficacy of Levetiracetam in Huntington Disease. *Clin Neuropharmacol* 28, no. 6 (2005): 280–84.

de Tommaso, M., O. Di Fruscolo, V. Sciruicchio, N. Specchio, and P. Livrea. Two Years' Follow-up of Rivastigmine Treatment in Huntington Disease. *Clin Neuropharmacol* 30, no. 1 (2007): 43–46.

Deng, Y. P., R. L. Albin, J. B. Penney, A. B. Young, K. D. Anderson, and A. Reiner. Differential Loss of Striatal Projection Systems in Huntington's Disease: A Quantitative Immunohistochemical Study. *J Chem Neuroanat* 27, no. 3 (2004): 143–64.

Di Prospero, N. A., and K. H. Fischbeck. Therapeutics Development for Triplet Repeat Expansion Diseases. *Nat Rev Genet* 6, no. 10 (2005): 756–65.

DiFiglia, M., E. Sapp, K. O. Chase, S. W. Davies, G. P. Bates, J. P. Vonsattel, and N. Aronin. Aggregation of Huntingtin in Neuronal Intranuclear Inclusions and Dystrophic Neurites in Brain. *Science* 277, no. 5334 (1997): 1990–93.

Dubinsky, R., and C. Gray. Cyte-I-Hd: Phase I Dose Finding and Tolerability Study of Cysteamine (Cystagon) in Huntington's Disease. *Mov Disord* 21, no. 4 (2006): 530–33.

Duff, K., J. S. Paulsen, L. J. Beglinger, D. R. Langbehn, and J. C. Stout. Psychiatric Symptoms in Huntington's Disease before Diagnosis: The Predict-HD Study. *Biol Psychiatry* 62, no. 12 (2007): 1341–46.

Duyao, M., C. Ambrose, R. Myers, A. Novelletto, F. Persichetti, M. Frontali, S. Folstein, C. Ross, M. Franz, M. Abbott, et al. Trinucleotide Repeat Length Instability and Age of Onset in Huntington's Disease. *Nat Genet* 4, no. 4 (1993): 387–92.

Falush, D., E. W. Almqvist, R. R. Brinkmann, Y. Iwasa, and M. R. Hayden. Measurement of Mutational Flow Implies Both a High New-Mutation Rate for Huntington Disease and Substantial Underascertainment of Late-Onset Cases. *Am J Hum Genet* 68, no. 2 (2001): 373–85.

Feigin, A., K. Kieburtz, K. Bordwell, P. Como, K. Steinberg, J. Sotack, C. Zimmerman, C. Hickey, C. Orme, and I. Shoulson. Functional Decline in Huntington's Disease. *Mov Disord* 10, no. 2 (1995): 211–14.

Feigin, A., and D. Zgaljardic. Recent Advances in Huntington's Disease: Implications for Experimental Therapeutics. *Curr Opin Neurol* 15, no. 4 (2002): 483–89.

Ferrante, R. J., M. F. Beal, N. W. Kowall, E. P. Richardson, Jr., and J. B. Martin. Sparing of Acetylcholinesterase-Containing Striatal Neurons in Huntington's Disease. *Brain Res* 411, no. 1 (1987): 162–66.

Ferrante, R. J., N. W. Kowall, M. F. Beal, E. P. Richardson, Jr., E. D. Bird, and J. B. Martin. Selective Sparing of a Class of Striatal Neurons in Huntington's Disease. *Science* 230, no. 4725 (1985): 561–63.

Gardian, G., S. E. Browne, D. K. Choi, P. Klivenyi, J. Gregorio, J. K. Kubilus, H. Ryu, B. Langley, R. R. Ratan, R. J. Ferrante, et al. Neuroprotective Effects of Phenylbutyrate in the N171-82q Transgenic Mouse Model of Huntington's Disease. *J Biol Chem* 280, no. 1 (2005): 556–63.

Gatchel, J. R., and H. Y. Zoghbi. Diseases of Unstable Repeat Expansion: Mechanisms and Common Principles. *Nat Rev Genet* 6, no. 10 (2005): 743–55.

Ginovart, N., A. Lundin, L. Farde, C. Halldin, L. Backman, C. G. Swahn, S. Pauli, and G. Sedvall. Pet Study of the Pre- and Post-Synaptic Dopaminergic Markers for the Neurodegenerative Process in Huntington's Disease. *Brain* 120 (Pt 3) (1997): 503–14.

Giorgini F, Guidetti P, Nguyen Q, Bennett SC, and Muchowski PJ. A Genomic Screen in Yeast Implicates Kynurenine 3-Monooxygenase as a Therapeutic Target for Huntington Disease. *Nat Genet* 37, no. 5 (2005): 526–31.

Gomez-Anson, B., M. Alegret, E. Munoz, A. Sainz, G. C. Monte, and E. Tolosa. Decreased Frontal Choline and Neuropsychological Performance in Preclinical Huntington Disease. *Neurology* 68, no. 12 (2007): 906–10.

Gutekunst, C. A., S. H. Li, H. Yi, J. S. Mulroy, S. Kuemmerle, R. Jones, D. Rye, R. J. Ferrante, S. M. Hersch, and X. J. Li. Nuclear and Neuropil Aggregates in Huntington's Disease: Relationship to Neuropathology. *J Neurosci* 19, no. 7 (1999): 2522–34.

Halliday, G. M., D. A. McRitchie, V. Macdonald, K. L. Double, R. J. Trent, and E. McCusker. Regional Specificity of Brain Atrophy in Huntington's Disease. *Exp Neurol* 154, no. 2 (1998): 663–72.

Harper, Peter S, ed. *Huntington's Disease, Major Problems in Neurology.* Philadelphia: W.B. Saunders Company, Ltd., 1991.

Harper, S. Q., P. D. Staber, X. He, S. L. Eliason, I. H. Martins, Q. Mao, L. Yang, R. M. Kotin, H. L. Paulson, and B. L. Davidson. RNA Interference Improves Motor and Neuropathological Abnormalities in a Huntington's Disease Mouse Model. *Proc Natl Acad Sci U S A* 102, no. 16 (2005): 5820–25.

Hedreen, J. C., C. E. Peyser, S. E. Folstein, and C. A. Ross. Neuronal Loss in Layers V and VI of Cerebral Cortex in Huntington's Disease. *Neurosci Lett* 133, no. 2 (1991): 257–61.

Hersch, S. M., and R. J. Ferrante. Translating Therapies for Huntington's Disease from Genetic Animal Models to Clinical Trials. *NeuroRx* 1, no. 3 (2004): 298–306.

Hersch, S. M., S. Gevorkian, K. Marder, C. Moskowitz, A. Feigin, M. Cox, P. Como, C. Zimmerman, M. Lin, L. Zhang, et al. Creatine in Huntington Disease Is Safe, Tolerable, Bioavailable in Brain and Reduces Serum 8oh2'dg. *Neurology* 66, no. 2 (2006): 250–52.

Hoth, K. F., J. S. Paulsen, D. J. Moser, D. Tranel, L. A. Clark, and A. Bechara. Patients with Huntington's Disease Have Impaired Awareness of Cognitive, Emotional, and Functional Abilities. *J Clin Exp Neuropsychol* 29, no. 4 (2007): 365–76.

Huntington, G. On Chorea. *J Neuropsychiatry Clin Neurosci* 15, no. 1 (2003): 109–12.

Huntington, G. On Chorea. *Medical and Surgical Reporter: A Weekly Journal* 26, no. 15 (1872): 317–21.

Huntington's Disease Collaborative Research Group. A Novel Gene Containing a Trinucleotide Repeat That Is Expanded and Unstable on Huntington's Disease Chromosomes. The Huntington's Disease Collaborative Research Group. *Cell* 72, no. 6 (1993): 971–83.

Huntington Study Group. Tetrabenazine as Antichorea Therapy in Huntington Disease: A Randomized Controlled Trial. *Neurology* 66, no. 3 (2006): 366–72.

Huntington Study Group. Minocycline Safety and Tolerability in Huntington Disease. *Neurology* 63, no. 3 (2004): 547–49.

Huntington Study Group. Dosage Effects of Riluzole in Huntington's Disease: A Multicenter Placebo-Controlled Study. *Neurology* 61, no. 11 (2003): 1551–56.

Huntington Study Group. A Randomized, Placebo-Controlled Trial of Coenzyme Q10 and Remacemide in Huntington's Disease. *Neurology* 57, no. 3 (2001): 397–404.

Huntington Study Group. Unified Huntington's Disease Rating Scale: Reliability and Consistency. *Mov Disord* 11, no. 2 (1996): 136–42.

Huntington Study Group PHAROS. At Risk for Huntington Disease: The Pharos (Prospective Huntington at Risk Observational Study) Cohort Enrolled. *Arch Neurol* 63, no. 7 (2006): 991–96.

Jongen, P. J., W. O. Renier, and F. J. Gabreels. Seven Cases of Huntington's Disease in Childhood and Levodopa Induced Improvement in the Hypokinetic–Rigid Form. *Clin Neurol Neurosurg* 82, no. 4 (1980): 251–61.

Keene, C. D., J. A. Sonnen, P. D. Swanson, O. Kopyov, J. B. Leverenz, T. D. Bird, and T. J. Montine. Neural Transplantation in Huntington Disease: Long-Term Grafts in Two Patients. *Neurology* 68, no. 24 (2007): 2093–98.

Kieburtz, K., M. MacDonald, C. Shih, A. Feigin, K. Steinberg, K. Bordwell, C. Zimmerman, J. Srinidhi, J. Sotack, J. Gusella, et al. Trinucleotide Repeat Length and Progression of Illness in Huntington's Disease. *J Med Genet* 31, no. 11 (1994): 872–74.

Kirkwood, S. C., E. Siemers, R. Viken, M. E. Hodes, P. M. Conneally, J. C. Christian, and T. Foroud. Longitudinal Personality Changes among Presymptomatic Huntington Disease Gene Carriers. *Neuropsychiatry Neuropsychol Behav Neurol* 15, no. 3 (2002): 192–97.

Kremer, B., P. Goldberg, S. E. Andrew, J. Theilmann, H. Telenius, J. Zeisler, F. Squitieri, B. Lin, A. Bassett, E. Almqvist, et al. A Worldwide Study of the Huntington's Disease Mutation. The Sensitivity and Specificity of Measuring Cag Repeats. *N Engl J Med* 330, no. 20 (1994): 1401–06.

Kremer, B., C. M. Clark, E. W. Almqvist, L. A. Raymond, P. Graf, C. Jacova, M. Mezei, M. A. Hardy, B. Snow, W. Martin, et al. Influence of Lamotrigine on Progression of Early Huntington Disease: A Randomized Clinical Trial. *Neurology* 53, no. 5 (1999): 1000–11.

Kuemmerle, S., C. A. Gutekunst, A. M. Klein, X. J. Li, S. H. Li, M. F. Beal, S. M. Hersch, and R. J. Ferrante. Huntington Aggregates May Not Predict Neuronal Death in Huntington's Disease. *Ann Neurol* 46, no. 6 (1999): 842–49.

Langbehn, D. R., R. R. Brinkman, D. Falush, J. S. Paulsen, and M. R. Hayden. A New Model for Prediction of the Age of Onset and Penetrance for Huntington's Disease Based on Cag Length. *Clin Genet* 65, no. 4 (2004): 267–77.

Lasker, A. G., D. S. Zee, T. C. Hain, S. E. Folstein, and H. S. Singer. Saccades in Huntington's Disease: Initiation Defects and Distractibility. *Neurology* 37, no. 3 (1987): 364–70.

Levesque, M., and A. Parent. The Striatofugal Fiber System in Primates: A Reevaluation of Its Organization Based on Single-Axon Tracing Studies. *Proc Natl Acad Sci U S A* 102, no. 33 (2005): 11888–93.

Li, J. L., M. R. Hayden, S. C. Warby, A. Durr, P. J. Morrison, M. Nance, C. A. Ross, R. L. Margolis, A. Rosenblatt, F. Squitieri, et al. Genome-Wide Significance for a Modifier of Age at Neurological Onset in Huntington's Disease at 6q23-24: The HD Maps Study. *BMC Med Genet* 7 (2006): 71.

Lipe, H., A. Schultz, and T. D. Bird. Risk Factors for Suicide in Huntington's Disease: A Retrospective Case Controlled Study. *Am J Med Genet* 48, no. 4 (1993): 231–33.

Lobo, M. K., S. L. Karsten, M. Gray, D. H. Geschwind, and X. W. Yang. Facs-Array Profiling of Striatal Projection Neuron Subtypes in Juvenile and Adult Mouse Brains. *Nat Neurosci* 9, no. 3 (2006): 443–52.

Lodi, R., A. H. Schapira, D. Manners, P. Styles, N. W. Wood, D. J. Taylor, and T. T. Warner. Abnormal in Vivo Skeletal Muscle Energy Metabolism in Huntington's Disease and Dentatorubropallidoluysian Atrophy. *Ann Neurol* 48, no. 1 (2000): 72–76.

Louis, E. D., K. E. Anderson, C. Moskowitz, D. Z. Thorne, and K. Marder. Dystonia-Predominant Adult-Onset Huntington Disease: Association between Motor Phenotype and Age of Onset in Adults. *Arch Neurol* 57, no. 9 (2000): 1326–30.

Macdonald, V., and G. Halliday. Pyramidal Cell Loss in Motor Cortices in Huntington's Disease. *Neurobiol Dis* 10, no. 3 (2002): 378–86.

Mangiarini, L., K. Sathasivam, M. Seller, B. Cozens, A. Harper, C. Hetherington, M. Lawton, Y. Trottier, H. Lehrach, S. W. Davies, et al. Exon 1 of the HD Gene with an Expanded Cag Repeat Is Sufficient to Cause a Progressive Neurological Phenotype in Transgenic Mice. *Cell* 87, no. 3 (1996): 493–506.

Marder, K., S. Sandler, A. Lechich, J. Klager, and S. M. Albert. Relationship between Cag Repeat Length and Late-Stage Outcomes in Huntington's Disease. *Neurology* 59, no. 10 (2002): 1622–24.

Marder, K., H. Zhao, R. H. Myers, M. Cudkowicz, E. Kayson, K. Kieburtz, C. Orme, J. Paulsen, J. B. Penney, Jr., E. Siemers, et al. Rate of Functional Decline in Huntington's Disease. Huntington Study Group. *Neurology* 54, no. 2 (2000): 452–58.

Marshall, F. J. Clinical Features and Treatment of Huntington's Disease. In *Movement Disorders: Neurologic Principles & Practice*, edited by R. L. Watts and W. C. Koller, 589–601. New York: McGraw-Hill, 2004.

Nakamura, K., and M. J. Aminoff. Huntington's Disease: Clinical Characteristics, Pathogenesis and Therapies. *Drugs Today (Barc)* 43, no. 2 (2007): 97–116.

Nance, M. A. Comprehensive Care in Huntington's Disease: A Physician's Perspective. *Brain Res Bull* 72, no. 2–3 (2007): 175–78.

O'Suilleabhain, P., and R. B. Dewey, Jr. A Randomized Trial of Amantadine in Huntington Disease. *Arch Neurol* 60, no. 7 (2003): 996–98.

Paulsen, J. S., N. Butters, J. R. Sadek, S. A. Johnson, D. P. Salmon, N. R. Swerdlow, and M. R. Swenson. Distinct Cognitive Profiles of Cortical and Subcortical Dementia in Advanced Illness. *Neurology* 45, no. 5 (1995): 951–56.

Paulsen, J. S., M. Hayden, J. C. Stout, D. R. Langbehn, E. Aylward, C. A. Ross, M. Guttman, M. Nance, K. Kieburtz, D. Oakes, et al. Preparing for Preventive Clinical Trials: The PREDICT-HD Study. *Arch Neurol* 63, no. 6 (2006): 883–90.

Paulsen, J. S., D. R. Langbehn, J. C. Stout, E. Aylward, C. A. Ross, M. Nance, M. Guttman, S. Johnson, M. MacDonald, L. J. Beglinger, et al. Detection of Huntington's Disease Decades before Diagnosis: The Predict-HD Study. *J Neurol Neurosurg Psychiatry* 79, no. 8 (2008): 874–80.

Paulsen, J. S., R. E. Ready, J. M. Hamilton, M. S. Mega, and J. L. Cummings. Neuropsychiatric Aspects of Huntington's Disease. *J Neurol Neurosurg Psychiatry* 71, no. 3 (2001): 310–14.

Quaid, K. A., and M. Morris. Reluctance to Undergo Predictive Testing: The Case of Huntington Disease. *Am J Med Genet* 45 (1993): 41–45.

Quarrell, O. W., A. S. Rigby, L. Barron, Y. Crow, A. Dalton, N. Dennis, A. E. Fryer, F. Heydon, E. Kinning, A. Lashwood, et al. Reduced Penetrance Alleles for Huntington's Disease: A Multi-Centre Direct Observational Study. *J Med Genet* 44, no. 3 (2007): e68.

Reiner, A., R. L. Albin, K. D. Anderson, C. J. D'Amato, J. B. Penney, and A. B. Young. Differential Loss of Striatal Projection Neurons in Huntington Disease. *Proc Natl Acad Sci U S A* 85, no. 15 (1988): 5733–37.

Ribai, P., K. Nguyen, V. Hahn-Barma, I. Gourfinkel-An, M. Vidailhet, A. Legout, C. Dode, A. Brice, and A. Durr. Psychiatric and Cognitive Difficulties as Indicators of Juvenile Huntington Disease Onset in 29 Patients. *Arch Neurol* 64, no. 6 (2007): 813–19.

Riley, B. E., and H. T. Orr. Polyglutamine Neurodegenerative Diseases and Regulation of Transcription: Assembling the Puzzle. *Genes Dev* 20, no. 16 (2006): 2183–92.

Rohrer, D., D. P. Salmon, J. T. Wixted, and J. S. Paulsen. The Disparate Effects of Alzheimer's Disease and Huntington's Disease on Semantic Memory. *Neuropsychology* 13, no. 3 (1999): 381–88.

Rosas, H. D., N. D. Hevelone, A. K. Zaleta, D. N. Greve, D. H. Salat, and B. Fischl. Regional Cortical Thinning in Preclinical Huntington Disease and Its Relationship to Cognition. *Neurology* 65, no. 5 (2005): 745–47.

Rosas, H. D., A. K. Liu, S. Hersch, M. Glessner, R. J. Ferrante, D. H. Salat, A. van der Kouwe, B. G. Jenkins, A. M. Dale, and B. Fischl. Regional and Progressive Thinning of the Cortical Ribbon in Huntington's Disease. *Neurology* 58, no. 5 (2002): 695–701.

Rosenblatt, A., N. G. Ranen, D. C. Rubinsztein, O. C. Stine, R. L. Margolis, M. V. Wagster, M. W. Becher, A. E. Rosser, J. Leggo, J. R. Hodges, et al. Patients with Features Similar to Huntington's Disease, without Cag Expansion in Huntingtin. *Neurology* 51, no. 1 (1998): 215–20.

Ross, C. A., and M. Poirier. Opinion: What is the Role of Protein Aggregation in Neurodegeneration? *Nat Rev Mol Cell Biol* 6, no. 11 (2005): 891–98.

Rubinsztein, D. C., J. Leggo, M. Chiano, A. Dodge, G. Norbury, E. Rosser, and D. Craufurd. Genotypes at the Glur6 Kainate Receptor Locus Are Associated with Variation in the Age of Onset of Huntington Disease. *Proc Natl Acad Sci U S A* 94, no. 8 (1997): 3872–76.

Saft, C., T. Lauter, P. H. Kraus, H. Przuntek, and J. E. Andrich. Dose-Dependent Improvement of Myoclonic Hyperkinesia Due to Valproic Acid in Eight Huntington's Disease Patients: A Case Series. *BMC Neurol* 6 (2006): 11.

Saft, C., J. Zange, J. Andrich, K. Muller, K. Lindenberg, B. Landwehrmeyer, M. Vorgerd, P. H. Kraus, H. Przuntek, and L. Schols. Mitochondrial Impairment in Patients and Asymptomatic Mutation Carriers of Huntington's Disease. *Mov Disord* 20, no. 6 (2005): 674–79.

Sarkar, S., E. O. Perlstein, S. Imarisio, S. Pineau, A. Cordenier, R. L. Maglathlin, J. A. Webster, T. A. Lewis, C. J. O'Kane, S. L. Schreiber, et al. Small Molecules Enhance Autophagy and Reduce Toxicity in Huntington's Disease Models. *Nat Chem Biol* 3, no. 6 (2007): 331–38.

Sawa, A., G. W. Wiegand, J. Cooper, R. L. Margolis, A. H. Sharp, J. F. Lawler, Jr., J. T. Greenamyre, S. H. Snyder, and C. A. Ross. Increased Apoptosis of Huntington Disease Lymphoblasts Associated with Repeat Length-Dependent Mitochondrial Depolarization. *Nat Med* 5, no. 10 (1999): 1194–98.

Schneider, S. A., R. H. Walker, and K. P. Bhatia. The Huntington's Disease-Like Syndromes: What to Consider in Patients with a Negative Huntington's Disease Gene Test. *Nat Clin Prac Neurol* 3, no. 9 (2007): 517–25.

Sieradzan, K. A., and D. M. Mann. The Selective Vulnerability of Nerve Cells in Huntington's Disease. *Neuropathol Appl Neurobiol* 27, no. 1 (2001): 1–21.

Siesling, S., J. P. van Vugt, K. A. Zwinderman, K. Kieburtz, and R. A. Roos. Unified Huntington's Disease Rating Scale: A Follow Up. *Mov Disord* 13, no. 6 (1998): 915–19.

Sorensen, S. A., and K. Fenger. Causes of Death in Patients with Huntington's Disease and in Unaffected First Degree Relatives. *J Med Genet* 29, no. 12 (1992): 911–14.

Sotrel, A., P. A. Paskevich, D. K. Kiely, E. D. Bird, R. S. Williams, and R. H. Myers. Morphometric Analysis of the Prefrontal Cortex in Huntington's Disease. *Neurology* 41, no. 7 (1991): 1117–23.

Sotrel, A., R. S. Williams, W. E. Kaufmann, and R. H. Myers. Evidence for Neuronal Degeneration and Dendritic Plasticity in Cortical Pyramidal Neurons of Huntington's Disease: A Quantitative Golgi Study. *Neurology* 43, no. 10 (1993): 2088–96.

Stevenson, Charles S. A Biography of George Huntington, M.D. In *Bulletin of the Institute of the History of Medicine*, edited by Henry E. Sigerist, 53–76. Baltimore: The Johns Hopkins University, 1934.

Stine, O. C., N. Pleasant, M. L. Franz, M. H. Abbott, S. E. Folstein, and C. A. Ross. Correlation between the Onset Age of Huntington's Disease and Length of the Trinucleotide Repeat in It-15. *Hum Mol Genet* 2, no. 10 (1993): 1547–49.

Strand, A. D., A. K. Aragaki, D. Shaw, T. Bird, J. Holton, C. Turner, S. J. Tapscott, S. J. Tabrizi, A. H. Schapira, C. Kooperberg, et al. Gene Expression in Huntington's Disease Skeletal Muscle: A Potential Biomarker. *Hum Mol Genet* 14, no. 13 (2005): 1863–76.

Tippett, L. J., H. J. Waldvogel, S. J. Thomas, V. M. Hogg, W. van Roon-Mom, B. J. Synek, A. M. Graybiel, and R. L. Faull. Striosomes and Mood Dysfunction in Huntington's Disease. *Brain* 130 (Pt 1) (2007): 206–21.

Todi, S. V., A. J. Williams, and H. L. Paulson. Polyglutamine Disorders Including Huntington's Disease. In *Molecular Neurology*, edited by Stephen G. Waxman, 257–75. Burlington, VT: Elsevier Academic Press, 2007.

Tsuang, D., E. W. Almqvist, H. Lipe, F. Strgar, L. DiGiacomo, D. Hoff, C. Eugenio, M. R. Hayden, and T. D. Bird. Familial Aggregation of Psychotic Symptoms in Huntington's Disease. *Am J Psychiatry* 157, no. 12 (2000): 1955–59.

van Oostrom, J. C., R. P. Maguire, C. C. Verschuuren-Bemelmans, L. Veenma-van der Duin, J. Pruim, R. A. Roos, and K. L. Leenders. Striatal Dopamine D2 Receptors, Metabolism, and Volume in Preclinical Huntington Disease. *Neurology* 65, no. 6 (2005): 941–43.

Van Raamsdonk, J. M., Z. Murphy, D. M. Selva, R. Hamidizadeh, J. Pearson, A. Petersen, M. Bjorkqvist, C. Muir, I. R. Mackenzie, G. L. Hammond, et al. Testicular Degeneration in Huntington Disease. *Neurobiol Dis* 26, no. 3 (2007): 512–20.

Verhagen Metman, L., M. J. Morris, C. Farmer, M. Gillespie, K. Mosby, J. Wuu, and T. N. Chase. Huntington's Disease: A Randomized, Controlled Trial Using the NMDA-Antagonist Amantadine. *Neurology* 59, no. 5 (2002): 694–99.

Vonsattel, J. P., and M. DiFiglia. Huntington Disease. *J Neuropathol Exp Neurol* 57, no. 5 (1998): 369–84.

Vonsattel, J. P., R. H. Myers, T. J. Stevens, R. J. Ferrante, E. D. Bird, and E. P. Richardson, Jr. Neuropathological Classification of Huntington's Disease. *J Neuropathol Exp Neurol* 44, no. 6 (1985): 559–77.

Weeks, R. A., P. Piccini, A. E. Harding, and D. J. Brooks. Striatal D1 and D2 Dopamine Receptor Loss in Asymptomatic Mutation Carriers of Huntington's Disease. *Ann Neurol* 40, no. 1 (1996): 49–54.

Wei, H., Z. H. Qin, V. V. Senatorov, W. Wei, Y. Wang, Y. Qian, and D. M. Chuang. Lithium Suppresses Excitotoxicity-Induced Striatal Lesions in a Rat Model of Huntington's Disease. *Neuroscience* 106, no. 3 (2001): 603–12.

WFN Research Group on Huntington's Chorea. Guidelines for the Molecular Genetics Predictive Test in Huntington's Disease. International Huntington Association (IHA) and the World Federation of Neurology (WFN) Research Group on Huntington's Chorea. *Neurology* 44, no. 8 (1994): 1533–36.

Zeng, W., T. Gillis, M. Hakky, L. Djousse, R. H. Myers, M. E. MacDonald, and J. F. Gusella. Genetic Analysis of the Grik2 Modifier Effect in Huntington's Disease. *BMC Neurosci* 7 (2006): 62.

2 Huntington's Disease Pathogenesis
Mechanisms and Pathways

Albert R. La Spada, Patrick Weydt, and Victor V. Pineda

CONTENTS

INTRODUCTION

The discovery in 1993 of the gene responsible for Huntington's disease (HD) represented a crucial turning point in the HD research field. At the time of the discovery, no one could predict that HD would belong to a large class of inherited neurological diseases all caused by the same type of genetic mutation (i.e., polyglutamine [polyQ] expansion) or that the mechanistic basis of HD (i.e., protein misfolding) would emerge as a common theme linking together all the major neurodegenerative disorders, including Alzheimer's disease (AD), Parkinson's disease (PD), and the prion diseases. The study of how the mutant HD gene product, an unusually large 3,144 amino acid protein (huntingtin [htt]) with few recognizable motifs or obvious functional domains that results in the degeneration and death of neurons in the striatum and cortex, has been an enormous undertaking. Indeed, a PubMed search using the term "huntingtin" yields 1,124 hits at the time of writing this chapter. Suffice it to say that dozens of theories of pathogenesis have been proposed and studied. The goal of this chapter will be to present some of the most enduring lines of investigation, with an emphasis on the latest developments, and to highlight emerging notions likely to drive basic research on HD in the future.

HD displays the genetic feature of anticipation, defined as earlier disease onset and more rapid disease progression in successive generations of a pedigree segregating the disease gene. This feature was an important clue for discovery of the causal mutation, as a trinucleotide repeat expansion encoding an elongated glutamine tract in the htt protein was determined to be responsible for HD in 1993, and a relationship between the length of the expanded glutamine tract and the severity of the HD phenotype was uncovered at that time [1]. HD is one of nine inherited neurodegenerative disorders caused by CAG trinucleotide repeats that expand to produce disease by encoding elongated polyQ tracts in their respective protein products. Included in this CAG/polyQ repeat disease class are spinal and bulbar muscular atrophy (SBMA), dentatorubral-pallidoluysian atrophy (DRPLA), and six forms of spinocerebellar ataxia (SCA1, SCA2, SCA3, SCA6, SCA7, and SCA17) [2]. Based on work done on all these disorders, investigators have learned that once glutamine tracts exceed the mid-30s, the polyQ tract adopts a novel conformation that is pathogenic. An anti-polyQ antibody (1C2) can specifically detect this structural transformation, as it will only bind to disease-length polyQ tracts from patients with different polyQ diseases [3]. The transition of polyQ-expanded proteins into this misfolded conformer is the crux of the molecular pathology in these disorders. Once in this conformation, however, it is unclear how polyQ tract expansions mediate the patterns of neuronal cell loss seen in each disease, as most of the polyQ disease gene products show overlapping patterns of expression within the central nervous system (CNS) but restricted pathology. In the case of HD, molecular explanations for disease pathogenesis must account for the selective vulnerability of the medium spiny neurons of the striatum and certain neuron subsets in the cortex.

PROTEIN AGGREGATION AND DEGRADATION

A major turning point in the HD and polyQ disease field came in 1997 when independent groups detected visible proteinaceous aggregates (or inclusion bodies) in the nuclei of neurons from patients with SCA3 and from patients and mice with HD [4–6]. These neuronal intranuclear inclusions (first abbreviated "NIIs" and then later "NIs") appear before the onset of disease in mouse models of HD, suggesting a primary role in pathogenesis. Because the NIs displayed immunoreactivity to antibodies directed against the polyQ disease protein and the expanded polyQ tract epitope, the NIs were thought to be "aggregates" of the mutant disease protein. This led to the "aggregation theory" of polyQ disease pathogenesis that posited aggregation of expanded polyQ tracts as the crucial step in the cascade of events that leads to neurodegeneration in these diseases. As the kinetics of polyQ tract self-aggregation increases with the lengthening of the glutamine repeat [7], paralleling the genotype-phenotype relationship documented in HD, the aggregation theory accumulated a number of strong proponents. However, at the same time, other lines of investigation began to suggest otherwise. When the Orr and Zoghbi groups crossed their SCA1 transgenic mice with mice lacking E6-AP ubiquitin ligase function, they observed limited aggregate formation but with an earlier onset of SCA1 neurodegeneration [8]. HD yeast artificial chromosome (YAC) transgenic mice were then noted to develop a motoric phenotype and obvious neurodegeneration in the absence of protein aggregates [9]. Thus, a contentious debate ensued over the role of aggregate formation in polyQ disease pathogenesis—with some workers espousing the view that aggregates were responsible for disease pathology, others suggesting that the aggregation process was a protective coping mechanism of the cell and thereby beneficial, and still others insisting that aggregates were incidental and irrelevant. This debate was complicated by the fact that the absence of aggregates at the light microscope level could occur in the presence of so-called "microaggregates" at the electron microscope level [10].

To deconstruct the nature of the microaggregates, investigators have used a variety of biophysical approaches—including transmission electron microscopy, Fourier transform infrared spectroscopy, and atomic force microscopy—to dissect the process of htt exon 1 peptide aggregation, and have found evidence for a number of sequential morphological and structural intermediates [11, 12]. Many have proposed that misfolding of expanded polyglutamine tracts into insoluble aggregates involves transition through a number of steps, including the formation of oligomers, then assembly of oligomers into protofibrils, and followed by protofibril assembly into fibrils (Figure 2.1). The importance of such a model is that it accounts for how aggregation can at the same time be toxic and protective because views of aggregation had been dramatically oversimplified. To differentiate the toxicity of oligomeric precursor forms from the ultimate visible aggregates in an unbiased fashion, one group developed an automated microscope system for temporally tracking polyQ-htt-expressing cells over time and found that neuron cell death could not be attributed to visible inclusion formation [13]. Rather, levels of diffuse polyQ-htt expression were a significant negative predictor of neuron survival, and polyQ-htt neurons *lacking* visible inclusions had a higher cumulative risk of cell death, suggesting that visible aggregate formation can be protective. Hence work done on HD and other polyQ diseases suggests that aggregates may not be toxic *per se*, but rather may signify the presence of misfolded proteins whose toxic action is playing out in the soluble phase and/or at the level of oligomers or protofibril structures ("microaggregates").

Initial studies of protein aggregates in HD and the other polyQ diseases documented the presence of molecular chaperones and components of the ubiquitin-proteasome system (UPS) in polyQ inclusions by demonstrating intense immunoreactivity of the aggregates with antibodies directed against such factors [14]. As accurate folding of proteins is essential for the proper functioning of all cells, eukaryotic cells possess a highly efficient multistep protein quality control system that can eliminate misfolded proteins. Molecular chaperones are small scaffolding proteins that can facilitate proper folding of their client proteins or tag them for degradation [15]. Many molecular chaperones are heat shock proteins (Hsp), as their expression is induced by increased temperature [16], an environmental stress

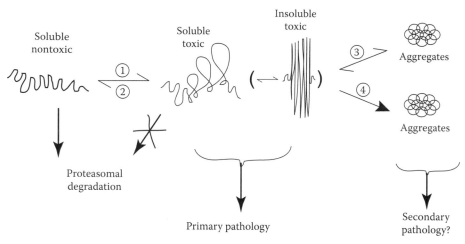

FIGURE 2.1 Model for polyglutamine neurotoxicity. Soluble (nontoxic) polyglutamine-expanded protein exists in equilibrium with a soluble toxic conformer/oligomer. Upon transition to the misfolded conformer, the polyglutamine-expanded protein can no longer be readily degraded. Note that certain processes, such as proteolytic cleavage (1), may favor transition to the misfolded conformer, whereas other processes, such as molecular chaperone interaction (2), likely favor refolding into a soluble, nontoxic, degradable conformation. Soluble toxic species become insoluble and ultimately coalesce into visible aggregates. Formation of visible aggregates may be protective (3), because they sequester soluble/insoluble toxic oligomers away; however, it is possible that excessive aggregate formation may cause secondary pathology. Current hypotheses of polyglutamine disease pathogenesis postulate that the microaggregates (i.e., the soluble and insoluble toxic species) are principally responsible for the neurotoxicity. (From Grote, S. K. and La Spada, A. R., *Cytogenet Genome Res* 100, 164, 2003. With permission.)

that elevates the likelihood of protein misfolding. The UPS is the main intracellular degradation pathway to remove short-lived proteins and to eliminate misfolded proteins [14]. The proteasome component of the UPS consists of a 19S entry ring where peptide unfolding occurs to permit delivery of the degradable substrate to a 20S barrel core with peptidase activity. A three-step conjugation system for ubiquitination of intended substrates is also required for the proper operation of this protein degradation pathway. Numerous studies have shown that inhibition of the UPS with pharmacological agents predisposes neuronal and non-neuronal cells to polyQ toxicity, whereas enhanced molecular chaperone activity (especially the Hsp40–Hsp70 combination) significantly ameliorates polyQ neurotoxicity [14]. However, although molecular chaperone and UPS function are important factors in countering misfolded polyQ protein toxicity, many aggregate-prone proteins, such as polyQ proteins, are inefficiently degraded by the proteasome [17–19]. Failure of adequate degradation of aggregate-prone proteins activates alternative protein turnover pathways in the cell, particularly macroautophagy (typically referred to simply as "autophagy"). Autophagy is a degradative process that begins with engulfment of cytosolic materials and/or organelles and progresses through a series of steps involving production of a double membrane-bound structure, culminating in the delivery of the engulfed material to lysosomes [20]. In the CNS, basal levels of autophagy are required for the continued health and normal function of neurons, as conditional inactivation of the autophagy pathway in neural cells in mice yields neuronal dysfunction and neurodegeneration characterized by the accumulation of proteinaceous material [21, 22]. A series of studies from the Rubinzstein laboratory has strongly implicated autophagy activation as an important compensatory pathway for countering htt toxicity in cell culture, *Drosophila*, and transgenic mouse models [23, 24]. Whether pharmacological induction of autophagy can be achieved in the CNS as a therapeutic intervention for HD and related neurological proteinopathies remains to be determined.

PROTEOLYTIC CLEAVAGE

Studies of HD suggest that proteolytic cleavage of the htt protein is a key step in the neurotoxicity pathway. The *htt* gene encodes a protein of 3,144 amino acids with the glutamine tract beginning at codon 18. An analysis of protein aggregates from *in vitro* models, *in vivo* models, and HD patients indicated that glutamine- and amino-terminal epitopes are present in nuclear, cytosolic, and axonal aggregates [25]. In a landmark study, the Bates group used a 1.6-kb fragment of the *huntingtin* gene, containing only the first 2% of the huntingtin coding region, to derive lines of transgenic mice (R6/1 and R6/2) that showed a neurological phenotype that resembled HD [26]. This study demonstrated that a tiny amino-terminal portion of the htt protein (including the polyQ tract) was sufficient to produce an HD-like illness in mice. Other studies of the htt protein have shown that htt is a substrate for proteolytic cleavage by caspases and calpains [27–29]. The elaboration of an amino-terminal truncation product in HD and in other polyQ diseases has led to the "toxic fragment hypothesis" [30]. According to this hypothesis, cleavage of the polyQ disease protein yields a polyQ-containing peptide that represents the principal toxic species at the molecular level. As the toxic fragment hypothesis is based on proteolytic cleavage of polyQ disease proteins by enzymes such as caspases and calpains, investigators have sought to define specific sites of proteolytic cleavage and the cleavage enzymes. Studies of HD mouse models indicate that not all amino-terminal proteolytic cleavage fragments are toxic, as one HD YAC mouse model ("shortstop") expresses a polyQ-expanded amino-terminal truncation fragment that yields pronounced aggregates in the CNS but no neuronal toxicity [31]. In 2006, the Hayden group convincingly demonstrated that polyQ-expanded huntingtin protein with a cleavage site mutation (*putatively* for caspase-6) was incapable of causing neurotoxicity in HD YAC transgenic mice [32]. This work strongly supports a role for a specific proteolytic cleavage of htt as a required step in the HD pathogenic cascade. Identification of the enzyme mediating this cleavage will be an important goal toward a potential treatment for HD.

TRANSCRIPTIONAL DYSREGULATION

The necessity of nuclear localization for HD disease pathogenesis highlighted nuclear pathology as a likely early step in the neurotoxicity cascade [33]. As glutamine tracts and glutamine-rich regions often occur in transcription factors and permit functional protein–protein interactions to produce transcription activation complexes in species as diverse as yeast, fruit flies, chicken, and humans, a hypothesis of "transcription interference" or "transcription dysregulation" was formulated. According to this hypothesis, polyQ-expanded disease proteins (or peptides), accumulating in the nucleus inappropriately, interact with transcription factors and regulators to disrupt normal transcriptional functions [34]. Studies of htt have implicated a number of important transcription factors and coactivators. One of the most studied and strongly implicated transcription factors is CREB-binding protein (CBP), a transcriptional coactivator involved in the regulation of multiple genes through its intrinsic histone acetyltransferase activity (that remodels chromatin to allow the transcription machinery to access target genes) [35, 36]. A number of studies have shown htt interference of CBP-mediated transcription in a polyQ length-dependent fashion [37–39]. Consequently, drugs that block histone deacetylation, and thereby favor the outcome of CBP action (i.e., histone acetylation), are therapeutically beneficial in *Drosophila* and mouse models of HD [40–42]. Sp1, a ubiquitously expressed DNA binding factor that recruits the transcription factor IID (TFIID) complex, and TAFII130, a factor that mediates transcription activation complex assembly, have also been identified as targets of mutant htt protein [43]. Interaction of polyQ-expanded htt with CBP, Sp1, and TAFII130 has been shown to occur in the nucleus and to involve the amino-terminal region of htt. Roles for CREB-CBP and Sp1 gene targets in maintenance of normal neuronal function are suggested by other studies [44–47], supporting the conclusion that interference with CBP and Sp1 action could have deleterious effects on neuronal health and survival. Loss of brain-derived neurotrophic factor (BDNF) expression, via aberrant repressor protein localization, may also contribute to the transcription dysregulation in HD [48]. Finally, we and others have documented interference of htt with the peroxisome proliferator-activated receptor γ coactivator 1α (PGC-1α) transcription factor [49, 50], a regulatory protein with a crucial role in modulating mitochondrial number and function [51, 52].

MECHANISMS OF HUNTINGTIN-MEDIATED TRANSCRIPTIONAL INTERFERENCE

Remodeling of chromatin permits the RNA polymerase II complex to bind and initiate transcript synthesis at actively expressed gene loci. This occurs primarily through the covalent addition of acetyl moieties to lysine residues in the tails of core histone proteins—a process that is mediated by enzymes with histone acetyltransferase (HAT) activity [53]. CBP has intrinsic HAT activity, whereas other transcription factors recruit multiprotein coactivator complexes that contain one or more HAT components. PolyQ-expanded htt inhibits the HAT activities of CBP and the p300/CBP-associated factor (P/CAF). This effectively disrupts transcription of CBP targets and any other genes under the control of P/CAF-dependent transcription factors [54]. Transcriptional interference by mutant htt can also occur when key elements of the transcriptional machinery are not properly reconstituted at the site of transcription. Sp1 binds to specific *cis*-elements and directs core components of transcriptional complexes, such as TFIID, TATA box-binding protein (TBP), and other TBP-associated factors (TAFs), to initiate transcription at the Sp1 target gene [43]. Mutant htt disrupts the interaction between Sp1 and components of the TFIID and TFIIF transcriptional complex, thus causing transcriptional dysregulation of Sp1-dependent genes [55]. Finally, a different mechanism is observed when polyQ-expanded htt disrupts BDNF transcription. Normally, cytosolic sequestration of the RE1-silencing transcription factor/neuron-restrictive silencing factor (REST/NRSF) by full-length htt prevents REST/NRSF from binding to the neuron-restrictive silencer element (NRSE) found in the regulatory region of the *BDNF* gene [56]. Therefore, normal htt should sequester REST in the cytosol and allow transcription of *BDNF*; however, mutant htt, as it accumulates in the nucleus, may trap REST there, allowing it to repress NRSE-containing genes such as *BDNF* (Figure 2.2).

EVIDENCE FOR MITOCHONDRIAL ABNORMALITIES AND DEFECTIVE ENERGY METABOLISM IN HD

Neurons in the brain have enormous demands for continued production of high-energy phosphate-bonded compounds such as ATP. In 1993, Beal et al. [57] reported that long-term administration of a mitochondrial toxin, 3-nitropropionic acid, resulted in a selective loss of medium spiny neurons in the striatum. This provocative finding suggested that mitochondrial dysfunction may underlie HD disease pathogenesis and perhaps account for the cell-type specificity in this neurodegenerative disorder. Follow-up studies performed on HD patient material have documented significant

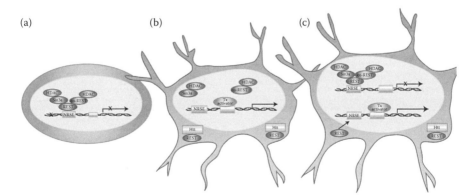

FIGURE 2.2 (See color insert following page 172.) Model for polyglutamine-expanded huntingtin transcription interference of *BDNF* gene expression. Promoters containing a neuron-restrictive silencing element (a) are bound by transcription repressors that prevent expression of downstream genes, such as *BDNF*, in non-neural cell types. In normal neurons (b), normal huntingtin protein is localized to the cytosol and binds the transcription repressor REST there, allowing *BDNF* gene expression to occur. However, in HD neurons (c), polyglutamine-expanded huntingtin protein accumulates in the nucleus and does not sequester REST in the cytosol. REST thus inappropriately enters neuronal nuclei and represses neuronal expression of *BDNF* in HD. (From Thompson, L. M., *Nat Genet* 35, 13, 2003. With permission.)

reductions in the enzymatic activities of complexes II, III, and IV of the mitochondrial oxidative phosphorylation pathway in caudate and putamen [58, 59] but have not detected such alterations in HD cerebella or fibroblasts [60]. Additional work has documented striatal-specific decreases in aconitase activity, a likely target of Ca^{++}-dependent, free radical-producing intramitochondrial enzymes [60]. Positron emission tomography (PET) scan analysis of HD patients also strongly supports the hypothesis of defective energy metabolism, as decreased rates of cerebral glucose metabolism are apparent in certain regions of the cortex and throughout the striatum [61]. Magnetic resonance spectroscopy corroborates such findings, revealing elevated lactate levels in striata of HD patients [62].

As mitochondrial energy production and metabolic pathways supply energy for ion exchange pumps, whose function is to maintain an electrochemical gradient across the mitochondrial membrane, defective energy metabolism could translate into an enhanced susceptibility of HD mitochondria to undergo depolarization. A number of studies have evaluated this and have indeed found that mitochondria from HD patients are exquisitely sensitive to depolarizing stresses. In one study, treatment of HD lymphoblasts with complex IV inhibitors resulted in mitochondrial depolarization and apoptotic cell death involving caspase activation [63]. In an independent study, electrical measurements of HD lymphoblast mitochondria yielded lower than normal membrane potentials and depolarization in response to modest Ca^{++} loads [64]. As mitochondrial membrane depolarization results in caspase activation and cleavage of htt protein appears to be mediated in part by caspases (and Ca^{2+}-activated calpains), mitochondrial dysfunction may represent an early step in the HD neurotoxicity cascade.

LINKING TRANSCRIPTION INTERFERENCE WITH MITOCHONDRIAL BIOENERGETIC ABNORMALITIES

Although more than a decade of study of HD has consistently implicated mitochondrial dysfunction as a central feature of disease pathogenesis, the molecular basis of the mitochondrial abnormality has remained elusive. At the same time, evidence continues to accumulate that nuclear pathology is likely preeminent in the polyQ diseases and that polyQ diseases displaying nuclear accumulation of mutant peptides are in reality "transcriptionopathies" [34]. Recent work on PGC-1α has suggested a connection between htt-mediated transcription dysregulation and mitochondrial abnormalities. PGC-1α is a transcription coactivator that lacks any apparent chromatin-remodeling enzymatic activity such as HAT activity [52]. It was cloned from a brown fat library and subsequently found to be a master regulator of complex transcriptional programs involved in the response to cold temperatures and to high caloric intake through its coactivation of peroxisome proliferator-activated receptor γ (PPARγ)-mediated gene expression. PGC-1α stimulates mitochondrial oxidative phosphorylation respiration and mitochondrial uncoupled respiration in brown fat and skeletal muscle cells [65]. Interestingly, PGC-1α also coordinates mitochondrial biogenesis by up-regulating the expression of the nuclear respiratory factors (NRF)-1 and NRF-2 [66]. After boosting the expression level of NRF-1,

PGC-1α directly interacts with NRF-1 to coactivate expression of mitochondrial transcription factor A (mtTFA), whose function is to transcribe and replicate the mitochondrial genome, permitting the production of increased numbers of mitochondria [51]. PPARγ also participates in mitochondrial biogenesis with PGC-1α by driving the expression of mitochondrial fatty acid oxidation enzymes [67]. All these findings indicate that PGC-1α is the key regulatory node in a complex network of transcription programs that culminate in adaptive thermogenesis or mitochondrial biogenesis. PGC-1α is very highly expressed in brain, where its role in mitochondrial biogenesis and uncoupling protein expression may be critical to neuron health and survival.

To determine the role of PGC-1α in metabolism and thermoregulation, the Spiegelman laboratory generated PGC-1α knockout mice [68]. Although these workers anticipated that PGC-1α$^{-/-}$ mice would display a predisposition to obesity, they instead noted that the mice were lean. The explanation for their enigmatic leanness turned out to be their phenotype of pronounced hyperactivity. Further analysis of the PGC-1α$^{-/-}$ mice revealed neurological abnormalities, including myoclonus, dystonia, exaggerated startle responses, and clasping (which is a stereotypical finding in all polyQ and HD mouse models). Neuropathology examination of the PGC-1α$^{-/-}$ mice yielded evidence of degeneration in cortex, thalamus, basal ganglia, and hippocampus, with the most pronounced degeneration in the striatum. The striatal degeneration was spongiform in nature and resulted from a significant drop-out of neurons. Interestingly, real-time reverse transcription-polymerase chain reaction (RT-PCR) analysis of hyperactive PGC-1α$^{-/-}$ mice documented significant reductions in the expression of mitochondrial genes. In addition to their phenotype of hyperactivity and striatal neurodegeneration, the PGC-1α$^{-/-}$ mice displayed reduced thermogenic capacity as a result of a failure of induction of uncoupling protein 1 (UCP1) gene expression. An independently generated PGC-1α knockout mouse model also developed a neurological phenotype with degeneration of the striatum, although this PGC-1α knockout model could properly regulate its body temperature when subjected to cold challenge for most of its lifespan [69]. In 2006, we reported that HD N171-82Q transgenic mice display profound thermoregulatory and metabolic defects [50]. Our discovery of deranged thermoregulation in HD mice led us to evaluate the PGC-1α pathway in the brain and the periphery of these HD mice and to survey PGC-1α-regulated target genes in the striatum of HD patients. We documented altered mitochondrial function in brown adipose tissue from HD N171-82Q mice and noted that the expression levels of PGC-1α target genes, whose protein products mediate oxidative metabolism in the mitochondria, were significantly reduced [50]. When we analyzed the expression levels of PGC-1α target genes in the striatum of HD patients by gene set enrichment analysis (GSEA) of microarray data, we observed significant reductions in 24 of these 26 PGC-1α target genes and confirmed these findings by RT-PCR (Figure 2.3). Thus, reduced expression and function of PGC-1α and its targets may be central to HD striatal degeneration, and PGC-1α transcription interference may provide a crucial link between two established aspects of HD molecular pathology: transcription dysregulation and mitochondrial dysfunction.

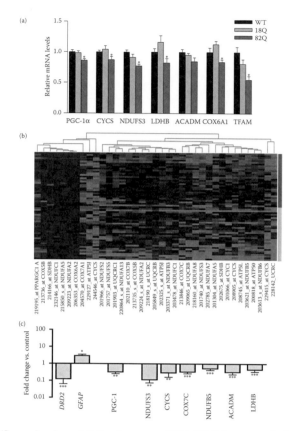

FIGURE 2.3 **(See color insert following page 172.)** PGC-1α transcription interference in HD mice and HD patients. (a) Real-time RT-PCR analysis of striatal RNAs from HD 82Q mice (red), 18Q mice (gray), and wild-type mice (black) reveals decreased mitochondrial gene expression in the HD mouse model. (b) Microarray expression analysis of PGC-1α-regulated genes in human caudate. Here we see a heat map comparing the caudate nucleus expression of 26 PGC-1α target genes for 32 Grade 0–2 HD patients (adjacent to gold bar) and 32 matched controls (adjacent to blue bar). Most PGC-1α target genes are down-regulated. (c) Confirmation of expression reduction of PGC-1α-regulated genes in human caudate. We measured RNA expression levels for six PGC-1α targets (NDUFS3, CYCS, COX7C, NDUFB5, ACADM, and LDHB), PGC-1α, and two control genes (*GFAP* and *DRD2*). In this way, we confirmed significant expression reductions in PGC-1α targets and detected reduced PGC-1α in human HD striatum from early-grade patients. Statistical comparisons were performed with the *t*-test (*, $p < 0.05$; **, $p < 0.005$; ***, $p < 0.0005$). (From Weydt, P., *Cell Metab* 4, 349, 2006. With permission.)

NEUROTROPHIC FACTORS

Neurotrophic factors are signaling molecules that mediate important physiological processes in the CNS and peripheral nervous system (PNS). The paracrine or autocrine effects are transduced through membrane receptor tyrosine kinases that mediate a number of phosphorylation events, culminating in *de novo* transcription in the nucleus. These neurotrophic factor-induced changes in gene expression regulate calcium homeostasis, modulate synaptic efficiency, and promote neuron survival. BDNF is a member of the neurotrophic factor family that has been strongly implicated in the pathogenesis of HD. In postmortem HD brains, levels of this neurotrophic factor are decreased in the striatum but not in all the cortical samples analyzed [70, 71]. A recent report also noted that serum levels of BDNF are lower in HD patients [72]. Loss of BDNF has been noted in a number of mouse and cellular models for HD [73]. In a study where the level of BDNF was genetically modulated in mice, strong evidence for a role of BDNF in HD pathogenesis emerged, as earlier disease onset was observed in HD mice that were heterozygous-null for BDNF gene dosage [74]. Indeed, when BDNF expression was ablated in cortical pyramidal cells, age-associated dendrite degeneration, followed by loss of medium-sized spiny neurons, occurred [75]—a pattern of results that closely mimics HD striatal degeneration. Microarray analysis of HD transgenic mice yields a pattern of gene expression alterations that closely parallels the gene expression alterations observed in BDNF knockout mice [76], lending further support to the importance of BDNF loss of function in HD pathogenesis.

Although BDNF immunoreactivity is rather intense in the striatum, the BDNF transcript is barely detectable there, especially compared with other regions of the CNS (such as the hippocampus and neocortex), where both BDNF mRNA and protein levels are very high. Striatal BDNF originates in corticostriatal projection neurons that deliver the signaling molecule to medium-sized spiny neurons via anterograde transport (Figure 2.4). Experimental evidence points to two possible molecular mechanisms that impair BDNF production in the corticostriatal pathway: (1) dysregulation of BDNF transcription may result from mutant htt-mediated neural sequestration of REST/NRSF, a repressor that resides in the cytosol with normal htt [75]; and (2) a decrease in striatal BDNF could reflect the disruption of anterograde axonal transport by polyQ-expanded htt. Normal htt has been shown to bind HAP1 and p150^{Glued} to mediate axonal transport of BDNF, but this process may be disrupted when mutant htt is present [71]. Whatever the mechanism, there can be little doubt that impaired cortical BDNF release would deleteriously impact striatal neuron survival. However, other pathological factors may compromise the prosurvival function of this neurotrophic factor. For example, reduced corticostriatal expression of TrkB, the BDNF receptor tyrosine kinase, has been observed in the brains of HD patients and was also found in two different HD mouse models [77]. Loss of this signal transduction receptor would negatively impact the target cell and the presynaptic neuron because BDNF has both autocrine and paracrine effects (Figure 2.4).

CYTOSKELETAL DEFECTS AND AXONAL TRANSPORT

Neurons have a unique problem because of their unusual cellular geometry and specialized cellular morphology. The cell body, where gene transcription and most protein translation occur, is usually a considerable distance from the synaptic terminals, as a single axon can be up to a meter long in humans and extend much further in larger mammals. Axons not only propagate electrical signals throughout the cell, they also serve as the main transport corridor for proteins and other metabolic

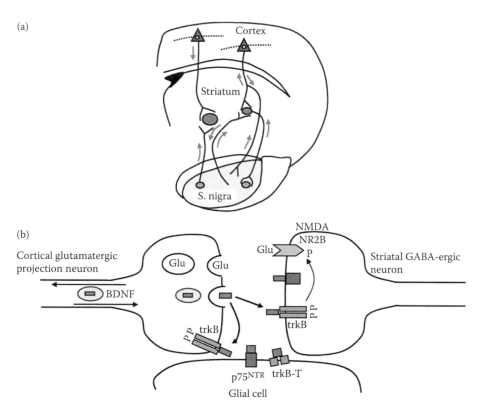

FIGURE 2.4 Cortico-striatal delivery and function of BDNF. (a) The cortico-striatal-nigral pathway. Anterograde and retrograde delivery of BDNF is indicated by the orange arrows. (b) Action of BDNF at the cortico-striatal synapse. BDNF vesicles fuse with the presynaptic membrane of the cortical projection neuron, causing BDNF to activate TrkB receptors on the postsynaptic membrane of striatal neurons, TrkB receptors on the presynaptic membrane, and TrkB-T or p75 receptors on glial cells. (From Zuccato, C. and Cattaneo, E., *Prog Neurobiol* 81, 294, 2007. With permission.)

components needed to maintain proper synaptic function. Hence a two-way transport system exists whose purpose is to shuttle ion channel components, membrane receptors, synaptic vesicle precursors, mitochondria, and signaling molecules, including neurotrophic factors and peptide neurotransmitters. The process is microtubule dependent and is powered by kinesin and dynein family members. Cargo is conveyed toward the synaptic terminal in an anterograde direction by kinesins, whereas transport toward the soma by dyneins occurs in a retrograde orientation [78, 79]. Disruption of axonal transport leads to aggregation of accumulated cargo, resulting in neuronal dysfunction and degeneration [reviewed in 78–80]. Charcot-Marie-Tooth type 2A1 (CMT2A1) and hereditary spastic paraplegia type 10 (SPG10) result from mutations in kinesin subunits, whereas familial and sporadic forms of amyotrophic lateral sclerosis (ALS) and lower motor neuron disease have been linked to mutations in the dynactin subunit p150^{Glued} [81, 82]. Defects in axonal transport have also been implicated in AD and in familial ALS type 1 as a result of Cu/Zn superoxide dismutase 1 (SOD1) mutations. The pathogenic protein associated with these neuro-degenerative disorders is posited to interact with the transport machinery and perturb normal axonal transport in the disease state.

In HD, two mechanisms for the axonal trafficking dysfunction have been proposed [83]. Based on studies in invertebrate models, normal htt was shown to play a role in fast axonal transport. Corollary studies of axonal transport in the face of polyQ-expanded htt or on reduced expression of an htt orthologue revealed marked reductions in the fast axonal transport pathway. The role of htt in axonal transport may depend on a presumed direct physical interaction with the axonal transport machinery. HAP1, a huntingtin interacting protein that associates with BDNF-containing vesicles in the cytosol, facilitates the interaction of htt with p150^{Glued} [71]. The resulting htt/p150^{Glued}/HAP1 protein complex interacts with microtubules to facilitate transport of the tethered vesicle along the axon (Figure 2.5). RNA interference-mediated knockdown of htt *or* polyQ expansion of full-length htt can disrupt binding of the HAP1/p150^{Glued} complex to microtubules and motor complexes and thereby can depress BDNF axonal transport and release [71]. The second proposed mechanism of polyQ-expanded htt axonal transport dysfunction posits titration of axonal transport components and/or physical disruption of microtubule-dependent movement of cargo secondary to cytosolic aggregation of mutant htt [84, 85]. This steric blockage of axonal transport function in HD could

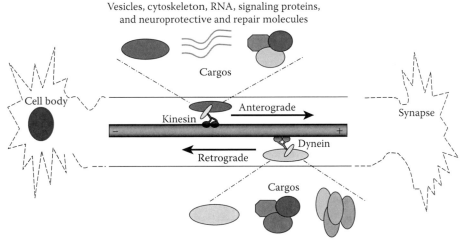

Vesicles, cytoskeleton, RNA, signaling proteins, and neuroprotective and repair molecules

Cargos

Cell body

Kinesin

Anterograde

Synapse

Dynein

Retrograde

Cargos

Vesicles, protein complexes, signaling proteins, and neuroprotective and repair molecules

FIGURE 2.5 Microtubule-based axonal transport pathways. The plus-end motor protein kinesin transports Golgi-derived vesicles, cytosolic proteins, RNAs, and other molecules anterogradely, whereas the minus-end motor protein dynein is principally responsible for retrograde transport. The huntingtin protein may directly interact with the dynein motor protein complex, suggesting that polyglutamine-expanded huntingtin protein could interfere with the normal functioning of this axonal transport pathway. Alternatively, misfolded huntingtin protein may form aggregates and create blockages in the pathway. (From Gunawardena S. and Goldstein, L. S., *Arch Neurol* 62, 46, 2005. With permission.)

involve the accumulation of autophagosomes whose maturation is blocked during dysfunctional autophagy.

A ROLE FOR THE LOSS OF HUNTINGTIN NORMAL FUNCTION IN HD?

Although polyQ expansion mutations produce a dominant gain-of-function toxicity, gain-of-function and loss-of-function mechanisms are not mutually exclusive in these diseases. There is considerable evidence for a pathogenic role of decreased normal function of disease proteins containing polyQ tract expansions [71, 86–90], including especially the htt protein. In a seminal study in 2000, postnatal elimination of htt protein expression yielded striatal degeneration in conditional knockout mice [91]. In the HD YAC128 mouse, the absence of endogenous huntingtin expression was achieved by crossing the HD YAC128 transgene onto an htt-null background, and this was shown to accelerate HD neuropathology [92, 93]. Similar studies of SBMA, a polyQ disorder with obvious disease protein-dependent non-neural loss-of-function phenotypes, have demonstrated that androgen receptor (AR) YAC100 mice display a more severe neuromuscular disease phenotype when placed on an AR-null background [94].

Given the evidence for a likely contribution of decreased htt normal function to HD pathogenesis, a crucial question is what is the normal function of htt, whose partial loss factors into the corticostriatal degeneration in HD? Different investigators have reported different potential normal htt functions [reviewed in 95]. Various *in vitro* studies have found that cells with depressed levels of htt expression are more susceptible to polyQ toxicity and have argued that htt is an important antiapoptotic factor [96, 97]. Dissection of htt's potential antiapoptotic actions has suggested that htt may inhibit procaspase-9 processing, perhaps by preventing the interaction of a proapoptotic initiator with the htt interacting protein-1 [98]. As noted previously, htt may be involved in the transcription regulation of BDNF production [95], and as htt possesses a functional nuclear export signal and has been shown to shuttle into and out of the nucleus [99, 100], htt could be regulating a process/factor that occurs in or originates in the nucleus. That htt is involved in the TFIID and TFIIF transcription factor complex through its interaction with TAFII130 indicates that one of its functions is to be a transcription cofactor [55]. Yet another function of htt is to mediate vesicular transport of BDNF [71], as discussed in the axonal transport section of this chapter. Htt has also been implicated in synaptic function, as it is highly enriched in synaptic terminals and may be involved in synaptic neurotransmission through its interaction with postsynaptic density protein-95 [101]. Although such divergent roles for htt may seem implausible, one must remember that the htt protein is very large (3,144 amino acids) and is subject to proteolytic cleavage; hence, different protein isoforms of htt may exist in neurons and carry out distinct functions in diverse subcellular compartments. The presence of three clusters of HEAT repeats indicates that one very likely normal function of htt is to serve as a scaffold protein, on which other sets of protein–protein interactions take place. Thus, it is reasonable to envision that htt (and peptide fragments thereof) are performing diverse tasks in neurons and other cell types; however, determining which function(s) are biologically relevant to HD pathogenesis remains a daunting challenge.

NON-CELL AUTONOMOUS DEGENERATION IN HD

The major neurodegenerative diseases, including HD, are defined by the neuronal population that is preferentially vulnerable: motor neurons in ALS; cortical neurons in AD; neurons of the substantia nigra in PD; and medium spiny neurons of the striatum (as well as cortical neurons) in HD [102]. In HD and in other neurodegenerative diseases, such as familial ALS, this pronounced selective vulnerability of neurons is an important conundrum, as expression of the offending protein occurs not only in neurons throughout the CNS but also in non-neural cells in the CNS and often throughout the rest of the body [103–105]. A fundamental insight that has emerged from the study of transgenic mouse models of neurodegenerative diseases in general and HD in particular is that the degeneration of neurons can be *non-cell autonomous*—meaning that cell types other than the dying neurons themselves will be critically involved in the degenerative process [106].

In HD, there is compelling evidence that both projecting neurons and surrounding glial cells are essential mediators of mutant htt toxicity. Studies of conditional transgene expression, using the Cre-lox system, have revealed that widespread expression of mutant htt in the brain, extending well beyond the primarily affected corticostriatal neurons, is necessary to produce motor dysfunction and brain pathology in HD mice [106]. Because mutant htt expression results in reduced BDNF gene transcription [48], one specific non-cell autonomous effect of the HD mutation is to deprive medium spiny neurons of their trophic input from cortical neurons. The relative importance of this scenario is underscored by a recent study from Strand et al. [76], in which targeted knockdown of BDNF expression in the forebrain yielded gene expression changes in the striatum that resemble the gene expression changes in human HD. However, non-neural cells are also impacted by the expression of mutant htt. The best evidence for this thesis comes from studies of astrocytes and microglia. (The role of microglia in HD is discussed under "Neuroinflammation in HD: How Important Are Microglia?"). In HD, a role for glutamate excitotoxicity has long been postulated and is supported by extensive literature [50–55]. Most recently, astrocytes derived from R6/2 HD transgenic mice or astrocytes infected with viral vectors encoding mutant huntingtin protein displayed a notable reduction in the glutamate transporter GLT-1/Slc1a2, as well as a limited ability to protect cocultured neurons from glutamatergic insult [56]. In *Drosophila*, the presence of polyQ-expanded peptides prevented the reactive up-regulation of glial glutamate transporters [107]. That polyQ expansions can impair the glutamate transport capacity of astrocytes *in vivo* has also been shown for SCA7, another polyQ repeat disease [108]. A likely contribution of impaired glial glutamate uptake to polyglutamine neurodegeneration implies that modulators of glutamate transporter expression could be vetted as potential therapeutic agents in these diseases. Although impaired glutamate uptake may primarily result from glial dysfunction in HD, it is also worth noting that cell autonomous factors may determine which types of neurons in the striatum are susceptible to the excitotoxic stress, as medium spiny striatal neurons in HD YAC transgenic mice display greater sensitivity to NMDA receptor activation, apparently because of preferential expression of the NR2B subunit in this neuronal population [109]. In the striatum of patients with early HD, evidence also exists for deficient *in vivo* glycolysis, a predominant function of astrocytes, suggesting yet another pathway by which glial dysfunction may contribute to the metabolic abnormalities occurring in HD [110].

For oligodendrocytes, the situation is less clear. White matter changes are a well-recognized feature of HD pathology. In postmortem HD brains, oligodendrocyte densities are increased independent of the presence of manifest astrocytosis [111, 112]. Also, white matter changes are found in imaging studies of presymptomatic HD patients [113]. Finally, PGC-1$\alpha^{-/-}$ mice, which recapitulate many aspects of the mouse HD phenotype, display significant oligodendrocyte abnormalities [68, 114]. Studies of the effect of mutant htt expression on oligodendrocyte function and their interactions with neurons are lacking; therefore, the role of oligodendrocyte dysfunction in HD pathogenesis is yet to be addressed.

NEUROINFLAMMATION IN HD: HOW IMPORTANT ARE MICROGLIA?

Microglia, resembling peripheral tissue macrophages, are the resident immune cells of the CNS and are the primary mediators of neuroinflammation. The past two decades have brought compelling evidence that microglia are important determinants of the microenvironment of the brain and are involved in many acute and chronic neurological diseases, including neurodegeneration [115]. In the unperturbed adult brain, microglia exist as so-called "resting" or "quiescent" microglia [116]. In this state, they have a small cell body with fine, ramified processes and minimal expression of surface antigens. Far from what the terminology of "quiescent" and "resting" would suggest, however, microglia in the healthy CNS are in fact busy "patrolling" the brain for lesions and intruders [117]. In the event of CNS injury, these cells are swiftly activated and therefore heavily involved in the pathology of almost all neurological disorders. The net result of neuroinflammation reflects the outcome of a delicate balance between the neurotoxic and neuroprotective factors that microglia release into their immediate environment [115]. Microglial effects on neurons and glia (astrocytes and oligodendrocytes) are mediated by the release of toxic substances such as nitric oxide, oxygen radicals, glutamate, proteases, and neurotoxic cytokines, as well as protective agents such as growth factors and neuroprotective cytokines [118]. These effects are modulated by cytokines and neurotransmitters released from astrocytes and neurons, giving rise to complex interactions between microglia, neurons, and astrocytes.

A large and growing body of evidence implicates microglia in the pathogenesis of the major sporadic neurodegenerative diseases: AD, PD, and ALS [115, 119]. For example, in AD, activated microglia are found near amyloid plaques and neurofibrillary tangles—abnormalities that are central to the pathogenesis of the disease [115]. In ALS mouse models, microglia are an important determinant of disease progression [120]. Recent evidence also suggests that microglial activation is detrimental for the generation of endogenous stem cells in the brain [121]. Microglial activation involves the up-regulation of the peripheral benzodiazepine binding site on the outer mitochondrial membrane. Through recent advances in brain imaging, it is possible to readily detect and visualize this process *in vivo* in experimental animals and in humans using the ^{11}C-labeled benzodiazepine receptor ligand PK-11195 [122]. Pathological imaging studies in HD show that microglial activation is an integral and remarkably early event in the disease process [123, 124]. Postmortem studies have revealed activated microglia mainly in the striatum and cortex of HD brains. The level of activation is a function of the degree of neuronal pathology [123]. As the distribution of the activated microglia extends well into the white matter, axonal pathology rather than neuronal loss may trigger and sustain neuroinflammation in HD. Indeed, PK-11195 imaging studies indicate that microglial activation is present not only in symptomatic HD patients but also in presymptomatic gene carriers [124, 125]. In presymptomatic gene carriers, microglial activation was closely associated anatomically with subclinical striatal dysfunction as measured by raclopide-PET. Striatal PK-11195 binding was also significantly correlated with a shorter "predicted time to symptomatic onset" of HD. In a gene array study of HD brain, mRNA expression revealed generalized activation of inflammatory pathways [126]. Using proteomics approaches, a systemic inflammatory response is also detectable in plasma and CSF of HD patients [127], whereas the transcriptome of peripheral blood did not show consistent inflammation [128]. Although microglial activation is also a feature of transgenic mouse models of HD at the histological level [129, 130], experimental treatments with inhibitors of microglial activation (namely, minocycline) have yielded conflicting results [131–134].

Another pathway central to neurodegeneration is oxidative stress and the concomitant production of reactive oxygen species (ROS). Glial cells likely play a role in propping up the antioxidant defenses of adjacent neurons. In PD, the drug rasagiline was identified based on its ability to block monoamine oxidase B metabolism of dopamine neurotransmitter, but it actually prevents the accumulation of iron in glial cells [135], suggesting that it reduces ROS production. Although glial cells are capable of relieving ROS stress, they could also conceivably be the source of it. In HD, kynurenine 3-monooxygenase was identified as a potent suppressor of huntingtin toxicity in yeast [136]. As the kynurenine pathway drives production of metabolites (3-hydroxy-kynurenine and quinolinic acid) known to increase ROS, and the kynurenine pathway operates in microglia, a model for microglia-induced non-cell autonomous degeneration of HD has been proposed. The availability of small molecules to inhibit the kynurenine 3-monooxygenase enzyme will permit investigators to evaluate efficacy in preclinical trials in HD mice.

In summary, there is compelling evidence from histopathological and imaging studies in human patients and transgenic mouse models that microglial activation is part of HD. In theory, this represents an alluring therapeutic target; however, further research is needed to determine how the complex neuroinflammatory response can be best tweaked to improve the clinical course of HD. Independently of the therapeutic implications, neuroinflammation and microglial activation may provide an opportunity to track disease progression with imaging and other biomarker techniques.

CONCLUSIONS

There can be little doubt that we have come a long way since interesting transcript 15 was elevated to the status of the HD gene in 1993. Through considerable funding support from patient groups and many governments, we have attacked the molecular basis of HD pathogenesis with an onslaught of basic research approaches that cut across an amazingly diverse range of disciplines. The net result of this truly global research enterprise has been to provide important insights into how polyQ expansions cause pathology. An important step has been the realization that HD presents a fascinating paradox: the protein misfolding process that is the crux of the pathology in HD is a fundamental feature of practically all neurodegenerative diseases, whereas the explanation for the selective loss of corticostriatal neurons in HD almost assuredly reflects some HD-specific process, most likely stemming from the normal biology of the huntingtin protein. The challenge now is to delineate which cellular pathologies and molecular abnormalities are central to HD onset *or* progression so as to better guide therapeutic efforts. Nuclear transcription interference remains a strong candidate for a principal role in HD pathogenesis, as does proteolytic cleavage. The linkage of transcription interference with mitochondrial dysfunction may prove fundamental. However, axonal transport and neurotrophic factor abnormalities also seem to be strong candidates for contributing to the ultimate demise of the affected neurons in HD and thus could make very valuable therapeutic targets. The relatively recent ascendancy of a role for astrocytes and microglia in HD has become widely accepted in the HD field and more broadly in the entire field of neurodegeneration. This paradigm shift has deepened our understanding of HD pathogenesis and has important implications for devising therapeutic strategies, especially stem cell–based approaches. The progress of the past 15 years has been significant enough that we now stand poised to apply translational tools to identified target pathways and players, in the hope that some of these endeavors will yield meaningful new treatments for this devastating disorder.

ACKNOWLEDGMENTS

We apologize to the many investigators whose work could not be discussed in this chapter because of space limitations. We also thank the Hereditary Disease Foundation and CHDI Foundation, Inc. for their support of our HD research program, as well as for their encouragement to all of us in the HD and polyglutamine disease field.

REFERENCES

1. Huntington's Disease Collaborative Research Group. 1993. A novel gene containing a trinucleotide repeat that is expanded and unstable on Huntington's disease chromosomes. *Cell* 72:971–83.

2. Zoghbi HY, Orr HT. 2000. Glutamine repeats and neurodegeneration. *Annu Rev Neurosci* 23:217–47.
3. Trottier Y, Lutz Y, Stevanin G, Imbert G, Devys D, Cancel G, et al. 1995. Polyglutamine expansion as a pathological epitope in Huntington's disease and four dominant cerebellar ataxias. *Nature* 378:403–06.
4. Paulson HL, Perez MK, Trottier Y, Trojanowski JQ, Subramony SH, Das SS, et al. 1997. Intranuclear inclusions of expanded polyglutamine protein in spinocerebellar ataxia type 3. *Neuron* 19:333–44.
5. Davies SW, Turmaine M, Cozens BA, DiFiglia M, Sharp AH, Ross CA, et al. 1997. Formation of neuronal intranuclear inclusions underlies the neurological dysfunction in mice transgenic for the HD mutation. *Cell* 90:537–48.
6. DiFiglia M, Sapp E, Chase KO, Davies SW, Bates GP, Vonsattel JP, et al. 1997. Aggregation of huntingtin in neuronal intranuclear inclusions and dystrophic neurites in brain. *Science* 277:1990–93.
7. Scherzinger E, Lurz R, Turmaine M, Mangiarini L, Hollenbach B, Hasenbank R, et al. 1997. Huntingtin-encoded polyglutamine expansions form amyloid-like protein aggregates in vitro and in vivo. *Cell* 90:549–58.
8. Cummings CJ, Reinstein E, Sun Y, Antalffy B, Jiang Y, Ciechanover A, et al. 1999. Mutation of the E6-AP ubiquitin ligase reduces nuclear inclusion frequency while accelerating polyglutamine-induced pathology in SCA1 mice. *Neuron* 24:879–92.
9. Hodgson JG, Agopyan N, Gutekunst CA, Leavitt BR, LePiane F, Singaraja R, et al. 1999. A YAC mouse model for Huntington's disease with full-length mutant huntingtin, cytoplasmic toxicity, and selective striatal neurodegeneration. *Neuron* 23:181–92.
10. Ross CA, Poirier MA. 2004. Protein aggregation and neurodegenerative disease. *Nat Med* 10 Suppl:S10–17.
11. Poirier MA, Li H, Macosko J, Cai S, Amzel M, Ross CA. 2002. Huntingtin spheroids and protofibrils as precursors in polyglutamine fibrilization. *J Biol Chem* 277:41032–37.
12. Wacker JL, Zareie MH, Fong H, Sarikaya M, Muchowski PJ. 2004. Hsp70 and Hsp40 attenuate formation of spherical and annular polyglutamine oligomers by partitioning monomer. *Nat Struct Mol Biol* 11:1215–22.
13. Arrasate M, Mitra S, Schweitzer ES, Segal MR, Finkbeiner S. 2004. Inclusion body formation reduces levels of mutant huntingtin and the risk of neuronal death. *Nature* 431:805–10.
14. Sherman MY, Goldberg AL. 2001. Cellular defenses against unfolded proteins: a cell biologist thinks about neurodegenerative diseases. *Neuron* 29:15–32.
15. Bukau B, Weissman J, Horwich A. 2006. Molecular chaperones and protein quality control. *Cell* 125:443–51.
16. Sherman MY, Goldberg AL. 2001. Cellular defenses against unfolded proteins: a cell biologist thinks about neurodegenerative diseases. *Neuron* 29:15–32.
17. Bence NF, Sampat RM, Kopito RR. 2001. Impairment of the ubiquitin-proteasome system by protein aggregation. *Science* 292:1552–55.
18. Holmberg CI, Staniszewski KE, Mensah KN, Matouschek A, Morimoto RI. 2004. Inefficient degradation of truncated polyglutamine proteins by the proteasome. *Embo J* 23:4307–18.
19. Venkatraman P, Wetzel R, Tanaka M, Nukina N, Goldberg AL. 2004. Eukaryotic proteasomes cannot digest polyglutamine sequences and release them during degradation of polyglutamine-containing proteins. *Mol Cell* 14:95–104.
20. Shintani T, Klionsky DJ. 2004. Autophagy in health and disease: a double-edged sword. *Science* 306:990–95.
21. Hara T, Nakamura K, Matsui M, Yamamoto A, Nakahara Y, Suzuki-Migishima R, et al. 2006. Suppression of basal autophagy in neural cells causes neurodegenerative disease in mice. *Nature* 441:885–89.

22. Komatsu M, Waguri S, Chiba T, Murata S, Iwata J, Tanida I, et al. 2006. Loss of autophagy in the central nervous system causes neurodegeneration in mice. *Nature* 441:880–84.
23. Ravikumar B, Acevedo-Arozena A, Imarisio S, Berger Z, Vacher C, O'Kane CJ, et al. 2005. Dynein mutations impair autophagic clearance of aggregate-prone proteins. *Nat Genet* 37:771–76.
24. Ravikumar B, Vacher C, Berger Z, Davies JE, Luo S, Oroz LG, et al. 2004. Inhibition of mTOR induces autophagy and reduces toxicity of polyglutamine expansions in fly and mouse models of Huntington disease. *Nat Genet* 36:585–95.
25. DiFiglia M. 2002. Huntingtin fragments that aggregate go their separate ways. *Mol Cell* 10:224–25.
26. Mangiarini L, Sathasivam K, Seller M, Cozens B, Harper A, Hetherington C, et al. 1996. Exon 1 of the HD gene with an expanded CAG repeat is sufficient to cause a progressive neurological phenotype in transgenic mice. *Cell* 87:493–506.
27. Gafni J, Ellerby LM. 2002. Calpain activation in Huntington's disease. *J Neurosci* 22:4842–49.
28. Goldberg YP, Nicholson DW, Rasper DM, Kalchman MA, Koide HB, Graham RK, et al. 1996. Cleavage of huntingtin by apopain, a proapoptotic cysteine protease, is modulated by the polyglutamine tract. *Nat Genet* 13:442–49.
29. Wellington CL, Ellerby LM, Gutekunst CA, Rogers D, Warby S, Graham RK, et al. 2002. Caspase cleavage of mutant huntingtin precedes neurodegeneration in Huntington's disease. *J Neurosci* 22:7862–72.
30. Ross CA. 2002. Polyglutamine pathogenesis: emergence of unifying mechanisms for Huntington's disease and related disorders. *Neuron* 35:819–22.
31. Slow EJ, Graham RK, Osmand AP, Devon RS, Lu G, Deng Y, et al. 2005. Absence of behavioral abnormalities and neurodegeneration in vivo despite widespread neuronal huntingtin inclusions. *Proc Natl Acad Sci U S A* 102:11402–07.
32. Graham RK, Deng Y, Slow EJ, Haigh B, Bissada N, Lu G, et al. 2006. Cleavage at the caspase-6 site is required for neuronal dysfunction and degeneration due to mutant huntingtin. *Cell* 125:1179–91.
33. Sisodia SS. 1998. Nuclear inclusions in glutamine repeat disorders: are they pernicious, coincidental, or beneficial? *Cell* 95:1–4.
34. La Spada AR, Taylor JP. 2003. Polyglutamines placed into context. *Neuron* 38:681–84.
35. Bannister AJ, Kouzarides T. 1996. The CBP co-activator is a histone acetyltransferase. *Nature* 384:641–43.
36. Ogryzko VV, Schiltz RL, Russanova V, Howard BH, Nakatani Y. 1996. The transcriptional coactivators p300 and CBP are histone acetyltransferases. *Cell* 87: 953–59.
37. Cong SY, Pepers BA, Evert BO, Rubinsztein DC, Roos RA, van Ommen GJ, et al. 2005. Mutant huntingtin represses CBP, but not p300, by binding and protein degradation. *Mol Cell Neurosci* 30:560–71.
38. Nucifora FC Jr., Sasaki M, Peters MF, Huang H, Cooper JK, Yamada M, et al. 2001. Interference by huntingtin and atrophin-1 with cbp-mediated transcription leading to cellular toxicity. *Science* 291:2423–28.
39. Steffan JS, Kazantsev A, Spasic-Boskovic O, Greenwald M, Zhu YZ, Gohler H, et al. 2000. The Huntington's disease protein interacts with p53 and CREB-binding protein and represses transcription. *Proc Natl Acad Sci U S A* 97:6763–68.
40. Ferrante RJ, Kubilus JK, Lee J, Ryu H, Beesen A, Zucker B, et al. 2003. Histone deacetylase inhibition by sodium butyrate chemotherapy ameliorates the neurodegenerative phenotype in Huntington's disease mice. *J Neurosci* 23:9418–27.
41. Hockly E, Richon VM, Woodman B, Smith DL, Zhou X, Rosa E, et al. 2003. Suberoylanilide hydroxamic acid, a histone deacetylase inhibitor, ameliorates motor deficits in a mouse model of Huntington's disease. *Proc Natl Acad Sci U S A* 100:2041–46.

42. Steffan JS, Bodai L, Pallos J, Poelman M, McCampbell A, Apostol BL, et al. 2001. Histone deacetylase inhibitors arrest polyglutamine-dependent neurodegeneration in Drosophila. *Nature* 413:739–43.
43. Dunah AW, Jeong H, Griffin A, Kim YM, Standaert DG, Hersch SM, et al. 2002. Sp1 and TAFII130 transcriptional activity disrupted in early Huntington's disease. *Science* 296:2238–43.
44. Li SH, Cheng AL, Zhou H, Lam S, Rao M, Li H, et al. 2002. Interaction of Huntington disease protein with transcriptional activator Sp1. *Mol Cell Biol* 22:1277–87.
45. Mantamadiotis T, Lemberger T, Bleckmann SC, Kern H, Kretz O, Martin Villalba A, et al. 2002. Disruption of CREB function in brain leads to neurodegeneration. *Nat Genet* 31:47–54.
46. Qiu Z, Norflus F, Singh B, Swindell MK, Buzescu R, Bejarano M, et al. 2006. Sp1 is up-regulated in cellular and transgenic models of Huntington disease, and its reduction is neuroprotective. *J Biol Chem* 281:16672–80.
47. Ryu H, Lee J, Zaman K, Kubilis J, Ferrante RJ, Ross BD, et al. 2003. Sp1 and Sp3 are oxidative stress-inducible, antideath transcription factors in cortical neurons. *J Neurosci* 23:3597–606.
48. Zuccato C, Ciammola A, Rigamonti D, Leavitt BR, Goffredo D, Conti L, et al. 2001. Loss of huntingtin-mediated BDNF gene transcription in Huntington's disease. *Science* 293:493–98.
49. Cui L, Jeong H, Borovecki F, Parkhurst CN, Tanese N, Krainc D. 2006. Transcriptional repression of PGC-1alpha by mutant huntingtin leads to mitochondrial dysfunction and neurodegeneration. *Cell* 127:59–69.
50. Weydt P, Pineda VV, Torrence AE, Libby RT, Satterfield TF, Lazarowski ER, et al. 2006. Thermoregulatory and metabolic defects in Huntington's disease transgenic mice implicate PGC-1alpha in Huntington's disease neurodegeneration. *Cell Metab* 4:349–62.
51. Kelly DP, Scarpulla RC. 2004. Transcriptional regulatory circuits controlling mitochondrial biogenesis and function. *Genes Dev* 18:357–68.
52. Puigserver P, Wu Z, Park CW, Graves R, Wright M, Spiegelman BM. 1998. A cold-inducible coactivator of nuclear receptors linked to adaptive thermogenesis. *Cell* 92:829–39.
53. Fry CJ, Peterson CL. 2002. Transcription. Unlocking the gates to gene expression. *Science* 295:1847–48.
54. Landles C, Bates GP. 2004. Huntingtin and the molecular pathogenesis of Huntington's disease. Fourth in molecular medicine review series. *EMBO Rep* 5:958–63.
55. Zhai W, Jeong H, Cui L, Krainc D, Tjian R. 2005. In vitro analysis of huntingtin-mediated transcriptional repression reveals multiple transcription factor targets. *Cell* 123:1241–53.
56. Thompson LM. 2003. An expanded role for wild-type huntingtin in neuronal transcription. *Nat Genet* 35:13–14.
57. Beal MF, Brouillet E, Jenkins BG, Ferrante RJ, Kowall NW, Miller JM, et al. 1993. Neurochemical and histologic characterization of striatal excitotoxic lesions produced by the mitochondrial toxin 3-nitropropionic acid. *J Neurosci* 13:4181–92.
58. Browne SE, Bowling AC, MacGarvey U, Baik MJ, Berger SC, Muqit MM, et al. 1997. Oxidative damage and metabolic dysfunction in Huntington's disease: selective vulnerability of the basal ganglia. *Ann Neurol* 41:646–53.
59. Gu M, Gash MT, Mann VM, Javoy-Agid F, Cooper JM, Schapira AH. 1996. Mitochondrial defect in Huntington's disease caudate nucleus. *Ann Neurol* 39:385–89.
60. Tabrizi SJ, Cleeter MW, Xuereb J, Taanman JW, Cooper JM, Schapira AH. 1999. Biochemical abnormalities and excitotoxicity in Huntington's disease brain. *Ann Neurol* 45:25–32.

61. Stoessl AJ, Martin WR, Clark C, Adam MJ, Ammann W, Beckman JH, et al. 1986. PET studies of cerebral glucose metabolism in idiopathic torticollis. *Neurology* 36:653–57.
62. Harms L, Meierkord H, Timm G, Pfeiffer L, Ludolph AC. 1997. Decreased N-acetyl-aspartate/choline ratio and increased lactate in the frontal lobe of patients with Huntington's disease: a proton magnetic resonance spectroscopy study. *J Neurol Neurosurg Psychiatry* 62:27–30.
63. Sawa A, Wiegand GW, Cooper J, Margolis RL, Sharp AH, Lawler JF Jr., et al. 1999. Increased apoptosis of Huntington disease lymphoblasts associated with repeat length-dependent mitochondrial depolarization. *Nat Med* 5:1194–98.
64. Panov AV, Gutekunst CA, Leavitt BR, Hayden MR, Burke JR, Strittmatter WJ, et al. 2002. Early mitochondrial calcium defects in Huntington's disease are a direct effect of polyglutamines. *Nat Neurosci* 5:731–36.
65. Puigserver P, Spiegelman BM. 2003. Peroxisome proliferator-activated receptor-gamma coactivator 1 alpha (PGC-1 alpha): transcriptional coactivator and metabolic regulator. *Endocr Rev* 24:78–90.
66. Wu Z, Puigserver P, Andersson U, Zhang C, Adelmant G, Mootha V, et al. 1999. Mechanisms controlling mitochondrial biogenesis and respiration through the thermogenic coactivator PGC-1. *Cell* 98:115–24.
67. Vega RB, Huss JM, Kelly DP. 2000. The coactivator PGC-1 cooperates with peroxisome proliferator-activated receptor alpha in transcriptional control of nuclear genes encoding mitochondrial fatty acid oxidation enzymes. *Mol Cell Biol* 20:1868–76.
68. Lin J, Wu PH, Tarr PT, Lindenberg KS, St-Pierre J, Zhang CY, et al. 2004. Defects in adaptive energy metabolism with CNS-linked hyperactivity in PGC-1alpha null mice. *Cell* 119:121–35.
69. Leone TC, Lehman JJ, Finck BN, Schaeffer PJ, Wende AR, Boudina S, et al. 2005. PGC-1alpha deficiency causes multi-system energy metabolic derangements: muscle dysfunction, abnormal weight control and hepatic steatosis. *PLoS Biol* 3:e101.
70. Ferrer I, Goutan E, Marin C, Rey MJ, Ribalta T. 2000. Brain-derived neurotrophic factor in Huntington disease. *Brain Res* 866:257–61.
71. Gauthier LR, Charrin BC, Borrell-Pages M, Dompierre JP, Rangone H, Cordelieres FP, et al. 2004. Huntingtin controls neurotrophic support and survival of neurons by enhancing BDNF vesicular transport along microtubules. *Cell* 118:127–38.
72. Ciammola A, Sassone J, Cannella M, Calza S, Poletti B, Frati L, et al. 2007. Low brain-derived neurotrophic factor (BDNF) levels in serum of Huntington's disease patients. *Am J Med Genet B Neuropsychiatr Genet* 144:574–77.
73. Zuccato C, Cattaneo E. 2007. Role of brain-derived neurotrophic factor in Huntington's disease. *Prog Neurobiol* 81:294–330.
74. Canals JM, Pineda JR, Torres-Peraza JF, Bosch M, Martin-Ibanez R, Munoz MT, et al. 2004. Brain-derived neurotrophic factor regulates the onset and severity of motor dysfunction associated with enkephalinergic neuronal degeneration in Huntington's disease. *J Neurosci* 24:7727–39.
75. Baquet ZC, Gorski JA, Jones KR. 2004. Early striatal dendrite deficits followed by neuron loss with advanced age in the absence of anterograde cortical brain-derived neurotrophic factor. *J Neurosci* 24:4250–58.
76. Strand AD, Baquet ZC, Aragaki AK, Holmans P, Yang L, Cleren C, et al. 2007. Expression profiling of Huntington's disease models suggests that brain-derived neurotrophic factor depletion plays a major role in striatal degeneration. *J Neurosci* 27:11758–68.
77. Gines S, Bosch M, Marco S, Gavalda N, Diaz-Hernandez M, Lucas JJ, et al. 2006. Reduced expression of the TrkB receptor in Huntington's disease mouse models and in human brain. *Eur J Neurosci* 23:649–58.
78. Duncan JE, Goldstein LS. 2006. The genetics of axonal transport and axonal transport disorders. *PLoS Genet* 2:e124.

79. Roy S, Zhang B, Lee VM, Trojanowski JQ. 2005. Axonal transport defects: a common theme in neurodegenerative diseases. *Acta Neuropathol* 109:5–13.
80. Gerdes JM, Katsanis N. 2005. Microtubule transport defects in neurological and ciliary disease. *Cell Mol Life Sci* 62:1556–70.
81. Munch C, Sedlmeier R, Meyer T, Homberg V, Sperfeld AD, Kurt A, et al. 2004. Point mutations of the p150 subunit of dynactin (DCTN1) gene in ALS. *Neurology* 63:724–26.
82. Schymick JC, Talbot K, Traynor BJ. 2007. Genetics of sporadic amyotrophic lateral sclerosis. *Hum Mol Genet* 16 Spec No. 2:R233–42.
83. Gunawardena S, Goldstein LS. 2005. Polyglutamine diseases and transport problems: deadly traffic jams on neuronal highways. *Arch Neurol* 62:46–51.
84. Lee WC, Yoshihara M, Littleton JT. 2004. Cytoplasmic aggregates trap polyglutamine-containing proteins and block axonal transport in a Drosophila model of Huntington's disease. *Proc Natl Acad Sci U S A* 101:3224–29.
85. Trushina E, Dyer RB, Badger JD 2nd, Ure D, Eide L, Tran DD, et al. 2004. Mutant huntingtin impairs axonal trafficking in mammalian neurons in vivo and in vitro. *Mol Cell Biol* 24:8195–209.
86. Li F, Macfarlan T, Pittman RN, Chakravarti D. 2002. Ataxin-3 is a histone-binding protein with two independent transcriptional corepressor activities. *J Biol Chem* 277:45004–12.
87. Palhan VB, Chen S, Peng GH, Tjernberg A, Gamper AM, Fan Y, et al. 2005. Polyglutamine-expanded ataxin 7 inhibits STAGA histone acetyltransferase activity to produce retinal degeneration. *Proc Natl Acad Sci U S A* 102:8472–77.
88. Chai Y, Berke SS, Cohen RE, Paulson HL. 2004. Poly-ubiquitin binding by the polyglutamine disease protein ataxin-3 links its normal function to protein surveillance pathways. *J Biol Chem* 279:3605–11.
89. Donaldson KM, Li W, Ching KA, Batalov S, Tsai CC, Joazeiro CA. 2003. Ubiquitin-mediated sequestration of normal cellular proteins into polyglutamine aggregates. *Proc Natl Acad Sci U S A* 100:8892–97.
90. Satterfield TF, Jackson SM, Pallanck LJ. 2002. A Drosophila homolog of the polyglutamine disease gene SCA2 is a dosage-sensitive regulator of actin filament formation. *Genetics* 162:1687–702.
91. Dragatsis I, Levine MS, Zeitlin S. 2000. Inactivation of Hdh in the brain and testis results in progressive neurodegeneration and sterility in mice. *Nat Genet* 26:300–06.
92. Leavitt BR, Guttman JA, Hodgson JG, Kimel GH, Singaraja R, Vogl AW, et al. 2001. Wild-type huntingtin reduces the cellular toxicity of mutant huntingtin in vivo. *Am J Hum Genet* 68:313–24.
93. Van Raamsdonk JM, Pearson J, Rogers DA, Bissada N, Vogl AW, Hayden MR, et al. 2005. Loss of wild-type huntingtin influences motor dysfunction and survival in the YAC128 mouse model of Huntington disease. *Hum Mol Genet* 14:1379–92.
94. Thomas PS Jr., Fraley GS, Damien V, Woodke LB, Zapata F, Sopher BL, et al. 2006. Loss of endogenous androgen receptor protein accelerates motor neuron degeneration and accentuates androgen insensitivity in a mouse model of X-linked spinal and bulbar muscular atrophy. *Hum Mol Genet* 15:2225–38.
95. Cattaneo E, Zuccato C, Tartari M. 2005. Normal huntingtin function: an alternative approach to Huntington's disease. *Nat Rev Neurosci* 6:919–30.
96. Leavitt BR, Raamsdonk JM, Shehadeh J, Fernandes H, Murphy Z, Graham RK, et al. 2006. Wild-type huntingtin protects neurons from excitotoxicity. *J Neurochem* 96:1121–29.
97. Rigamonti D, Bauer JH, De-Fraja C, Conti L, Sipione S, Sciorati C, et al. 2000. Wild-type huntingtin protects from apoptosis upstream of caspase-3. *J Neurosci* 20:3705–13.

98. Gervais FG, Singaraja R, Xanthoudakis S, Gutekunst CA, Leavitt BR, Metzler M, et al. 2002. Recruitment and activation of caspase-8 by the Huntingtin-interacting protein Hip-1 and a novel partner Hippi. *Nat Cell Biol* 4:95–105.

99. Truant R, Atwal RS, Burtnik A. 2007. Nucleocytoplasmic trafficking and transcription effects of huntingtin in Huntington's disease. *Prog Neurobiol* 83:211–27.

100. Xia J, Lee DH, Taylor J, Vandelft M, Truant R. 2003. Huntingtin contains a highly conserved nuclear export signal. *Hum Mol Genet* 12:1393–403.

101. Sun Y, Savanenin A, Reddy PH, Liu YF. 2001. Polyglutamine-expanded huntingtin promotes sensitization of N-methyl-D-aspartate receptors via post-synaptic density 95. *J Biol Chem* 276:24713–18.

102. Martin JB. 1999. Molecular basis of the neurodegenerative disorders. *N Engl J Med* 340:1970–80.

103. Li SH, Schilling G, Young WS 3rd, Li XJ, Margolis RL, Stine OC, et al. 1993. Huntington's disease gene (IT15) is widely expressed in human and rat tissues. *Neuron* 11:985–93.

104. Rosen DR, Siddique T, Patterson D, Figlewicz DA, Sapp P, Hentati A, et al. 1993. Mutations in Cu/Zn superoxide dismutase gene are associated with familial amyotrophic lateral sclerosis. *Nature* 362:59–62.

105. Strong TV, Tagle DA, Valdes JM, Elmer LW, Boehm K, Swaroop M, et al. 1993. Widespread expression of the human and rat Huntington's disease gene in brain and nonneural tissues. *Nat Genet* 5:259–65.

106. Lobsiger CS, Cleveland DW. 2007. Glial cells as intrinsic components of non-cell-autonomous neurodegenerative disease. *Nat Neurosci* 10:1355–60.

107. Lievens JC, Rival T, Iche M, Chneiweiss H, Birman S. 2005. Expanded polyglutamine peptides disrupt EGF receptor signaling and glutamate transporter expression in Drosophila. *Hum Mol Genet* 14:713–24.

108. Custer SK, Garden GA, Gill N, Rueb U, Libby RT, Schultz C, et al. 2006. Bergmann glia expression of polyglutamine-expanded ataxin-7 produces neurodegeneration by impairing glutamate transport. *Nat Neurosci* 9:1302–11.

109. Zeron MM, Hansson O, Chen N, Wellington CL, Leavitt BR, Brundin P, et al. 2002. Increased sensitivity to N-methyl-D-aspartate receptor-mediated excitotoxicity in a mouse model of Huntington's disease. *Neuron* 33:849–60.

110. Powers WJ, Videen TO, Markham J, McGee-Minnich L, Antenor-Dorsey JV, Hershey T, et al. 2007. Selective defect of in vivo glycolysis in early Huntington's disease striatum. *Proc Natl Acad Sci U S A* 104:2945–49.

111. Gomez-Tortosa E, MacDonald ME, Friend JC, Taylor SA, Weiler LJ, Cupples LA, et al. 2001. Quantitative neuropathological changes in presymptomatic Huntington's disease. *Ann Neurol* 49:29–34.

112. Myers RH, Vonsattel JP, Paskevich PA, Kiely DK, Stevens TJ, Cupples LA, et al. 1991. Decreased neuronal and increased oligodendroglial densities in Huntington's disease caudate nucleus. *J Neuropathol Exp Neurol* 50:729–42.

113. Thieben MJ, Duggins AJ, Good CD, Gomes L, Mahant N, Richards F, et al. 2002. The distribution of structural neuropathology in pre-clinical Huntington's disease. *Brain* 125:1815–28.

114. Leone TC, Lehman JJ, Finck BN, Schaeffer PJ, Wende AR, Boudina S, et al. 2005. PGC-1alpha deficiency causes multi-system energy metabolic derangements: muscle dysfunction, abnormal weight control and hepatic steatosis. *PLoS Biol* 3:e101.

115. Wyss-Coray T, Mucke L. 2002. Inflammation in neurodegenerative disease—a double-edged sword. *Neuron* 35:419–32.

116. Kreutzberg GW. 1996. Microglia: a sensor for pathological events in the CNS. *Trends Neurosci* 19:312–18.

117. Nimmerjahn A, Kirchhoff F, Helmchen F. 2005. Resting microglial cells are highly dynamic surveillants of brain parenchyma in vivo. *Science* 308:1314–18.

118. Hanisch UK. 2002. Microglia as a source and target of cytokines. *Glia* 40:140–55.
119. Weydt P, Moller T. 2005. Neuroinflammation in the pathogenesis of amyotrophic lateral sclerosis. *Neuroreport* 16:527–31.
120. Boillee S, Yamanaka K, Lobsiger CS, Copeland NG, Jenkins NA, Kassiotis G, et al. 2006. Onset and progression in inherited ALS determined by motor neurons and microglia. *Science* 312:1389–92.
121. Monje ML, Toda H, Palmer TD. 2003. Inflammatory blockade restores adult hippocampal neurogenesis. *Science* 302:1760–65.
122. Banati RB. 2002. Visualising microglial activation in vivo. *Glia* 40:206–17.
123. Sapp E, Kegel KB, Aronin N, Hashikawa T, Uchiyama Y, Tohyama K, et al. 2001. Early and progressive accumulation of reactive microglia in the Huntington disease brain. *J Neuropathol Exp Neurol* 60:161–72.
124. Tai YF, Pavese N, Gerhard A, Tabrizi SJ, Barker RA, Brooks DJ, et al. 2007. Microglial activation in presymptomatic Huntington's disease gene carriers. *Brain* 130:1759–66.
125. Pavese N, Gerhard A, Tai YF, Ho AK, Turkheimer F, Barker RA, et al. 2006. Microglial activation correlates with severity in Huntington disease: a clinical and PET study. *Neurology* 66:1638–43.
126. Hodges A, Strand AD, Aragaki AK, Kuhn A, Sengstag T, Hughes G, et al. 2006. Regional and cellular gene expression changes in human Huntington's disease brain. *Hum Mol Genet* 15:965–77.
127. Dalrymple A, Wild EJ, Joubert R, Sathasivam K, Bjorkqvist M, Petersen A, et al. 2007. Proteomic profiling of plasma in Huntington's disease reveals neuroinflammatory activation and biomarker candidates. *J Proteome Res* 6:2833–40.
128. Runne H, Kuhn A, Wild EJ, Pratyaksha W, Kristiansen M, Isaacs JD, et al. 2007. Analysis of potential transcriptomic biomarkers for Huntington's disease in peripheral blood. *Proc Natl Acad Sci U S A* 104:14424–29.
129. Ma L, Morton AJ, Nicholson LF. 2003. Microglia density decreases with age in a mouse model of Huntington's disease. *Glia* 43:274–80.
130. Simmons DA, Casale M, Alcon B, Pham N, Narayan N, Lynch G. 2007. Ferritin accumulation in dystrophic microglia is an early event in the development of Huntington's disease. *Glia* 55:1074–84.
131. Chen M, Ona VO, Li M, Ferrante RJ, Fink KB, Zhu S, et al. 2000. Minocycline inhibits caspase-1 and caspase-3 expression and delays mortality in a transgenic mouse model of Huntington disease. *Nat Med* 6:797–801.
132. Hersch S, Fink K, Vonsattel JP, Friedlander RM. 2003. Minocycline is protective in a mouse model of Huntington's disease. *Ann Neurol* 54:841; author reply 842–43.
133. Smith DL, Woodman B, Mahal A, Sathasivam K, Ghazi-Noori S, Lowden PA, et al. 2003. Minocycline and doxycycline are not beneficial in a model of Huntington's disease. *Ann Neurol* 54:186–96.
134. Stack EC, Smith KM, Ryu H, Cormier K, Chen M, Hagerty SW, et al. 2006. Combination therapy using minocycline and coenzyme Q10 in R6/2 transgenic Huntington's disease mice. *Biochim Biophys Acta* 1762:373–80.
135. Youdim MB, Fridkin M, Zheng H. 2005. Bifunctional drug derivatives of MAO-B inhibitor rasagiline and iron chelator VK-28 as a more effective approach to treatment of brain ageing and ageing neurodegenerative diseases. *Mech Ageing Dev* 126:317–26.
136. Giorgini F, Guidetti P, Nguyen Q, Bennett SC, Muchowski PJ. 2005. A genomic screen in yeast implicates kynurenine 3-monooxygenase as a therapeutic target for Huntington disease. *Nat Genet* 37:526–31.

3 Protein Interactions and Target Discovery in Huntington's Disease

John P. Miller and Robert E. Hughes

CONTENTS

INTRODUCTION

Huntington's disease (HD) is a devastating neurological disorder for which we currently have no effective treatments. Although patients are typically treated with drugs that can modify symptoms, none of these current drug regimens are thought to modify the onset, progress, or ultimate fatality of HD. A major step forward in terms of understanding the mechanism underlying HD and thus toward developing rational approaches to drug development occurred in 1993 with the cloning of the

HD gene (Willard, 1993). From a drug development perspective, one great outcome of this advancement was to create the ability to express the mutant HD gene in cell-based and transgenic models that can experimentally recapitulate aspects of HD pathology (see Chapters 5, 6, and 7, this volume). Such assays are invaluable tools for characterizing pathogenic mechanisms and discovering targets and small molecule modifiers of toxicity mediated by mutant Htt expression. The identification of the protein also allows us to determine its interacting partners and thereby place the pathogenesis of the disease in a proteomic context. However, despite early enthusiasm suggesting that the discovery of the precise genetic cause of HD could provide a fast track to an effective treatment, disease-modifying small molecule interventions for HD remain to be fully developed.

Target discovery and target validation are key early steps in the drug discovery process (see Chapter 4, this volume). There are a number of approaches to target discovery that include nomination and testing of candidate targets based on consideration of biological and molecular features of specific diseases. Candidate targets can also be inferred from genetic modifier screens in model organisms such as *Caenorhabditis elegans* and *Drosophila* (see Chapter 6, this volume). More recently, high-throughput RNA interference (RNAi) screening in cell-based models of disease has provided an opportunity to use unbiased genome-wide screens to identify potential targets capable of modifying *in vitro* models of disease phenotypes (Cronin et al., 2009; Krishnan et al., 2008; Luo et al., 2009). Candidate genes identified through RNAi-mediated phenotypes can be further validated in higher content models, such as crossing transgenic HD mice with a strain that bears a genetic modification of a candidate target. Another powerful method for target identification is the use of protein interaction studies. This chapter will focus on the role of protein interactions in HD and specifically on how knowledge of protein interaction networks can inform target discovery and validation processes for HD drug development.

PROTEIN INTERACTIONS IMPLICATE MECHANISMS AND TARGET PATHWAYS IN HD

The precise pathogenic mechanisms underlying HD remain relatively unclear. An interesting phenomenon contributing to this uncertainty may be the great number of pathways that have been reported to be dysfunctional in cell- and organism-based models of this disease. Pathways that have been implicated as impaired in HD include transcriptional regulation, vesicle transport, mitochondrial energy production, synaptic function, and protein homeostasis (Landles and Bates, 2004). Despite the fact that there have been a number of cellular dysfunctions associated with expanded Htt expression, components of these dysfunctional pathways should not be assumed to be *de facto* targets for HD. Candidate targets are typically defined as gene products whose activities can modify the progress or severity of a disease state.

In many cases of dysfunctional pathways linked to HD, there have been specific protein-interaction partners discovered for huntingtin that either initially implicated a particular pathway, offered corroborative evidence, and/or provided a specific target within a pathway of interest as being mechanistically important. Htt protein interactions and their putative roles in specific pathogenic processes have been reviewed elsewhere (Harjes and Wanker, 2003; Li and Li, 2004).

TECHNOLOGIES FOR DETECTION OF PROTEIN INTERACTIONS AND PROTEIN COMPLEXES

There are a number of established techniques for the discovery and characterization of protein interactions. Currently, the two most commonly used techniques for the high-throughput identification of protein-interacting partners and complexes are yeast two-hybrid (Y2H) screening and tandem affinity purification followed by mass spectrometric analysis. Each of these techniques will be described briefly.

An understanding of protein interactions at the level of individual proteins and in the context of the cellular network is now considered to be a key step toward elucidation of biological functions and processes. Traditionally, the discovery and characterization of protein interactions through classic biochemical methods such as chromatography and chemical sequencing was a slow and laborious process. The invention of Y2H screening technology represented a powerful leap forward in our ability to rapidly and unambiguously identify binary interactions for a given protein at the genome-scale level (Fields and Song, 1989; Miller and Stagljar, 2004). This genetic technology is based on the modular nature of transcription factors that allows for DNA-binding domains and transcriptional activation domains to be noncovalently reassociated through interactions by fusion protein partners. Although initially practiced on a single protein basis, the automation of Y2H allowed this technique to become a major contributor to proteome-scale protein-interaction mapping (Rual et al., 2005; Schwikowski et al., 2000; Stelzl et al., 2005; Uetz et al., 2000). The application of high-throughput Y2H technology to questions relevant to HD research is discussed below.

A complementary biochemical technology for discovery of protein interactions is tandem affinity purification (TAP) of protein complexes followed by protein identification using mass spectrometry (MS) (Collins and Choudhary, 2008; Kaiser et al., 2008). This technology often uses bait proteins of interest fused to TAP tags that allow for high levels of protein complex purity. In contrast to Y2H technology, which provides information about binary interactions, TAP-MS analysis provides information about protein complexes. It is important to note the output of TAP-MS does not provide information about potential pair-wise connectivities between proteins detected as being present in a complex. Like Y2H, TAP-MS has been scaled to a level that can approach genome-wide analysis, at least in model systems such as yeast (Gavin et al., 2002; Ho et al., 2002; Krogan et al., 2006).

LARGE-SCALE INTERACTION NETWORKS AND NEXT-GENERATION DATA MINING

The conventional application of protein-interaction discovery as a "one-off" experiment is gradually giving way to high-throughput methods that allow for parallel screening and complex network generation. To date, two high-throughput protein interaction studies centered on huntingtin protein interactions have been published. In the study of Goehler et al. (2004), a combination of library-based and matrix-based screening was used to generate a huntingtin protein interaction network containing 186 protein interaction pairs. Further analysis and validation of proteins contained in this network revealed that GIT1, a protein interacting directly with huntingtin in the network, could modify huntingtin aggregation in cultured mammalian cells and is colocalized with huntingtin-containing aggregates in the brains of mouse models of HD and in postmortem human HD brain tissue. It is not clear how GIT1, a G protein-coupled receptor kinase-interacting protein involved in cytoskeletal and membrane functions, may modify HD, but the discovery that this protein does associate with huntingtin in HD brain underscores the values of mining large-scale protein interaction networks to identify candidate targets.

Another large-scale interaction network centered on huntingtin was generated using the complementary approaches of Y2H and affinity purification/MS (Kaltenbach et al., 2007; Li and Li, 2004). This study involved exhaustive two-hybrid searches using huntingtin as a bait to screen multiple human cDNA libraries. In parallel, purified amino-terminal huntingtin protein fragments were used to probe soluble protein tissue lysates prepared from mouse brain, mouse muscle, and postmortem human brain samples. Overall, these protein interaction screens identified 104 proteins by Y2H and 130 proteins by TAP. An intriguing feature of this study was the functional validation observed by testing a large number of the interacting proteins for their ability to modify neurodegeneration in a fly model of HD. *Drosophila* orthologues of genes encoding interacting proteins were tested for their ability to act as modifiers of retinal degeneration in transgenic flies expressing mutant huntingtin. Of the 60 genes tested, 80% were able to enhance and/or suppress the toxic effects of mutant huntingtin through either overexpression or a partial loss of function. Modifier genes validated in this manner encoded proteins involved in functions such as transcription, signal transduction, and synaptic vesicle fusion. Proteins with functions in synaptic vesicle fusion and neurotransmission (e.g., STX1A, SNAP, and CACNA2D1) were further validated as having a role in mutant huntingtin toxicity in follow-up studies in HD transgenic *Drosophila* (Romero et al., 2008).

TRANSCRIPTION PROTEIN INTERACTIONS WITH HUNTINGTIN

In this chapter, we will focus on one area of biology that has been the subject of a significant amount of investigation in HD research: transcriptional dysregulation (Butler and Bates, 2006; Cha, 2000, 2007; Hughes, 2002; Sugars and Rubinsztein, 2003). Protein interactions likely play an important role in this facet of the disease as a number of transcription factors contain polyglutamine tracts, and the propensity for these sequences to self-assemble suggests a mechanism by which mutant Htt influences transcription. Furthermore, polyglutamine tracts and polyproline tracts, both of which Htt possesses, can function as transcription-activating domains when fused to a DNA-binding domain (Gerber et al., 1994). Beyond interactions based on polyglutamine tract annealing, Htt also interacts with transcriptional machinery constituents that contain no polyglutamine stretch, or, as in the case of CREB-binding protein (CBP), interacts even after the polyglutamine stretch is removed from the protein (Steffan et al., 2001).

From a therapeutic standpoint, transcription factors that interact with Htt present a class of targets with two opposing intervention strategies. First, in the case where an interaction is seen with normal Htt, but not with mutant Htt, the interaction may be important for cell survival, and stabilizing/enhancing the interaction may be therapeutic. In the alternative situation, wherein mutant Htt interacts with a protein but wild-type Htt does not, inhibition of the interaction, or even levels of the protein itself, is the objective of intervention. This latter circumstance is the more expected case from a theoretical perspective; i.e., the gain of an inappropriate interaction confers the detrimental consequences of the disease protein. However, prominent examples indicate the contribution of the loss of at least some normal interactions is also important in disease etiology (Zuccato et al., 2003).

A number of the transcriptional proteins found to interact with Htt also interact with at least one other Htt-interacting transcription protein (Figure 3.1). Intriguingly, 32 of the transcription-related Htt-interacting proteins form an interconnected group with each other. This suggests that Htt could potentially influence a number of the proteins via its direct interaction with just one of them. On the one hand, wild-type Htt may have a prosurvival role that depends on an interaction with one or more of these proteins. Such an interaction may be reduced or lost with the mutant protein, which additionally interacts inappropriately with other transcription factor proteins, with cytotoxic results.

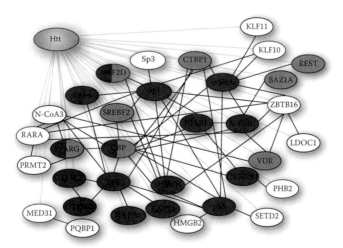

FIGURE 3.1   (See color insert following page 172.) The interactions among transcription-related proteins that interact with Htt. As discussed in the text, evidence suggests the interaction of Htt with some partners is likely to be beneficial to the cell (green), whereas other interactions are most likely to be detrimental (red). Conflicting data exist for some of the proteins (half red, half green). (From the "Interactions" section of the online Entrez Gene database [http://www. ncbi.nlm.nih.gov/entrez/query.fcgi?db=gene; Maglott et al., 2005] of the National Center for Biotechnology Information website. The original publications describing each interaction can be viewed using the "Interactions" section "PubMed" links next to the appropriate interaction partner. Other sources for interactions not listed in Entrez Gene are referenced in the text.)

TABLE 3.1

Transcription-Related Proteins That Interact with Huntingtin

Protein	Gene ID	Description	PolyQ Tract	References for Htt Interaction	Beneficial/Detrimental
BAZ1A (hACF1)	11177	Nuclear hormone receptor repressor	No	Kaltenbach et al. (2007)	Beneficial
CBP	1387	CREB-binding protein	Yes	Kazantsev et al. (1999); Nucifora et al. (2001); Steffan et al. (2000, 2001)	Both
				Jeong et al. (2009); Kaltenbach et al. (2007)	
CTBP1	1487	Transcriptional repressor	No	Kegel et al. (2002)	Beneficial
CTNNB1	1499	β-catenin	No	Kaltenbach et al. (2007)	Detrimental
EHMT1	79813	Part of transcription Repressor complex	No	Kaltenbach et al. (2007)	Detrimental
ENOA	2023	Represses c-myc Promoter	No	Kaltenbach et al. (2007)	?
GTF2I	2969	Phosphoprotein with roles in transcription and signal transduction	No	Kaltenbach et al. (2007)	Detrimental
GTF3C2	2976	RNA PolIII transcription factor	No	Kaltenbach et al. (2007)	Detrimental
GTF3C3	9330	RNA PolIII transcription factor	No; Glu repeats	Kaltenbach et al. (2007)	Detrimental
HMGB2	3148	Nonhistone chromatin-binding protein	No; Glu repeats	Kaltenbach et al. (2007)	?
HTT	3064	Htt self-interacts and binds DNA on its own	Yes	Benn et al. (2008); Shaffar et al. (2004)	Detrimental
HYPA (PRPF40A)	55660	mRNA processing	No	Faber et al. (1998)	Detrimental

continued

HYPB (SETD2)	29072	Chromatin activating	4Q	Faber et al. (1998); Kaltenbach et al. (2007)	?
HYPC (PRPF40B)	25766	mRNA processing	4Q	Faber et al. (1998)	?
KLF10 (TIEG)	7071	Sp1-like, Kruppel-like factor 10	No	Kaltenbach et al. (2007)	?
KLF11 (TIEG2)	8462	Sp1-like, Kruppel-like factor 11	No	Kaltenbach et al. (2007)	?
LDOC1	23641	Negative regulator of NFkB	No	Kaltenbach et al. (2007)	?
LXRα (NR1H3)	10062	Nuclear receptor	No	Futter et al. (2009)	Beneficial
MED15	51586	Polymerase II transcriptional activator	Yes	Kaltenbach et al. (2007)	Beneficial
MED23	9439	Subunit of cofactor required for Sp1 activation	No	Kaltenbach et al. (2007)	Detrimental
MED31	51003	Pol II mediator component	4Q	Goehler et al. (2004)	?
MEF2D	4209	Transcriptional activator	QQP repeats	Kaltenbach et al. (2007)	Both
mSin3a	25942	Transcriptional repressor	No	Boutell et al. (1999); Steffan et al. (2000)	Detrimental
N-CoA3	8202	Nuclear coreceptor activator	Yes	Kaltenbach et al. (2007)	?
N-CoR1	9611	Unliganded nuclear receptor corepressor	7Q	Boutell et al. (1999); Kaltenbach et al. (2007)	Detrimental
NEUROD1	4760	Transcriptional activator of E-box promoters	No	Marcora et al. (2003)	Beneficial
NFKB1	4790	Transcription factor	No	Kaltenbach et al. (2007); Takano and Gusella (2002)	Detrimental
OPTN	10133	Coiled-coil protein that interacts with TFIIIA	No	Hattula and Peranen (2000)	?

TABLE 3.1 (continued)
Transcription-Related Proteins That Interact with Huntingtin

Protein	Gene ID	Description	PolyQ Tract	References for Htt Interaction	Beneficial/Detrimental
PHB2	11331	Estrogen-responsive corepressor	No	Kaltenbach et al. (2007)	?
PPARG	5468	Nuclear hormone receptor	No	Futter et al. (2009); Kaltenbach et al. (2007)	Both
PQBP1	10084	PolyQ-binding protein 1	No	Waragai et al. (1999)	?
PRMT2 (HRMT1L1)	3275	Coactivator with androgen receptor	No	Kaltenbach et al. (2007)	?
PTRF	284119	RNA POLI transcript release factor	No	Kaltenbach et al. (2007)	?
RAP30 (GTF2F2)	2963	TFIIF subunit	No	Zhai et al. (2005)	Detrimental
RAP74 (GTF2F1)	2962	TFIIF subunit	No	Zhai et al. (2005)	Detrimental
RARα (RARA)	5914	Retinoic acid receptor	No	Kaltenbach et al. (2007)	?
REST/NRSF	5978	Repressor of neuronal genes in non-neuronal cells	No	Zuccato et al. (2003)	Beneficial
RNF20	56254	Homologue to yeast BRE1 histone modifier	No	Kaltenbach et al. (2007)	?
RNF40	9810	Functions in complex with RNF20	No	Kaltenbach et al. (2007)	?
Sp1	6667	Specificity transcription factor	4Q, Q-rich	Dunah et al. (2002); Li et al. (2002); Zhai et al. (2005)	Detrimental
Sp3	6670	Specificity transcription factor	5Q on longer isoform	Kaltenbach et al. (2007)	?
Sp8	221833	Specificity transcription factor	No, Ser/Ala/Gly-rich	Kaltenbach et al. (2007)	?

Protein	Entrez Gene	Description	Q-rich	Reference	Effect
SREBF2 (SREBP)	6721	Sterol-regulated transcription factor	Q-rich	Kaltenbach et al. (2007)	Beneficial
SUMO1	7341	Small modifier that is covalently attached to proteins	No	Steffan et al. (2004); Subramaniam et al. (2009)	Detrimental
SUPT5H	6829	Transcription factor	No	Kaltenbach et al. (2007)	Detrimental
TAF4	6874	TFIID subunit	No	Dunah et al. (2002); Kaltenbach et al. (2007); Shimohata et al. (2000)	Detrimental
TBP	6908	TATA-binding protein	Yes	Huang et al. (1998); Shaffar et al. (2004)	Detrimental
TCERG1	10915	Regulates transcriptional elongation and pre-mRNA splicing	No, Glu/Ala repeats	Holbert et al. (2001); Kaltenbach et al. (2007)	Beneficial
P53	7157	Tumor suppressor	No	Bae et al. (2005); Steffan et al. (2000)	Detrimental
TRα1 (THRA)	7067	Thyroid hormone receptor	No	Futter et al. (2009)	Beneficial
VDR	7421	Vitamin D nuclear hormone receptor	No	Futter et al. (2009)	Beneficial
WBP4	11193	Spliceosome component	No	Kaltenbach et al. (2007)	Beneficial
ZBTB16	7704	Transcription factor	No	Kaltenbach et al. (2007)	?
ZNF133	7692	Transcriptional repressor	No	Kaltenbach et al. (2007)	Beneficial
ZNF537 (TSHZ3)	57616	Zinc finger protein	No	Kaltenbach et al. (2007)	Beneficial
ZNF655	79027	Zinc finger protein	No	Kaltenbach et al. (2007)	?
ZNF675	171392	Modulates TRAF6 signaling on NFkB transcription factor	No	Kaltenbach et al. (2007)	?
ZNF91	7644	Zinc finger protein	No	Kaltenbach et al. (2007)	Beneficial

Source: Entrez Gene entries for the protein unless the original publication is referenced in the table. With permission.

The transcription-related protein interactions with Htt will be summarized in this section (Table 3.1). Where there is sufficient evidence to do so, a description of whether the interaction appears beneficial or detrimental is presented. For several prominent examples, approaches that may develop a given interaction into an effective therapeutic target will be discussed. We will focus primarily on directed therapeutic strategies, as the screening for small molecules that remedy transcriptional dysregulation in HD has been explored elsewhere (Kazantsev and Hersch, 2007; Rigamonti et al., 2007), although such strategies have potential applicability to all the interactions discussed.

POTENTIAL PROSURVIVAL PROTEIN INTERACTIONS WITH NORMAL HUNTINGTIN

Of 58 transcription-related proteins (Table 3.1) reported to interact with Htt in the literature, evidence for 14 proteins suggests that the mutant form does not interact to the same extent as the wild type form. If these then represent "normal" interactions, they may contribute to pathology in that they become reduced in abundance or possibly converted to harmful interactions when Htt within the cell harbors an expanded polyglutamine tract. Hence, promotion of these interactions may provide a therapeutic benefit. This may come about by permitting mutant Htt to perform the implicated function normally provided by wild-type huntingtin. Alternatively, an interacting partner that functions to modulate stability or post-translational modifications on Htt may not be as effective in the context of the polyglutamine-expanded protein, and facilitation of the interaction may restore a therapeutic degree of function. Strategies for intervention center around increasing the amount or activity of the "normal" interaction partner, with the hope of thereby promoting the interaction.

BAZ1A (HACF1)

Bromodomain adjacent to zinc finger domain 1A (BAZ1A) is a component of an ATP-using chromatin assembly and remodeling factor complex (He et al., 2006) that interacts with nuclear receptor corepressor 1 (N-CoR-1) and is important in repressing transcription within unliganded nuclear hormone receptor complexes (Ewing et al., 2007). BAZ1A was found to interact more strongly with wild-type Htt fragments than with mutant Htt fragments in a Y2H study, suggesting a detrimental effect on the interaction of the expanded polyglutamine tract (Kaltenbach et al., 2007). Hence, the BAZ1A interaction with Htt may be part of the normal function for both proteins. One hypothesis is that in the presence of mutant Htt, reduced interaction allows BAZ1A accumulation in the nucleus, creating excessive repression at unliganded nuclear hormone receptor recognition sites. A potential therapeutic approach would then be administration of the appropriate hormones or vitamins (vitamin D_3 response sequences are one established N-CoR/SMRT [silencing mediator for retinoid and thyroid hormone receptors] recruiting element that is influenced by BAZ1A function [Ewing et al., 2007]) to reduce the impact of the nuclear BAZ1A repressive presence. Such an approach first requires confirmation of the changes of BAZ1A target genes/regions in HD, as well as an establishment of a role for BAZ1A in disease dysfunction.

CTBP1

The C-terminal binding protein 1 (CTBP1) is a transcriptional repressor that inter-acts with multiple components of the cellular repression machinery and provides a bridge between more general repressive factors (e.g., histone deacetylases [HDACs]) and DNA-binding transcriptional repressors. Based on the presence of a motif highly conserved in proteins that interact with CTBP1, Kegel et al. (2002) tested the ability of Htt to interact with this repressor. They found that Htt interacts with CTBP1 and that an expanded polyglutamine tract reduced this interaction. Furthermore, full-length wild-type and mutant Htt both behaved as transcriptional repressors for a luciferase reporter, but N-terminal fragments of Htt were only able to repress tran-scription if they contained a mutant polyglutamine stretch.

Based on these observations, we anticipate that the interaction of wild-type Htt with CTBP1 represents the appropriate interaction and that the reduced binding with mutant Htt results in transcriptional dysregulation. Because of the ability of only mutant Htt N-terminal fragments to repress transcription in the same experi-mental context (Kegel et al., 2002), it appears the reduced interaction with CTBP1 may allow inappropriate transcriptional repression through these mutant Htt frag-ments on their own. Strategies to promote the degradation of mutant Htt fragments would then be of therapeutic benefit. Alternatively, enhancement of nuclear export of mutant Htt fragments would be another way to suppress the defective presence of these molecules in the cell.

There is also evidence that CTBP1 may be a scaffold for sumoylation of substrates in repressor complexes that also contain the small ubiquitin-like modifier (SUMO)-conjugating enzyme UBC9 (Kuppuswamy et al., 2008). In this context, the interac-tion with Htt would be expected to have detrimental consequences, as it is reported that sumoylation of mutant Htt leads to its stabilization and enhances its transcrip-tional repressive activity toward the multidrug resistance gene (Steffan et al., 2004). Potentially, the reason for reduced interaction with CTBP1 (Kegel et al., 2002) is that the mutant protein is a better substrate for sumoylation, therefore causing it to be more rapidly released. Specific inhibition of Htt sumoylation may then be an effec-tive approach to help cells survive the burden of mutant Htt that disruptively impacts cellular transcriptional programs (see below).

SREBF2 (SREBP)

The sterol response element binding transcription factor 2 (SREBF2) is a transmem-brane protein that undergoes proteolytic release of its transcription factor domain from the membrane in response to reduced levels of sterols (Brown and Goldstein, 1997; Wang et al., 1994). In the nucleus, the soluble transcription factor domain activates the expression of genes encoding sterol biosynthetic enzymes. Thus, this protein senses the cellular levels of sterols and up-regulates their synthesis when low amounts are present in the cell.

Kaltenbach et al. (2007) identified an interaction of SREBF2 and Htt that was more robust when the nonexpanded polyglutamine protein was used in Y2H. This result is intriguing in light of the fact that studies have found that cholesterol biosynthetic enzyme mRNAs are reduced in cell and animal models of HD and in HD patients (Valenza et al., 2005, 2007a, 2007b). These results suggest that the presence of mutant huntingtin induces a down-regulation of mRNAs involved in cholesterol biosynthesis, and a weakened interaction with SREBF2 may be a part of the mechanism for this phenotype. One hypothesis is that normal Htt protein interaction with SREBF2 promotes the activity of SREBF2, and in cells containing mutant Htt a deficit in this interaction results in reduced expression of cholesterol-synthesizing gene products. Alternatively, some SREBF2 protein may be sequestered away from its normal function through interaction with mutant Htt.

Appropriate targeting of the impact of the defective interaction could be complicated. One approach to therapeutic intervention would be to increase cholesterol in the diets of HD patients. This has obvious caveats associated with atherosclerosis and inflammation risk and would require careful monitoring of cholesterol levels. However, circumvention of the reduced cholesterol biosynthesis in HD cells may be therapeutic if the ultimate sterol levels are the culprit.

REST/NRSF

Repressor element-1 transcription factor/neuron-restrictive silencing factor (REST/NRSF) is a transcriptional repressor that maintains neural-specific genes in a repressed state in non-neural cell types (Chong et al., 1995; Schoenherr and Anderson, 1995). In neurons, wild-type huntingtin sequesters REST in the cytoplasm, thus allowing the expression of genes that are held in REST-dependent repressed states in other cell types. Mutant Htt is defective in the interaction with REST, and in HD models and patient brains REST accumulates in the nucleus (Zuccato et al., 2003). The aberrant nuclear presence of REST results in repression of prosurvival genes such as brain-derived neurotrophic factor (BDNF) and represents a strong candidate for therapeutic intervention.

In this case, a reduction in the amount of REST to decrease its presence in the nucleus would be predicted to be beneficial. This could potentially be accomplished by the introduction of small hairpin RNA constructs that target the REST transcript for degradation by the RNAi pathway (see Chapter 9, this volume). This "gene therapy" approach for such constructs will require developments in that field before it can be used in HD patients, but establishing the benefits of the strategy in mouse models of the disease could be done with current methodologies. As opposed to direct intervention at the level of this transcriptional repressor, the downstream gene targets it impacts may offer more straightforward therapeutic solutions.

In this context, the evidence of the detrimental consequences of defective BDNF delivery to striatal cells in HD is compelling (Zuccato and Cattaneo, 2007; Zuccato et al., 2001, 2005). Interestingly, BDNF knockout mice recapitulate the transcriptional changes in HD brain more closely than any HD mouse model tested (Strand et al., 2007). The body of work in this area argues that restoration of BDNF levels in the striatum may counteract the fundamental defect in HD brains. Administration of BDNF alone or in combination with other neurotrophic factors repressed by REST may circumvent the elevated levels of REST in the nucleus. However, the blood–brain barrier provides an obstacle to effective delivery of neurotrophic factors into the central nervous system. Recent work transplanting bone marrow cells that are transduced with the BDNF gene has shown this to be an effective way to deliver detectable amounts of BDNF in the brain (Makar et al., 2004, 2009). The recent developments in stem cell and induced pluripotent stem cell research may offer alternative cell types with advantages over bone marrow cells with which to deliver therapeutic factors such as BDNF. Alternatively, fusion of a protein transduction domain to BDNF has been shown to successfully facilitate transport of the factor across the blood–brain barrier, resulting in improved spatial learning and memory in a mouse model of Alzheimer's disease (Zhou et al., 2008).

Importantly, Cattaneo and colleagues developed a BDNF promoter-based luciferase assay to identify small molecules that activate expression of BDNF (Rigamonti et al., 2007). They found three related compounds that at nanomolar levels could activate reporter and endogenous BDNF levels, as well as importantly increase expression of other REST/NRSF gene targets. This suggests that the compounds target REST/NRSF generally and not solely the BDNF promoter. This screening platform has thereby successfully identified compounds showing therapeutic promise, and indeed one of these compounds has been shown to be able to rescue toxicity in an HD cell model.

PPARγ

Peroxisome proliferator-activated receptor γ (PPARγ) is a nuclear receptor that is important for adipocyte differentiation (Tontonoz et al., 1995). Hypolipidemic drugs and other PPAR activators, including naturally occurring polyunsaturated fatty acids, promote the differentiation of PPARγ-expressing cells in a dose-dependent manner, and retroviral expression of PPARγ is sufficient to cause fibroblast lines to become preadipocytes (Tontonoz et al., 1994). PPARγ is also expressed in other tissues, including brain (Heneka and Landreth, 2007), and has been shown to have neuroprotective effects (Bordet et al., 2006; Heneka et al., 2007).

PPARγ has been shown to interact with Htt by two independent groups (Futter et al., 2009; Kaltenbach et al., 2007). In a fly model of degeneration, heterozygosity for a loss-of-function allele for PPARγ was a suppressor, suggesting that the interaction with mutant Htt is a detrimental one that can be rescued by decreasing the levels of PPARγ (Kaltenbach et al., 2007). However, coimmunoprecipitation experiments show wild-type Htt more strongly interacting with PPARγ (Futter et al., 2009), suggesting that the mutant protein is defective in its interaction with the receptor. This is more in line with multiple studies showing that PPARγ activation is important for mitochondrial biogenesis, and thus reductions in PPARγ may be expected to be toxic.

Two recent studies have provided additional evidence for an important contribution of PPARγ in HD. The PPARγ coactivator 1α (PGC-1α) transcriptional coactivator is observed to be transcriptionally repressed in cells expressing mutant Htt (Cui et al., 2006). Crossing mutant Htt mice with PGC-1α knockout mice led to enhanced neurodegeneration, whereas overexpression of PGC-1α in a mutant Htt striatal cell line partially suppressed toxicity. Furthermore, a second study showed that HD mice are hypothermic, and PGC-1α-regulated genes are reduced in both brown adipose tissue and the striatum (Weydt et al., 2006). This group additionally observed PGC-1α transcriptional deficits in HD patient striatum, suggesting this defect manifests in the disease as it does in mutant Htt model systems.

Based on its role as a coactivator for PPARγ, the reduced PGC-1α function suggests that activation of PPARγ may be of therapeutic benefit in HD. This is additionally supported by multiple studies describing beneficial effects of PPARγ agonists in a variety of neurodegenerative diseases, including Alzheimer's disease (Inestrosa et al., 2005; Watson et al., 2005), amyotrophic lateral sclerosis (Kiaei et al., 2005), multiple sclerosis (Watson and Craft, 2006), and Parkinson's disease (Breidert et al., 2002; Quinn et al., 2008; Schintu et al., 2009). The existence of agonists and antagonists for PPARγ allows for direct experiments into its role in mutant Htt toxicity.

An agonist of PPARγ, rosiglitazone, has been shown to have protective effects in a mutant Htt striatal cell line challenged by thapsigargin exposure (Quintanilla et al., 2008). Addition of thapsigargin, a disruptor of calcium homeostasis, to mutant Htt cells resulted in mitochondrial dysfunction as indicated by elevated reactive oxygen species production and a decrease in the mitochondrial membrane potential. This effect of thapsigargin was blocked by pretreatment with rosiglitazone, suggesting that PPARγ activation could protect cells from the mitochondrial dysfunction. Importantly, the authors showed that the effects were abrogated if an antagonist of PPARγ, GW9662, was also added.

Finally, the interaction of wild-type Htt with PPARγ is likely reduced in mutant Htt-containing cells. Because mutant Htt also represses the levels of PGC-1α, the activity of PPARγ is blunted on two fronts by the mutant protein. Use of activating PPARγ ligand mimetics should prove effective in overcoming both defects and thus may be promising as therapeutic molecules for the treatment of the disease.

NUCLEAR HORMONE RECEPTORS

Specific interactions between multiple nuclear hormone receptors and Htt have been reported. Kaltenbach et al. (2007) showed an interaction between the retinoic acid receptor α (RARα) and Htt using Y2H. Futter et al. (2009) demonstrated coimmunoprecipitation of the liver X receptor α (LXRα), vitamin D receptor (VDR), and thyroid hormone receptor α (THRα) with Htt. Interactions with these nuclear hormone receptors were stronger with wild-type Htt, and reporters and endogenous targets of LXRα transcriptional activity showed coactivation by wild-type Htt. Mutant Htt did not confer coactivation, leading to the conclusion that the polyglutamine tract expansion results in a loss of this activity.

These results are intriguing in combination with the demonstration of interactions between Htt and N-CoR (Boutell et al., 1999; Kaltenbach et al., 2007) and nuclear receptor coactivator 3 (N-CoA3) (Kaltenbach et al., 2007). N-CoR binds to unliganded nuclear hormone receptors and represses expression of their target genes, whereas N-CoA binds to liganded receptors and activates the expression of their target genes. This supports the model of mutant Htt impacting the expression of hormone-responsive genes by independently determined interactions with the specific and general factors involved in their regulation. The evidence suggests that wild-type Htt is a coactivator of multiple nuclear hormone receptors, whereas mutant Htt is at least defective in coactivation and also impacts the normal repression of these receptors by inducing the mislocalization of their main corepressor, N-CoR, to the cytoplasm.

Thus, this family of transcription-related interaction partners potentially provides one of the most amenable therapeutic targets for HD. This is because of readily available agonists (and cognate ligands) for these factors. Their ligand-binding mechanism of action makes the nuclear hormone receptors an inherently "druggable" group of transcription factors. Additionally, based on the results showing differential interaction and resultant consequences with wild-type and mutant Htt, the use of agonists is likely to restore some of the lost regulation that occurs in the presence of mutant Htt. For these reasons, the use of nuclear hormone receptor agonists may represent tractable therapeutic approaches based on identified protein interactions between Htt and transcription-related proteins.

POTENTIAL DETRIMENTAL INTERACTION PARTNERS OF MUTANT HUNTINGTIN

Of the 58 interactions between transcription-related proteins and Htt reported in the literature, 19 include evidence of a detrimental effect of the interaction. Included among these is the interaction of Htt with itself, based on the evidence for a direct transcriptional role of Htt in the cell (Benn et al., 2008; Zhai et al., 2005). The mechanisms by which mutant Htt can interact with a transcriptional protein with harmful consequences can be generally thought of as one of three nonexclusive possibilities. First, mutant Htt may sequester a prosurvival transcription factor in the cytoplasm or within an intranuclear inclusion where it is unavailable to perform its function. Second, mutant Htt may promote the activity of an apoptotic/cell death-inducing transcription factor. Third, the protein may promote the toxic impact of mutant Htt on the cell. In the following discussion we will attempt to suggest which of these is in play for a given interaction partner based on the evidence to date.

MEF2D

The myocyte enhancer factor 2D (MEF2D) is a transcriptional activator with an established role in commitment of myogenic lineages (Breitbart et al., 1993). Importantly, the MEF2 family of transcription factors has also been shown to have prosurvival roles in neurons (Gong et al., 2003; Ikeshima et al., 1995; Mao et al., 1999; Okamoto et al., 2000, 2002). The regulation of these transcription factors is mediated by a variety of inputs, including phosphorylation (Gong et al., 2003) and antagonistic binding of corepressor proteins that inhibit the expression of MEF2 transcriptional targets.

Using the Y2H method, it was observed that MEF2D interacts with Htt, and this interaction decreases with a longer polyglutamine tract Htt (Kaltenbach et al., 2007). Further, in the same study, heterozygous loss-of-function mutations in *Mef2* suppressed the phenotype caused by expression of a mutant Htt fragment in the *Drosophila* eye. This latter result argues that the interaction between mutant Htt and MEF2D is detrimental and that a moderate reduction of the Mef2 protein level is beneficial for mutant Htt-containing cells, presumably by decreasing the harmful effect of the mutant Htt interaction with Mef2.

To establish MEF2D as a therapeutic target, the mechanism of the harmful interaction with Htt will need to be elucidated. The suppression of mutant Htt toxicity by a heterozygous *Mef2* loss-of-function allele suggests that mutant Htt may promote elevated expression of Mef2 target genes. The neurons perhaps undergo apoptosis in response to inappropriately expressed factors that are necessary and beneficial during development or at particular levels of expression but trigger tumor suppressor pathways when maintained or supplemented by the action of mutant Htt in an inappropriate context. Alternatively, mutant Htt may mediate the assembly of complexes of transcriptional repressors at MEF2D sites, and the reduction of MEF2D levels reduces the recruitment of these to important prosurvival genes. Until the phenotypic consequences of mutant Htt on MEF2D are better understood, it is unclear what intervention approach would be beneficial.

Sp1 AND Sp3

The specificity transcription factor 1 (Sp1) and specificity transcription factor 3 (Sp3) are important transcriptional activators that target particular genes for expression (Cook et al., 1999). In the particular case of Sp1, it is observed to interact with multiple other transcription factors (see Figure 3.1) and plays a major role in gene expression by recruiting the basal transcription machinery in the form of transcription factor IID to its target genes (Pugh and Tjian, 1990). Sp1 also contains a glutamine-rich region (Courey and Tjian, 1988).

The evidence of an interaction between Sp1 and Htt came initially from two sources. Dunah et al. (2002) recognized that mRNA expression arrays of HD mice had changes in genes that contained recognition sites for Sp1 binding. They demonstrated an interaction between Htt and Sp1 and showed that overexpression of Sp1 rescues mutant Htt toxicity and repression at the dopamine D2 receptor gene. Additionally, they showed that mutant Htt prevents Sp1 from binding to DNA. Similarly, Li et al. (2002) initiated another study when they noticed that many genes that were differentially expressed in HD cell lines and mouse models harbored Sp1 recognition motifs. They observed that polyglutamine expansion enhances the interaction of N-terminal Htt with Sp1, and whereas soluble mutant Htt interacts with Sp1, aggregated mutant Htt does not. Using the nerve growth factor receptor gene promoter as a reporter, they found that mutant Htt represses Sp1 transcription, and Sp1 overexpression also suppressed toxicity in their models. Later work further demonstrated direct mutant Htt repression of Sp1-mediated transcription in an *in vitro* transcription system (Zhai et al., 2005).

Because of the central importance of Sp1 in the expression of a large number of genes, therapeutic targeting to relieve Sp1 of repression by mutant Htt should be an effective approach to alleviating the transcriptional dysregulation aspect of the disease. Activation of Sp1 or increasing its levels would be one approach that could counteract the influence of mutant Htt. Studies have demonstrated that Sp1 is activated by phosphorylation regulated by several signal transduction cascades

(Chuang et al., 2008; Horovitz-Fried et al., 2007; Merchant et al., 1999; Tan and Khachigian, 2009; Zheng et al., 2000, 2001), suggesting that growth factors or stimulants of these pathways could be used to increase Sp1 activity. Sp1 is also sumoylated, and this modification results in relocalization to the cytoplasm and enhanced degradation of Sp1 by the proteasome (Wang et al., 2008). Hence, inhibition of Sp1 sumoylation may increase its levels and help offset the problematic interaction of the protein with mutant Htt (see below).

CBP (CREBBP)

The CBP plays essential roles in homeostasis and other processes based on its bridging of transcription factor recognition with chromatin remodeling (Kalkhoven, 2004). The interaction of CBP with CREB, as well as the determination of its transcriptional activation function, established it as having a central role in the cell (Chrivia et al., 1993). CBP is a histone acetyltransferase (Bannister and Kouzarides, 1996; Ogryzko et al., 1996) whose modification of histones results in the transcriptional activation of genes.

The interaction of CBP with Htt has been demonstrated by numerous groups. CBP was found to be trapped in expanded polyglutamine aggregates in a transfected cell line only when its own 19-glutamine stretch was present in the protein (Kazantsev et al., 1999). Htt exon 1 interacts with CBP *in vitro*, the interaction is enhanced in expanded polyglutamine-containing exon 1, and CBP colocalizes with intranuclear Htt inclusions in HD transgenic mice (Steffan et al., 2000). Htt exon 1 can bind to and inhibit the acetyltransferase domain of CBP, as well as that of other acetyltransferases (Steffan et al., 2001). CBP is depleted from its normal nuclear location by mutant Htt in cell culture, HD transgenic mice, and in human HD postmortem brain; expanded polyglutamine Htt represses CBP-activated transcription; and overexpression of CBP rescues polyglutamine-induced neuronal toxicity (Nucifora et al., 2001). A high-throughput Y2H screen found CBP as an interactor of Htt (Kaltenbach et al., 2007). Finally, CBP was recently shown to acetylate mutant Htt, and this modification targets it for degradation by autophagy (Jeong et al., 2009). This last result, in contrast with the other lines of evidence, suggests that the interaction of CBP and mutant Htt may actually be beneficial (see below).

The preponderance of evidence suggests that the interaction between Htt and CBP is detrimental to cells because Htt inhibits CBP function. Thus, the prevention of the interaction of Htt with this critical transcriptional coactivator would most likely prove beneficial. However, this is difficult to accomplish by small molecule inhibitors of the interaction because both the polyQ region and the acetyltransferase domain of CBP can mediate seemingly independent interactions with Htt, suggesting the interaction motifs may include two (or more) surfaces of CBP. Activation of CBP or increasing its levels would be additional strategies to overcome the CBP deficiency caused by mutant Htt.

An alternative approach to activating CBP is to inhibit the enzymes that counteract its activity. Such enzymes, the HDACs, are amenable to targeting by small molecules. The inhibition of HDACs enriches the amount of acetylated histones and nonhistone proteins, accomplishing the same effect as would be expected for increasing the activity or levels of CBP or other acetyltransferase enzymes. HDAC inhibitors have been explored as antitumor therapeutics based on their differentiation-inducing effects (Bolden et al., 2006; Botrugno et al., 2009). More importantly, a number of studies have established a therapeutic benefit of HDAC inhibitors in cell and animal models of HD (Bates et al., 2006; Hockly et al., 2003; Hughes et al., 2001; McCampbell et al., 2001; Pallos et al., 2008; Parker et al., 2005; Steffan et al., 2001). Therefore, modulation of the acetylation state of key proteins is a therapeutic strategy with great potential in the treatment of HD that may restore the deficiency caused by lost CBP activity imparted by its interaction with mutant huntingtin.

Of note, a direct consequence of Htt on CBP has been described in multiple studies, but only one emphasizes a reciprocal impact of CBP on mutant Htt (Jeong et al., 2009). The authors show that CBP acetylates mutant Htt, and this modification causes autophagic clearance of the toxic protein. This suggests a beneficial effect of CBP binding to mutant Htt, with the interaction leading to reduced mutant Htt in the cell. Overexpression of CBP rescues toxicity in this model, as does RNAi knockdown of an HDAC enzyme whose overexpression conversely decreases mutant Htt acetylation. In HD, the binding of mutant Htt to CBP may be a directed process to cause its degradation, but endogenous levels of the two proteins result in CBP becoming sequestered and inactivated by the interaction. Therapeutic effects of deacetylase inhibition could rescue toxicity through multiple targets in a way that is independent of the biological role of the actual protein interaction—be it beneficial or detrimental.

N-CoR AND mSin3A

The N-CoR was identified by its binding to unliganded nuclear receptor heterodimers (Horlein et al., 1995). Unliganded nuclear hormone receptors had been observed to repress transcription, and binding of N-CoR (or the closely related SMRT [Chen and Evans, 1995]) was found to mediate this repression. This most often occurs through the recruitment of complexes containing HDAC activity (Jepsen and Rosenfeld, 2002). N-CoR has been shown to potently regulate such nuclear receptors as PPARγ and VDR (Abedin et al., 2009). Hence, the function of N-CoR in the cell appears to center around maintaining genes that are responsive to hormone/vitamin cues in a repressed state in the absence of their signaling molecules. However, it has recently been shown that at least one ligand can mediate corepression when binding to its receptor in a heterodimeric context, causing the unliganded partner receptor to bind N-CoR or another corepressor (Sanchez-Martinez et al., 2008).

In HD, the function of N-CoR is disrupted. A Y2H study identified a specific interaction of the C-terminal region of N-CoR with Htt that was polyglutamine length-dependent (Boutell et al., 1999). The authors went on to demonstrate that N-CoR was exclusively cytoplasmic in HD cortex and caudate, whereas in control brain the protein is present in both the cytoplasm and the nucleus. This suggests a relocalization of N-CoR in mutant Htt cells, presumably because of an inappropriate interaction with the mutant protein.

A second Y2H study independently identified an interaction between Htt and N-CoR (Kaltenbach et al., 2007). Intriguingly, using their mutant Htt-dependent fly retinal degeneration model, they identified N-CoR as a hemizygous loss-of-function suppressor of toxicity. One possibility that would explain this counterintuitive result is that the relocalization of N-CoR in HD brain confers toxicity potentially through trapping a partner protein in the cytoplasm, and reduction of N-CoR levels decreases this sequestration. Alternatively, N-CoR elicits some as yet unrealized toxic response in the cytoplasm. Further work will be required to establish whether either of these mechanisms may have a role in HD.

One of N-CoR's partner proteins, mammalian *SIN3* (yeast switch independent 3) homologue A (mSin3a), forms transcription-repressing complexes in combination with HDAC1/2 enzymes (Laherty et al., 1997). Boutell et al. (1999) saw the relocalization of mSin3a exclusively to the cytoplasmic compartment in HD patient samples, with the only exceptions being ~5% of neuronal intranuclear inclusions where mSin3a was also observed to be present. An independent study described a polyglutamine length-dependent interaction between mSin3a and Htt (Steffan et al., 2000). Studies to establish the effects of changes in the levels of N-CoR and mSin3a are required before the potential interplay of their redistribution in mutant Htt cells can be understood. However, based on the existing evidence, one can speculate that the trapping of mSin3a in the cytoplasm is a detrimental consequence of mutant huntingtin's enhanced interaction with the N-CoR/mSin3a complex. Reduction of N-CoR levels reduces mSin3a association with mutant Htt and thus maintains some of its presence in the nucleus, allowing it to carry out enough of its nuclear function to reduce the defect.

If this mechanism is correct, targeting of the consequences of this interaction by HDAC inhibition could have beneficial effects. As mentioned previously, the transcriptional repression complexes that contain N-CoR and mSin3a frequently contain HDAC enzymes, and thus their redistribution to the cytoplasm may either sequester HDACs in the cytoplasm or result in an accumulation of HDAC enzymes in the nucleus. As a result of the observation of the toxicity-rescuing effects of HDAC inhibition in a number of HD models (Bates et al., 2006; Hockly et al., 2003; McCampbell et al., 2001; Pallos et al., 2008; Parker et al., 2005; Steffan et al., 2001), the latter mechanism seems likely, but the inappropriate presence of HDACs in the cytoplasm could also have toxic consequences, such as the stabilization of mutant Htt (Jeong et al., 2009). Again, irrespective of the mechanism, HDAC inhibition has potential for therapeutic rescue based on these interactions with mutant Htt.

p53

The tumor suppressor p53 is mutated in a large number of cancer cells (Malkin et al., 1990; Srivastava et al., 1990; Vogelstein, 1990). The mutations largely prevent or reduce binding to DNA, leading to an inability of the protein to regulate transcription of its target genes (Kern et al., 1991a, 1991b). p53, through its transcriptional activity, also plays an important role in promoting neuronal apoptosis (Ra et al., 2009; Uo et al., 2007). The neuronal apoptotic role of p53 and its putative modulation by an interaction with mutant Htt is of particular interest.

An initial description of an interaction between Htt and p53 found *in vitro* that there was no preference of p53 for expanded versus nonexpanded polyglutamine (Steffan et al., 2000). The authors also observed the presence of p53 in mutant Htt aggregates, as well as the ability of Htt to repress transcription of p53-dependent genes using luciferase reporters. Interestingly, a gene that is repressed by p53 could, in a p53 knockout cell line, be repressed by a mutant Htt fragment to a much greater degree than by a wild-type Htt fragment. This suggests that Htt may exacerbate p53 function independently of their physical association.

A later study of the interaction between Htt and p53 established the consequences of their interaction more fully (Bae et al., 2005). The authors showed an interaction between mutant Htt and p53 that is not observed with a second expanded polyglutamine protein. Further, mutant Htt increases p53 nuclear abundance and its transcriptional activity. p53 levels are also increased in transgenic mice and HD patient brains. Most importantly, these authors show that p53 inhibition by genetic deletion, RNAi, and pfithrin-α treatment all lead to reduced cytotoxicity in HD cells.

The results of Bae et al. (2005) support the idea that p53 inhibition could be a therapeutic intervention in HD. Inhibition of a central tumor suppressor such as p53 is not an ideal approach for obvious reasons, but if neural-specific delivery of a small molecule inhibitor or RNAi-mediated knockdown can be accomplished, this may be a useful approach to not only HD but also to other neurodegenerative diseases as well. Chemical compounds that inhibit p53 have been developed and should be investigated in HD models (Nayak et al., 2009).

SUMOYLATION

The attachment of the SUMO to the lysine side chains of proteins can have varied consequences on the substrate protein (Hay, 2005; Johnson, 2004). In terms of transcription-related consequences, sumoylation of proteins can cause translocation into the nucleus, stabilization of the protein, or its degradation. These effects are likely mediated by one of three mechanisms. Attachment of SUMO to a protein permits its recognition by SUMO-binding proteins, causing the substrate protein to be targeted by the relevant activity. A second possibility is that the modification by SUMO competes with an alternative modification at a given lysine residue. For example, attachment of ubiquitin to a particular lysine of a protein may promote its degradation, whereas attachment of SUMO to the same residue prevents both the attachment of ubiquitin and the resultant degradation. Third, attachment of SUMO may modulate the post-translational modification of other sites on the protein, leading to their enhancement or inhibition.

The interaction of SUMO with Htt is through covalent attachment of SUMO to one or more of three N-terminal lysine residues of mutant Htt (Steffan et al., 2004; Subramaniam et al., 2009). This modification results in the stabilization of mutant Htt, and sumoylation promotes cytotoxicity. Steffan et al. (2004) also showed that sumoylation of mutant Htt enhanced repression of a luciferase reporter driven by a mutant Htt-responsive promoter and that hemizygosity of the SUMO homologue Smt3 reduced neurodegeneration as a result of mutant Htt in flies. These results suggest that direct prevention of Htt sumoylation may be an effective therapeutic target and, further, that a general reduction of the cellular SUMO1 pool may prove beneficial. Subsequent studies have confirmed that SUMO plays a key role in the cytotoxicity of mutant Htt (Subramaniam et al., 2009).

A strategy incorporating the general inhibition of sumoylation is also supported by the effect of sumoylation on several of the interaction partners of Htt presented here. coREST, a corepressor with REST, is sumoylated, and, in the absence of SUMO1, coREST repressive activity is decreased (Muraoka et al., 2008). CBP modification by SUMO1 reduces CBP transcriptional activity by promoting an interaction with a transcriptional corepressor (Kuo et al., 2005). One study (Spengler and Brattain, 2006) found that Sp1 is sumoylated, and this inhibits its transcriptional activity; a second study (Wang et al., 2008) showed that sumoylation of Sp1 led to its cytoplasmic localization and interaction with a proteasome component, resulting in its degradation. Sumoylation of Sp3 causes it to have transcriptional repressing activity by inducing proximal heterochromatic-based silencing (Stielow et al., 2008). SREBF2 is also sumoylated, and the modification results in reduced transcriptional activity of the factor (Hirano et al., 2003).

Based on the toxicity-promoting role of SUMO modification of mutant Htt and the likely exacerbating inhibitory effects it has on multiple protein partners of Htt, inhibition of sumoylation is a strong candidate for rescuing mutant Htt toxicity. Recently, an inhibitor of the initial step in the sumoylation enzymatic cascade has been identified. Ginkgolic acid and the analogous anacardic acid were shown to bind to the SUMO E1 enzyme and block the creation of the E1-SUMO intermediate (Fukuda et al., 2009). Therefore, these compounds may provide a way to reduce overall sumoylation within the cell. In cells with mutant Htt, therapeutic intervention working on multiple proteins that are SUMO substrates could potentially suppress multiple defects caused by mutant Htt.

TRANSCRIPTION PROTEIN INTERACTION SUMMARY

In conclusion, the interaction of Htt with transcription-related proteins consists of both normal cellular associations and inappropriate mutant-specific protein complexes. Strategies to promote the interactions that wild-type Htt engages in within the context of HD cells may restore lost functions of critical importance to the cell. On the other hand, the inhibition of interactions of proteins with mutant Htt may suppress harmful impacts of the mutant protein. Targeting of such interactions with small-molecule therapeutics is nontrivial because of the large relative size of the interacting surfaces of two proteins. Hence, the alternative approach of increasing the levels of a prosurvival interaction partner or inhibiting the activity or amounts of a detrimental protein interactor may more readily accomplish the same goal. Nuclear hormone receptor agonists, HDAC inhibitors, and sumoylation inhibitors are three strategies that appear to have great promise and will hopefully fulfill that potential in being developed into therapeutics for HD. Finally, the value of knowledge concerning protein-interacting partners such as those transcription factors discussed here goes beyond strategies designed to target these specific proteins. Ultimately, information about the specific pathways that lie downstream of these factors (e.g., sterol biosynthesis) can also provide inroads into therapeutic candidates.

In this chapter we have attempted to outline the ways in which protein interaction discovery can lead us not only into thinking about pathogenic mechanisms in HD but also how protein interactions can lead to specific targets and thereby

strategies for therapeutic intervention. Selecting from the breadth of protein interactions and pathogenic mechanisms involved in HD, we focused on one central
aspect of HD pathobiology, that of transcriptional dysregulation. However, the principles of using interacting proteins as a means of understanding disease pathways
in terms of drug discovery is generally applicable to other implicated processes.
In the past decade, the field of protein interaction studies has exploded from single
investigator/single target-driven activities to highly automated, massively parallel
proteome-scale research enterprises. The proliferation of complex interaction networks being reported in the literature for humans (as well as other organisms of
biological and medical interest) is exciting and accelerating. Many of these studies
are unbiased in nature in that they focus on the broad class of human proteins rather
than specific types of proteins, such as huntingtin. The two large-scale HD-specific
protein interaction networks constructed to date provide an opportunity to understand targets not just in the context of binary relationships but rather at a network
level. Computational methods that will allow us to properly mine the data contained
in these networks are just beginning to be developed, and they should soon help us
identify the signals that exist among the noise in these complex datasets. It is clear
that HD is highly complex in the number and nature of biological pathways that
have been demonstrated to be perturbed by the expression of the mutant huntingtin
protein. Ultimately, a comprehensive understanding of this complex disease and
the volumes of information being generated by its study will require increasingly
complex methods of analysis to realize fully the development of comprehensive
therapeutic strategies.

ACKNOWLEDGMENTS

We thank the Hereditary Disease Foundation, CHDI Foundation Inc., and the
National Institutes of Health for their support.

REFERENCES

Abedin, S.A., Thorne, J.L., Battaglia, S., Maguire, O., Hornung, L.B., Doherty, A.P., Mills,
    I.G. and Campbell, M.J. (2009) Elevated NCOR1 disrupts a network of dietary-sensing
    nuclear receptors in bladder cancer cells. *Carcinogenesis*, **30**, 449–456.
Bae, B.I., Xu, H., Igarashi, S., Fujimuro, M., Agrawal, N., Taya, Y., Hayward, S.D., Moran,
    T.H., Montell, C., Ross, C.A., Snyder, S.H. and Sawa, A. (2005) p53 mediates cellular
    dysfunction and behavioral abnormalities in Huntington's disease. *Neuron*, **47**, 29–41.
Bannister, A.J. and Kouzarides, T. (1996) The CBP co-activator is a histone acetyltransferase.
    Nature, **384**, 641–643.
Bates, E.A., Victor, M., Jones, A.K., Shi, Y. and Hart, A.C. (2006) Differential contributions
    of Caenorhabditis elegans histone deacetylases to huntingtin polyglutamine toxicity. *J
    Neurosci*, **26**, 2830–2838.
Benn, C.L., Sun, T., Sadri-Vakili, G., McFarland, K.N., DiRocco, D.P., Yohrling, G.J., Clark,
    T.W., Bouzou, B. and Cha, J.H. (2008) Huntingtin modulates transcription, occupies
    gene promoters in vivo, and binds directly to DNA in a polyglutamine-dependent man-
    ner. *J Neurosci*, **28**, 10720–10733.
Bolden, J.E., Peart, M.J. and Johnstone, R.W. (2006) Anticancer activities of histone deacety-
    lase inhibitors. *Nat Rev Drug Discov*, **5**, 769–784.

Bordet, R., Ouk, T., Petrault, O., Gele, P., Gautier, S., Laprais, M., Deplanque, D., Duriez, P., Staels, B., Fruchart, J.C. and Bastide, M. (2006) PPAR: a new pharmacological target for neuroprotection in stroke and neurodegenerative diseases. *Biochem Soc Trans*, **34**, 1341–1346.

Botrugno, O.A., Santoro, F. and Minucci, S. (2009) Histone deacetylase inhibitors as a new weapon in the arsenal of differentiation therapies of cancer. *Cancer Lett*, **280**, 134–144.

Boutell, J.M., Thomas, P., Neal, J.W., Weston, V.J., Duce, J., Harper, P.S. and Jones, A.L. (1999) Aberrant interactions of transcriptional repressor proteins with the Huntington's disease gene product, huntingtin. *Hum Mol Genet*, **8**, 1647–1655.

Breidert, T., Callebert, J., Heneka, M.T., Landreth, G., Launay, J.M. and Hirsch, E.C. (2002) Protective action of the peroxisome proliferator-activated receptor-gamma agonist pioglitazone in a mouse model of Parkinson's disease. *J Neurochem*, **82**, 615–624.

Breitbart, R.E., Liang, C.S., Smoot, L.B., Laheru, D.A., Mahdavi, V. and Nadal-Ginard, B. (1993) A fourth human MEF2 transcription factor, hMEF2D, is an early marker of the myogenic lineage. *Development*, **118**, 1095–1106.

Brown, M.S. and Goldstein, J.L. (1997) The SREBP pathway: regulation of cholesterol metabolism by proteolysis of a membrane-bound transcription factor. *Cell*, **89**, 331–340.

Butler, R. and Bates, G.P. (2006) Histone deacetylase inhibitors as therapeutics for polyglutamine disorders. *Nat Rev Neurosci*, **7**, 784–796.

Cha, J.H. (2000) Transcriptional dysregulation in Huntington's disease. *Trends Neurosci*, **23**, 387–392.

Cha, J.H. (2007) Transcriptional signatures in Huntington's disease. *Prog Neurobiol*, **83**, 228–248.

Chen, J.D. and Evans, R.M. (1995) A transcriptional co-repressor that interacts with nuclear hormone receptors. *Nature*, **377**, 454–457.

Chong, J.A., Tapia-Ramirez, J., Kim, S., Toledo-Aral, J.J., Zheng, Y., Boutros, M.C., Altshuller, Y.M., Frohman, M.A., Kraner, S.D. and Mandel, G. (1995) REST: a mammalian silencer protein that restricts sodium channel gene expression to neurons. *Cell*, **80**, 949–957.

Chrivia, J.C., Kwok, R.P., Lamb, N., Hagiwara, M., Montminy, M.R. and Goodman, R.H. (1993) Phosphorylated CREB binds specifically to the nuclear protein CBP. *Nature*, **365**, 855–859.

Chuang, J.Y., Wang, Y.T., Yeh, S.H., Liu, Y.W., Chang, W.C. and Hung, J.J. (2008) Phosphorylation by c-Jun NH2-terminal kinase 1 regulates the stability of transcription factor Sp1 during mitosis. *Mol Biol Cell*, **19**, 1139–1151.

Collins, M.O. and Choudhary, J.S. (2008) Mapping multiprotein complexes by affinity purification and mass spectrometry. *Curr Opin Biotechnol*, **19**, 324–330.

Cook, T., Gebelein, B. and Urrutia, R. (1999) Sp1 and its likes: biochemical and functional predictions for a growing family of zinc finger transcription factors. *Ann N Y Acad Sci*, **880**, 94–102.

Courey, A.J. and Tjian, R. (1988) Analysis of Sp1 in vivo reveals multiple transcriptional domains, including a novel glutamine-rich activation motif. *Cell*, **55**, 887–898.

Cronin, S.J., Nehme, N.T., Limmer, S., Liegeois, S., Pospisilik, J.A., Schramek, D., Leibbrandt, A., Simoes Rde, M., Gruber, S., Puc, U. et al. (2009) Genome-wide RNAi screen identifies genes involved in intestinal pathogenic bacterial infection. *Science*, **325**, 340–343.

Cui, L., Jeong, H., Borovecki, F., Parkhurst, C.N., Tanese, N. and Krainc, D. (2006) Transcriptional repression of PGC-1alpha by mutant huntingtin leads to mitochondrial dysfunction and neurodegeneration. *Cell*, **127**, 59–69.

Dunah, A.W., Jeong, H., Griffin, A., Kim, Y.M., Standaert, D.G., Hersch, S.M., Mouradian, M.M., Young, A.B., Tanese, N. and Krainc, D. (2002) Sp1 and TAFII130 transcriptional activity disrupted in early Huntington's disease. *Science*, **296**, 2238–2243.

Ewing, A.K., Attner, M. and Chakravarti, D. (2007) Novel regulatory role for human Acf1 in transcriptional repression of vitamin D3 receptor-regulated genes. *Mol Endocrinol*, **21**, 1791–1806.

Fields, S. and Song, O. (1989) A novel genetic system to detect protein-protein interactions. *Nature*, **340**, 245–246.

Fukuda, I., Ito, A., Hirai, G., Nishimura, S., Kawasaki, H., Saitoh, H., Kimura, K., Sodeoka, M. and Yoshida, M. (2009) Ginkgolic acid inhibits protein SUMOylation by blocking formation of the E1-SUMO intermediate. *Chem Biol*, **16**, 133–140.

Futter, M., Diekmann, H., Schoenmakers, E., Sadiq, O., Chatterjee, K. and Rubinsztein, D.C. (2009) Wild-type but not mutant huntingtin modulates the transcriptional activity of liver X receptors. *J Med Genet*, **46**, 438–446.

Gavin, A.C., Bosche, M., Krause, R., Grandi, P., Marzioch, M., Bauer, A., Schultz, J., Rick, J.M., Michon, A.M., Cruciat, C.M. et al. (2002) Functional organization of the yeast proteome by systematic analysis of protein complexes. *Nature*, **415**, 141–147.

Gerber, H.P., Seipel, K., Georgiev, O., Hofferer, M., Hug, M., Rusconi, S. and Schaffner, W. (1994) Transcriptional activation modulated by homopolymeric glutamine and proline stretches. *Science*, **263**, 808–811.

Goehler, H., Lalowski, M., Stelzl, U., Waelter, S., Stroedicke, M., Worm, U., Droege, A., Lindenberg, K.S., Knoblich, M., Haenig, C., Herbst, M. et al. (2004) A protein interaction network links GIT1, an enhancer of huntingtin aggregation, to Huntington's disease. *Mol Cell*, **15**, 853–865.

Gong, X., Tang, X., Wiedmann, M., Wang, X., Peng, J., Zheng, D., Blair, L.A., Marshall, J. and Mao, Z. (2003) Cdk5-mediated inhibition of the protective effects of transcription factor MEF2 in neurotoxicity-induced apoptosis. *Neuron*, **38**, 33–46.

Harjes, P. and Wanker, E.E. (2003) The hunt for huntingtin function: interaction partners tell many different stories. *Trends Biochem Sci*, **28**, 425–433.

Hay, R.T. (2005) SUMO: a history of modification. *Mol Cell*, **18**, 1–12.

He, X., Fan, H.Y., Narlikar, G.J. and Kingston, R.E. (2006) Human ACF1 alters the remodeling strategy of SNF2h. *J Biol Chem*, **281**, 28636–28647.

Heneka, M.T. and Landreth, G.E. (2007) PPARs in the brain. *Biochim Biophys Acta*, **1771**, 1031–1045.

Heneka, M.T., Landreth, G.E. and Hull, M. (2007) Drug insight: effects mediated by peroxisome proliferator-activated receptor-gamma in CNS disorders. *Nat Clin Pract Neurol*, **3**, 496–504.

Hirano, Y., Murata, S., Tanaka, K., Shimizu, M. and Sato, R. (2003) Sterol regulatory element-binding proteins are negatively regulated through SUMO-1 modification independent of the ubiquitin/26 S proteasome pathway. *J Biol Chem*, **278**, 16809–16819.

Ho, Y., Gruhler, A., Heilbut, A., Bader, G.D., Moore, L., Adams, S.L., Millar, A., Taylor, P., Bennett, K. Boutilier, K. et al. (2002) Systematic identification of protein complexes in Saccharomyces cerevisiae by mass spectrometry. *Nature*, **415**, 180–183.

Hockly, E., Richon, V.M., Woodman, B., Smith, D.L., Zhou, X., Rosa, E., Sathasivam, K., Ghazi-Noori, S., Mahal, A., Lowden, P.A. et al. (2003) Suberoylanilide hydroxamic acid, a histone deacetylase inhibitor, ameliorates motor deficits in a mouse model of Huntington's disease. *Proc Natl Acad Sci U S A*, **100**, 2041–2046.

Horlein, A.J., Naar, A.M., Heinzel, T., Torchia, J., Gloss, B., Kurokawa, R., Ryan, A., Kamei, Y., Soderstrom, M., Glass, C.K. et al. (1995) Ligand-independent repression by the thyroid hormone receptor mediated by a nuclear receptor co-repressor. *Nature*, **377**, 397–404.

Horovitz-Fried, M., Jacob, A.I., Cooper, D.R. and Sampson, S.R. (2007) Activation of the nuclear transcription factor SP-1 by insulin rapidly increases the expression of protein kinase C delta in skeletal muscle. *Cell Signal*, **19**, 556–562.

Hughes, R.E. (2002) Polyglutamine disease: acetyltransferases awry. *Curr Biol*, **12**, R141–143.

Hughes, R.E., Lo, R.S., Davis, C., Strand, A.D., Neal, C.L., Olson, J.M. and Fields, S. (2001) Altered transcription in yeast expressing expanded polyglutamine. *Proc Natl Acad Sci U S A*, **98**, 13201–13206.

Huntington's Disease Collaborative Research Group (1993) A novel gene containing a tri-nucleotide repeat that is expanded and unstable on Huntington's disease chromosomes. *Cell*, **72**, 971–983.

Ikeshima, H., Imai, S., Shimoda, K., Hata, J. and Takano, T. (1995) Expression of a MADS box gene, MEF2D, in neurons of the mouse central nervous system: implication of its binary function in myogenic and neurogenic cell lineages. *Neurosci Lett*, **200**, 117–120.

Inestrosa, N.C., Godoy, J.A., Quintanilla, R.A., Koenig, C.S. and Bronfman, M. (2005) Peroxisome proliferator-activated receptor gamma is expressed in hippocampal neurons and its activation prevents beta-amyloid neurodegeneration: role of Wnt signaling. *Exp Cell Res*, **304**, 91–104.

Jeong, H., Then, F., Melia, T.J. Jr., Mazzulli, J.R., Cui, L., Savas, J.N., Voisine, C., Paganetti, P., Tanese, N., Hart, A.C. et al. (2009) Acetylation targets mutant huntingtin to autopha-gosomes for degradation. *Cell*, **137**, 60–72.

Jepsen, K. and Rosenfeld, M.G. (2002) Biological roles and mechanistic actions of co-repressor complexes. *J Cell Sci*, **115**, 689–698.

Johnson, E.S. (2004) Protein modification by SUMO. *Annu Rev Biochem*, **73**, 355–382.

Kaiser, P., Meierhofer, D., Wang, X. and Huang, L. (2008) Tandem affinity purification combined with mass spectrometry to identify components of protein complexes. *Methods Mol Biol*, **439**, 309–326.

Kalkhoven, E. (2004) CBP and p300: HATs for different occasions. *Biochem Pharmacol*, **68**, 1145–1155.

Kaltenbach, L.S., Romero, E., Becklin, R.R., Chettier, R., Bell, R., Phansalkar, A., Strand, A., Torcassi, C., Savage, J., Hurlburt, A. et al. (2007) Huntingtin interacting proteins are genetic modifiers of neurodegeneration. *PLoS Genet*, **3**, e82.

Kazantsev, A., Preisinger, E., Dranovsky, A., Goldgaber, D. and Housman, D. (1999) Insoluble detergent-resistant aggregates form between pathological and nonpathological lengths of polyglutamine in mammalian cells. *Proc Natl Acad Sci U S A*, **96**, 11404–11409.

Kazantsev, A.G. and Hersch, S.M. (2007) Drug targeting of dysregulated transcription in Huntington's disease. *Prog Neurobiol*, **83**, 249–259.

Kegel, K.B., Meloni, A.R., Yi, Y., Kim, Y.J., Doyle, E., Cuiffo, B.G., Sapp, E., Wang, Y., Qin, Z.H., Chen, J.D. et al. (2002) Huntingtin is present in the nucleus, interacts with the transcriptional corepressor C-terminal binding protein, and represses transcription. *J Biol Chem*, **277**, 7466–7476.

Kern, S.E., Kinzler, K.W., Baker, S.J., Nigro, J.M., Rotter, V., Levine, A.J., Friedman, P., Prives, C. and Vogelstein, B. (1991a) Mutant p53 proteins bind DNA abnormally in vitro. *Oncogene*, **6**, 131–136.

Kern, S.E., Kinzler, K.W., Bruskin, A., Jarosz, D., Friedman, P., Prives, C. and Vogelstein, B. (1991b) Identification of p53 as a sequence-specific DNA-binding protein. *Science*, **252**, 1708–1711.

Kiaei, M., Kipiani, K., Chen, J., Calingasan, N.Y. and Beal, M.F. (2005) Peroxisome prolif-erator-activated receptor-gamma agonist extends survival in transgenic mouse model of amyotrophic lateral sclerosis. *Exp Neurol*, **191**, 331–336.

Krishnan, M.N., Ng, A., Sukumaran, B., Gilfoy, F.D., Uchil, P.D., Sultana, H., Brass, A.L., Adametz, R., Tsui, M., Qian, F. et al. (2008) RNA interference screen for human genes associated with West Nile virus infection. *Nature*, **455**, 242–245.

Krogan, N.J., Cagney, G., Yu, H., Zhong, G., Guo, X., Ignatchenko, A., Li, J., Pu, S., Datta, N., Tikuisis, A.P. et al. (2006) Global landscape of protein complexes in the yeast Saccharomyces cerevisiae. *Nature*, **440**, 637–643.

Kuo, H.Y., Chang, C.C., Jeng, J.C., Hu, H.M., Lin, D.Y., Maul, G.G., Kwok, R.P. and Shih, H.M. (2005) SUMO modification negatively modulates the transcriptional activity of CREB-binding protein via the recruitment of Daxx. *Proc Natl Acad Sci U S A*, **102**, 16973–16978.

Kuppuswamy, M., Vijayalingam, S., Zhao, L.J., Zhou, Y., Subramanian, T., Ryerse, J. and Chinnadurai, G. (2008) Role of the PLDLS-binding cleft region of CtBP1 in recruitment of core and auxiliary components of the corepressor complex. *Mol Cell Biol*, **28**, 269–281.

Laherty, C.D., Yang, W.M., Sun, J.M., Davie, J.R., Seto, E. and Eisenman, R.N. (1997) Histone deacetylases associated with the mSin3 corepressor mediate mad transcriptional repression. *Cell*, **89**, 349–356.

Landles, C. and Bates, G.P. (2004) Huntingtin and the molecular pathogenesis of Huntington's disease. Fourth in molecular medicine review series. *EMBO Rep*, **5**, 958–963.

Li, S.H., Cheng, A.L., Zhou, H., Lam, S., Rao, M., Li, H. and Li, X.J. (2002) Interaction of Huntington disease protein with transcriptional activator Sp1. *Mol Cell Biol*, **22**, 1277–1287.

Li, S.H. and Li, X.J. (2004) Huntingtin-protein interactions and the pathogenesis of Huntington's disease. *Trends Genet*, **20**, 146–154.

Luo, J., Emanuele, M.J., Li, D., Creighton, C.J., Schlabach, M.R., Westbrook, T.F., Wong, K.K. and Elledge, S.J. (2009) A genome-wide RNAi screen identifies multiple synthetic lethal interactions with the Ras oncogene. *Cell*, **137**, 835–848.

Maglott, D., Ostell, J., Pruitt, K.D. and Tatusova, T. (2005) Entrez Gene: gene-centered information at NCBI. *Nucleic Acids Res*, **33**, D54–58.

Makar, T.K., Bever, C.T., Singh, I.S., Royal, W., Sahu, S.N., Sura, T.P., Sultana, S., Sura, K.T., Patel, N., Dhib-Jalbut, S. and Trisler, D. (2009) Brain-derived neurotrophic factor gene delivery in an animal model of multiple sclerosis using bone marrow stem cells as a vehicle. *J Neuroimmunol*, **210**, 40–51.

Makar, T.K., Trisler, D., Eglitis, M.A., Mouradian, M.M. and Dhib-Jalbut, S. (2004) Brain-derived neurotrophic factor (BDNF) gene delivery into the CNS using bone marrow cells as vehicles in mice. *Neurosci Lett*, **356**, 215–219.

Malkin, D., Li, F.P., Strong, L.C., Fraumeni, J.F. Jr., Nelson, C.E., Kim, D.H., Kassel, J., Gryka, M.A., Bischoff, F.Z., Tainsky, M.A. et al. (1990) Germ line p53 mutations in a familial syndrome of breast cancer, sarcomas, and other neoplasms. *Science*, **250**, 1233–1238.

Mao, Z., Bonni, A., Xia, F., Nadal-Vicens, M. and Greenberg, M.E. (1999) Neuronal activity- dependent cell survival mediated by transcription factor MEF2. *Science*, **286**, 785–790.

McCampbell, A., Taye, A.A., Whitty, L., Penney, E., Steffan, J.S. and Fischbeck, K.H. (2001) Histone deacetylase inhibitors reduce polyglutamine toxicity. *Proc Natl Acad Sci U S A*, **98**, 15179–15184.

Merchant, J.L., Du, M. and Todisco, A. (1999) Sp1 phosphorylation by Erk 2 stimulates DNA binding. *Biochem Biophys Res Commun*, **254**, 454–461.

Miller, J. and Stagljar, I. (2004) Using the yeast two-hybrid system to identify interacting proteins. *Methods Mol Biol*, **261**, 247–262.

Muraoka, A., Maeda, A., Nakahara, N., Yokota, M., Nishida, T., Maruyama, T. and Ohshima, T. (2008) Sumoylation of CoREST modulates its function as a transcriptional repressor. *Biochem Biophys Res Commun*, **377**, 1031–1035.

Nayak, S.K., Panesar, P.S. and Kumar, H. (2009) p53-induced apoptosis and inhibitors of p53. *Curr Med Chem*, **16**, 2627–2640.

Nucifora, F.C. Jr., Sasaki, M., Peters, M.F., Huang, H., Cooper, J.K., Yamada, M., Takahashi, H., Tsuji, S., Troncoso, J., Dawson, V.L. et al. (2001) Interference by huntingtin and atrophin-1 with cbp-mediated transcription leading to cellular toxicity. *Science*, **291**, 2423–2428.

Ogryzko, V.V., Schiltz, R.L., Russanova, V., Howard, B.H. and Nakatani, Y. (1996) The transcriptional coactivators p300 and CBP are histone acetyltransferases. *Cell*, **87**, 953–959.

Okamoto, S., Krainc, D., Sherman, K. and Lipton, S.A. (2000) Antiapoptotic role of the p38 mitogen-activated protein kinase-myocyte enhancer factor 2 transcription factor pathway during neuronal differentiation. *Proc Natl Acad Sci U S A*, **97**, 7561–7566.

Okamoto, S., Li, Z., Ju, C., Scholzke, M.N., Mathews, E., Cui, J., Salvesen, G.S., Bossy-Wetzel, E. and Lipton, S.A. (2002) Dominant-interfering forms of MEF2 generated by caspase cleavage contribute to NMDA-induced neuronal apoptosis. *Proc Natl Acad Sci U S A*, **99**, 3974–3979.

Pallos, J., Bodai, L., Lukacsovich, T., Purcell, J.M., Steffan, J.S., Thompson, L.M. and Marsh, J.L. (2008) Inhibition of specific HDACs and sirtuins suppresses pathogenesis in a Drosophila model of Huntington's disease. *Hum Mol Genet*, **17**, 3767–3775.

Parker, J.A., Arango, M., Abderrahmane, S., Lambert, E., Tourette, C., Catoire, H. and Neri, C. (2005) Resveratrol rescues mutant polyglutamine cytotoxicity in nematode and mammalian neurons. *Nat Genet*, **37**, 349–350.

Pugh, B.F. and Tjian, R. (1990) Mechanism of transcriptional activation by Sp1: evidence for coactivators. *Cell*, **61**, 1187–1197.

Quinn, L.P., Crook, B., Hows, M.E., Vidgeon-Hart, M., Chapman, H., Upton, N., Medhurst, A.D. and Virley, D.J. (2008) The PPARgamma agonist pioglitazone is effective in the MPTP mouse model of Parkinson's disease through inhibition of monoamine oxidase B. *Br J Pharmacol*, **154**, 226–233.

Quintanilla, R.A., Jin, Y.N., Fuenzalida, K., Bronfman, M. and Johnson, G.V. (2008) Rosiglitazone treatment prevents mitochondrial dysfunction in mutant huntingtin-expressing cells: possible role of peroxisome proliferator-activated receptor-gamma (PPARgamma) in the pathogenesis of Huntington disease. *J Biol Chem*, **283**, 25628–25637.

Ra, H., Kim, H.L., Lee, H.W. and Kim, Y.H. (2009) Essential role of p53 in TPEN-induced neuronal apoptosis. *FEBS Lett*, **583**, 1516–1520.

Rigamonti, D., Bolognini, D., Mutti, C., Zuccato, C., Tartari, M., Sola, F., Valenza, M., Kazantsev, A.G. and Cattaneo, E. (2007) Loss of huntingtin function complemented by small molecules acting as repressor element 1/neuron restrictive silencer element silencer modulators. *J Biol Chem*, **282**, 24554–24562.

Romero, E., Cha, G.H., Verstreken, P., Ly, C.V., Hughes, R.E., Bellen, H.J. and Botas, J. (2008) Suppression of neurodegeneration and increased neurotransmission caused by expanded full-length huntingtin accumulating in the cytoplasm. *Neuron*, **57**, 27–40.

Rual, J.F., Venkatesan, K., Hao, T., Hirozane-Kishikawa, T., Dricot, A., Li, N., Berriz, G.F., Gibbons, F.D., Dreze, M., Ayivi-Guedehoussou, N. et al. (2005) Towards a proteome-scale map of the human protein-protein interaction network. *Nature*, **437**, 1173–1178.

Sanchez-Martinez, R., Zambrano, A., Castillo, A.I. and Aranda, A. (2008) Vitamin D-dependent recruitment of corepressors to vitamin D/retinoid X receptor heterodimers. *Mol Cell Biol*, **28**, 3817–3829.

Schintu, N., Frau, L., Ibba, M., Caboni, P., Garau, A., Carboni, E. and Carta, A.R. (2009) PPAR-gamma-mediated neuroprotection in a chronic mouse model of Parkinson's disease. *Eur J Neurosci*, **29**, 954–963.

Schoenherr, C.J. and Anderson, D.J. (1995) The neuron-restrictive silencer factor (NRSF): a coordinate repressor of multiple neuron-specific genes. *Science*, **267**, 1360–1363.

Schwikowski, B., Uetz, P. and Fields, S. (2000) A network of protein-protein interactions in yeast. *Nat Biotechnol*, **18**, 1257–1261.

Spengler, M.L. and Brattain, M.G. (2006) Sumoylation inhibits cleavage of Sp1 N-terminal negative regulatory domain and inhibits Sp1-dependent transcription. *J Biol Chem*, **281**, 5567–5574.

Srivastava, S., Zou, Z.Q., Pirollo, K., Blattner, W. and Chang, E.H. (1990) Germ-line trans- mission of a mutated p53 gene in a cancer-prone family with Li-Fraumeni syndrome. *Nature*, **348**, 747–749.

Steffan, J.S., Agrawal, N., Pallos, J., Rockabrand, E., Trotman, L.C., Slepko, N., Illes, K., Lukacsovich, T., Zhu, Y.Z., Cattaneo, E. et al. (2004) SUMO modification of Huntingtin and Huntington's disease pathology. *Science*, **304**, 100–104.

Steffan, J.S., Bodai, L., Pallos, J., Poelman, M., McCampbell, A., Apostol, B.L., Kazantsev, A., Schmidt, E., Zhu, Y.Z., Greenwald, M. et al. (2001) Histone deacetylase inhibi- tors arrest polyglutamine-dependent neurodegeneration in Drosophila. *Nature*, **413**, 739–743.

Steffan, J.S., Kazantsev, A., Spasic-Boskovic, O., Greenwald, M., Zhu, Y.Z., Gohler, H., Wanker, E.E., Bates, G.P., Housman, D.E. and Thompson, L.M. (2000) The Huntington's disease protein interacts with p53 and CREB-binding protein and represses transcrip- tion. *Proc Natl Acad Sci U S A*, **97**, 6763–6768.

Stelzl, U., Worm, U., Lalowski, M., Haenig, C., Brembeck, F.H., Goehler, H., Stroedicke, M., Zenkner, M., Schoenherr, A., Koeppen, S. et al. (2005) A human protein-protein interac- tion network: a resource for annotating the proteome. *Cell*, **122**, 957–968.

Stielow, B., Sapetschnig, A., Wink, C., Kruger, I. and Suske, G. (2008) SUMO-modified Sp3 represses transcription by provoking local heterochromatic gene silencing. *EMBO Rep*, **9**, 899–906.

Strand, A.D., Baquet, Z.C., Aragaki, A.K., Holmans, P., Yang, L., Cleren, C., Beal, M.F., Jones, L., Kooperberg, C., Olson, J.M. and Jones, K.R. (2007) Expression profiling of Huntington's disease models suggests that brain-derived neurotrophic factor depletion plays a major role in striatal degeneration. *J Neurosci*, **27**, 11758–11768.

Subramaniam, S., Sixt, K.M., Barrow, R. and Snyder, S.H. (2009) Rhes, a striatal specific protein, mediates mutant-huntingtin cytotoxicity. *Science*, **324**, 1327–1330.

Sugars, K.L. and Rubinsztein, D.C. (2003) Transcriptional abnormalities in Huntington dis- ease. *Trends Genet*, **19**, 233–238.

Tan, N.Y. and Khachigian, L.M. (2009) Sp1 phosphorylation and its regulation of gene tran- scription. *Mol Cell Biol*, **29**, 2483–2488.

Tontonoz, P., Hu, E. and Spiegelman, B.M. (1994) Stimulation of adipogenesis in fibroblasts by PPAR gamma 2, a lipid-activated transcription factor. *Cell*, **79**, 1147–1156.

Tontonoz, P., Hu, E. and Spiegelman, B.M. (1995) Regulation of adipocyte gene expression and differentiation by peroxisome proliferator activated receptor gamma. *Curr Opin Genet Dev*, **5**, 571–576.

Uetz, P., Giot, L., Cagney, G., Mansfield, T.A., Judson, R.S., Knight, J.R., Lockshon, D., Narayan, V., Srinivasan, M., Pochart, P. et al. (2000) A comprehensive analysis of pro- tein-protein interactions in Saccharomyces cerevisiae. *Nature*, **403**, 623–627.

Uo, T., Kinoshita, Y. and Morrison, R.S. (2007) Apoptotic actions of p53 require transcrip- tional activation of PUMA and do not involve a direct mitochondrial/cytoplasmic site of action in postnatal cortical neurons. *J Neurosci*, **27**, 12198–12210.

Valenza, M., Carroll, J.B., Leoni, V., Bertram, L.N., Bjorkhem, I., Singaraja, R.R., Di Donato, S., Lutjohann, D., Hayden, M.R. and Cattaneo, E. (2007a) Cholesterol biosynthesis pathway is disturbed in YAC128 mice and is modulated by huntingtin mutation. *Hum Mol Genet*, **16**, 2187–2198.

Valenza, M., Leoni, V., Tarditi, A., Mariotti, C., Bjorkhem, I., Di Donato, S. and Cattaneo, E. (2007b) Progressive dysfunction of the cholesterol biosynthesis pathway in the R6/2 mouse model of Huntington's disease. *Neurobiol Dis*, **28**, 133–142.

Valenza, M., Rigamonti, D., Goffredo, D., Zuccato, C., Fenu, S., Jamot, L., Strand, A., Tarditi, A., Woodman, B., Racchi, M. et al. (2005) Dysfunction of the cholesterol biosynthetic pathway in Huntington's disease. *J Neurosci*, **25**, 9932–9939.

Vogelstein, B. (1990) Cancer. A deadly inheritance. *Nature*, **348**, 681–682.

Wang, X., Sato, R., Brown, M.S., Hua, X. and Goldstein, J.L. (1994) SREBP-1, a membrane-bound transcription factor released by sterol-regulated proteolysis. *Cell*, **77**, 53–62.

Wang, Y.T., Chuang, J.Y., Shen, M.R., Yang, W.B., Chang, W.C. and Hung, J.J. (2008) Sumoylation of specificity protein 1 augments its degradation by changing the localization and increasing the specificity protein 1 proteolytic process. *J Mol Biol*, **380**, 869–885.

Watson, G.S., Cholerton, B.A., Reger, M.A., Baker, L.D., Plymate, S.R., Asthana, S., Fishel, M.A., Kulstad, J.J., Green, P.S., Cook, D.G. et al. (2005) Preserved cognition in patients with early Alzheimer disease and amnestic mild cognitive impairment during treatment with rosiglitazone: a preliminary study. *Am J Geriatr Psychiatry*, **13**, 950–958.

Watson, G.S. and Craft, S. (2006) Insulin resistance, inflammation, and cognition in Alzheimer's disease: lessons for multiple sclerosis. *J Neurol Sci*, **245**, 21–33.

Weydt, P., Pineda, V.V., Torrence, A.E., Libby, R.T., Satterfield, T.F., Lazarowski, E.R., Gilbert, M.L., Morton, G.J., Bammler, T.K., Strand, A.D. et al. (2006) Thermoregulatory and metabolic defects in Huntington's disease transgenic mice implicate PGC-1alpha in Huntington's disease neurodegeneration. *Cell Metab*, **4**, 349–362.

Willard, H.F. (1993) The needle found! Trinucleotide repeat expansion in the Huntington's disease gene. *Hum Mol Genet*, **2**, 497–498.

Zhai, W., Jeong, H., Cui, L., Krainc, D. and Tjian, R. (2005) In vitro analysis of huntingtin-mediated transcriptional repression reveals multiple transcription factor targets. *Cell*, **123**, 1241–1253.

Zheng, X.L., Matsubara, S., Diao, C., Hollenberg, M.D. and Wong, N.C. (2000) Activation of apolipoprotein AI gene expression by protein kinase A and kinase C through transcription factor, Sp1. *J Biol Chem*, **275**, 31747–31754.

Zheng, X.L., Matsubara, S., Diao, C., Hollenberg, M.D. and Wong, N.C. (2001) Epidermal growth factor induction of apolipoprotein A-I is mediated by the Ras-MAP kinase cascade and Sp1. *J Biol Chem*, **276**, 13822–13829.

Zhou, J.P., Feng, Z.G., Yuan, B.L., Yu, S.Z., Li, Q., Qu, H.Y. and Sun, M.J. (2008) Transduced PTD-BDNF fusion protein protects against beta amyloid peptide-induced learning and memory deficits in mice. *Brain Res*, **1191**, 12–19.

Zuccato, C. and Cattaneo, E. (2007) Role of brain-derived neurotrophic factor in Huntington's disease. *Prog Neurobiol*, **81**, 294–330.

Zuccato, C., Ciammola, A., Rigamonti, D., Leavitt, B.R., Goffredo, D., Conti, L., MacDonald, M.E., Friedlander, R.M., Silani, V., Hayden, M.R. et al. (2001) Loss of huntingtin-mediated BDNF gene transcription in Huntington's disease. *Science*, **293**, 493–498.

Zuccato, C., Liber, D., Ramos, C., Tarditi, A., Rigamonti, D., Tartari, M., Valenza, M. and Cattaneo, E. (2005) Progressive loss of BDNF in a mouse model of Huntington's disease and rescue by BDNF delivery. *Pharmacol Res*, **52**, 133–139.

Zuccato, C., Tartari, M., Crotti, A., Goffredo, D., Valenza, M., Conti, L., Cataudella, T., Leavitt, B.R., Hayden, M.R., Timmusk, T. et al. (2003) Huntingtin interacts with REST/NRSF to modulate the transcription of NRSE-controlled neuronal genes. *Nat Genet*, **35**, 76–83.

4 Target Validation for Huntington's Disease

Seung P. Kwak, James K. T. Wang, and David S. Howland

CONTENTS

INTRODUCTION

During the past several decades, pharmaceutical safety and efficacy have improved considerably with the elucidation of mechanisms of drug action in the disease context and a better understanding of the desired drug target profiles. The ability to fine-tune the target profile of many drugs has led to the premise that the optimal basis for the development of a safe and effective drug is the targeting of a single gene product. In the "omics" era of biology, the ability to rapidly identify and characterize the "druggable genome" in its entirety has created the opportunity to broadly test the validity of this premise. However, the torrent of potential drug targets resulting from this approach presents a new set of challenges, particularly the development of a robust and efficient target validation process. Such a process must be capable of screening a sizeable number of putative targets and identifying the best candidates for the lengthy, uncertain, and expensive process of drug development.

An effective target validation process depends on biological models or systems that exhibit the known properties of the disease and respond to intervention in a manner predictive of clinical outcome. Pre-existing benchmarks or "gold standard" treatments known to be effective in the clinic can be used to confirm the effectiveness of a target validation process by showing that the process is capable of "discovering" the benchmarks. The challenge is much greater for diseases without existing therapeutics. The absence of clinically successful benchmarks introduces significant uncertainty into all upstream drug discovery processes, as their validity cannot be confirmed. This is the difficult situation faced by therapeutic development for Huntington's disease (HD) and arguably for all neurodegenerative diseases. This chapter will describe the approaches adopted by the CHDI Foundation—an organization that is dedicated to the singular mission of developing HD therapeutics—to manage these limitations while moving forward with HD drug discovery and development (see also Chapter 8, this volume).

The chapter is divided into four sections. In the first section, we briefly review the challenges inherent in target validation for HD. In the second, we present our current target validation processes and their deployment for the existing list of potential targets. In the third, we discuss the operational bottlenecks and scientific challenges in the target validation process, as well as several methods for meeting these challenges. In the last section, we discuss our longer-term goal of developing new approaches to improve our current processes. The ultimate validity of our target validation processes can only be demonstrated by the development of an effective drug for HD.

THE CHALLENGE OF TARGET VALIDATION FOR HD

Because of the absence of clinically effective treatments for HD, research strategies and biological models for target validation are fraught with uncertainty and require constant re-evaluation and adjustment in light of new data. Moreover, the broad spectrum of biological processes that interact with Htt presents a multitude of potential biological approaches, none of which are supported by clinical evidence (reviewed in Borrell-Pages et al., 2006; Perez-De La Cruz and Santamaria, 2006; Ramaswamy et al., 2007). Furthermore, the slow, progressive neurodegeneration characteristic of HD, in which clinical onset (currently defined by motoric symptoms) lags behind molecular, biochemical, and cellular dysfunctions, also creates a higher standard, as optimal targets should be relevant both early and late in the disease process. Although we have tailored our current target validation process toward the identification of a single molecular target, it is not clear that a disease as complex as HD can be effectively treated with drugs that act on just one target. There is increasing evidence that the most effective treatments for complex diseases, such as cancer and HIV, target multiple steps in the mechanism of disease progression. It is possible, if not probable, that this will hold true for HD. Nevertheless, we believe that a robust target validation process to identify the best possible single molecular targets is a necessary first step. Despite its challenges, HD offers a critical advantage for drug discovery: it is a monogenic, dominant, and almost fully penetrant genetic disease with a high correlation between the size of the polyglutamine expansion and the age of disease onset. Ultimately, the optimal target validation process for HD will be defined by clinical successes and failures.

THE CHDI TARGET VALIDATION PROCESS

STEP-WISE TARGET VALIDATION—THE CHDI TARGET VALIDATION SCORING SYSTEM

To effectively assess the long list of potential HD targets, disease models, and intervention modalities that range from small molecules to proteins, we have established an organizational framework of standardized metrics and coherent yardsticks to evaluate the targets. This framework includes a scoring system (the Target Validation [TV] score) with a uniform set of scores or metrics that define a scale of increasing validity in HD targets. The scale ranges from a low of TV 0 to a high of TV 5 and progresses from the identification of targets to the demonstration of direct roles of targets in HD and finally to the development of target-based therapies with proven efficacy in the clinic (Figure 4.1). These definitions are summarized below:

TV 0 (target acquisition): Genes identified by any genome-wide screen with relevance to HD or any other neurodegenerative, polyglutamine repeat or brain disease.

TV 1.0 (target localization): Gene products expressed in HD-relevant brain regions (e.g., cortex or striatum) or otherwise linked to biological mechanisms of relevance to HD.

TV 2.0 (HD association): Gene products that interact with Htt or have altered expression or distribution in HD; targets at this level will also be used in our systems biology approach of building HD-relevant pathways and genetic networks.

TV 2.5 (functional association): Gene products and/or their pathways are altered or abnormal in HD or HD models.

TV 3.0 (causal relationship): The direct demonstration of a causal relationship between gene products and HD biology—a critical threshold in our target validation strategy; evidence consists of altered pathophysiology as a result of modulation of target function by genetic methods such as RNA interference (RNAi) and cDNA or by pharmacological probes and can be derived from derived from mammalian *in vitro/ex vivo* studies or nonmammalian whole organism disease models.

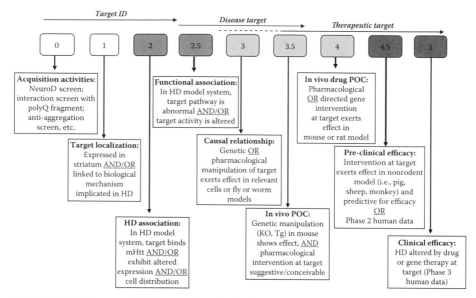

FIGURE 4.1 Target validation scoring criteria. Data collection for each target is done to progressively increase relevance in HD. Starting from the exploratory biology stages for acquisition of target and its association with HD, the target gene becomes a disease target when its causal relationship to HD is established. From this stage onward, research needs to take on a decidedly industrial "drug discovery" perspective and must exercise standardized experimental practices and decision-making processes to effectively promote therapeutic targets.

TV 3.5 (*in vivo* genetic proof-of-concept [POC]): To reach this level, alteration of the disease phenotype by modulation of the target gene product must be shown *in vivo* via genetic approaches in rodent HD models (e.g., viral gene delivery, RNAi, knockout [KO], knockin [KI], or transgenic rodents).

TV 4.0 (*in vivo* drug POC): The therapeutic potential of compounds or proteins is shown *in vivo* in rodent HD models; this stage marks the transition from disease target to therapeutic target.

TV 4.5 (pre- or Phase 2 clinical efficacy): At this level, the compound's efficacy at modulating the target must be shown in nonhuman primates, large animals, or Phase 2 human clinical trials.

TV 5.0 (Phase 3 clinical efficacy): At this level, the compound's efficacy at modulating the target must be shown in Phase 3 human clinical trials.

Although this scoring system appears to be a continuous spectrum, it is important to note discontinuities in the form of major gaps between some of the stages, for example, between TV 2.5 and 3.0 when the first causal functional relationship is made between a target gene and HD and between the preclinical target validation at TV 4.0 and clinical efficacy data at TV 4.5–5.0. These discontinuities also reflect the lack of a clinically validated drug discovery path for HD. Although this system defines a serial progression through the TV scores from low to high, some targets are initially assigned higher TV scores based on the strength of evidence appropriate for that particular stage, even in the absence of data supporting lower TV stages. For example, at TV 4.0 histone deacetylases (HDACs) lack some of the data demanded by TV 1.0 (localization in cortex and striatum) for many family members. However, HDACs as a class are supported by sufficient pharmacological efficacy data to be assigned to TV 4.0. Finally, the lack of clinically validated benchmarks (and thus no targets at stages TV 4.5 or 5.0) to guide the selection of appropriate assays and models to serve as the gatekeepers of each TV level leaves a degree of uncertainty in the target validation process that can only be resolved by success in clinical translation. In the meantime, as will be discussed later, the risks can be mitigated by applying multiple different assay systems and models, thereby increasing the chance of capturing the appropriate disease-relevant biological context. Therefore, the flow scheme of target advancement is not a rigid process but rather a work in progress. Opportunistic POC studies of advanced drug candidates, such as from drug companies, will more rapidly advance the pace of therapeutic development and attain a clinically successful drug to serve as the benchmark for guidance of future target validation efforts.

HD-Relevant Pathways and Mechanisms—Where Do Targets Come From?

Effective target validation depends on the proper acquisition of targets in appropriate disease-relevant systems. One might imagine that this is easier for a monogenic, dominant, and fully penetrant genetic disease such as HD, for which the mutation is known. However, as pointed out earlier, not only is a clinically proven benchmark drug lacking, but even basic questions on the pathophysiological mechanism of mutant Htt also remain unanswered, such as whether it is purely a gain of function associated with CAG expansion (Petersen et al., 1999; Tarlac and Storey, 2003) or also the result of loss of normal Htt function (Anne et al., 2007; Cattaneo, 2003). Moreover, Htt has no known intrinsic activity but is involved in myriad biological processes and serves as a scaffold for many proteins (Groves et al., 1999; Takano and Gusella, 2002; Truant et al., 2006) in various cellular compartments (Dorsman et al., 1999; Hackam et al., 1998, 1999; Petrasch-Parwez et al., 2007; Truant et al., 2006; Xia et al., 2003), making it difficult to sort out the specific mechanism of HD pathophysiology. Thus, even though Htt itself belongs at TV 4.0, the highest level achieved by any target thus far, it is difficult to envision how to modify mutant Htt to reduce its toxicity. The only avenue currently being pursued is the reduction of expression of Htt by antisense oligos or RNAi strategies. Although the many Htt-associated pathways or mechanisms (Hannan, 2005; Leegwater-Kim and Cha, 2004; Li and Li, 2004; Subba Rao, 2007) provide a bounty of potential targets (Table 4.1 shows a partial list of HD-associated pathways and mechanisms; see also Chapters 2 and 3, this volume, for further details), it is not clear which of these pathways will have an impact on the progression of HD. Thus, a robust system for prioritizing targets for further prosecution is clearly needed.

Observational data from HD clinical trials, particularly those that pertain to potential genetic modifiers or epigenetic phenomena that delay onset, clinical presentation, or course of the disease, are another valuable source of targets (Chattopadhyay et al., 2005; Di Maria et al., 2006; Djousse et al., 2004; Kishikawa et al., 2006; Li et al., 2006; Metzger et al., 2006b). These data take on particular significance in light of the current lack of clinically validated benchmarks for target validation. Targets

TABLE 4.1
HD-Relevant Biological Pathways

Direct Mechanisms	Indirect Mechanisms, Pathways
Htt structural modification	Selective vulnerability of MSN
Proteolysis of mutant Htt	Synaptic dysregulation
Toxic conformation adoption of mutant Huntingtin	Growth factor depletion
Nuclear/cytoplasmic aggregation	e.g., BDNF, IGF, and VEGF
Post-translational modification	Energy metabolism
Transglutaminase, sumoylation and ubiquitinylation, phosphorylation, palmitoylation, and acetylation	Mitochondrial permeability transition
Htt protein lifecycle	Reactive oxygen species and antioxidants
Protein production	Mitochondrial biogenesis
Protein clearance by UPS pathway	Nonmitochondrial energy metabolism
Protein clearance by autophagy	Dysfunction of vesicular trafficking/transport/release
Chaperone-mediated clearance	Transport motor dysfunction (anterograde, retrograde)
Intracellular localization	Vesicle endocytosis (clathrin-, caveolin-mediated)
Htt gene loci	Vesicle loading
Loss of normal huntingtin function	Vesicular fusion/NT release
Htt transcriptional modification	Membrane perturbation (cholesterol biosynthesis)
CREB, REST, HDACs, and chromatin remodeling	Neuroinflammation (astrocytic, microglial, peripheral)
Somatic expansion of CAG repeat	Intracellular calcium dysregulation
Htt interaction with HIPs, HAPs	Second messenger signaling
	Transcriptional dysregulation
	Proapoptotic, prosurvival pathways

HDAC, histone deacetylase; REST, RE1-silencing transcription factor; MSN, medium spiny neuron; BDNF, brain-derived neurotrophic factor; IGF, insulin-like growth factor; VEGF, vascular endothelial growth factor; UPS, ubiquitin proteosome system; NT, neurotrophin; HIP, huntingtin-interacting protein; HAP, huntingtin-associated proteins.

relevant to other neurodegenerative disorders should also be considered if they are not disease-specific and play more general roles in neurodegeneration, such as apoptosis (Ekshyyan and Aw, 2004; Pattison et al., 2006), aggregation (Ross and Poirier, 2004; Trzesniewska et al., 2004; Wanker, 2000), or glial dysfunction (Bonifati and Kishore, 2007; Guidetti and Schwarcz, 2003; Maragakis and Rothstein, 2006; Nemeth et al., 2006). Finally, mammalian cell-based HD models and lower organism models such as *Caenorhabditis elegans* and *Drosophila* are amenable for whole genome screens of modifiers because of their fast turnaround, well-defined phenotypes, and widespread availability of genetic tools. These unbiased and mechanism-agnostic screens are a powerful method in discovering novel targets and pathways that play functionally significant roles in HD. However, without any other known association with the disease, it will be important to build a more robust dossier of supportive evidence before assigning targets derived from such approaches to TV 3.0.

Our HD target validation process does not suffer from a lack of targets but rather faces the challenge of how to prioritize and triage many potential targets to generate a short list of worthy candidates for drug development. In the following sections, we will summarize our current TV list, the particular bottlenecks and challenges facing the validation process, and several methods under consideration for overcoming them.

SUMMARY OF CURRENT TV LIST

The current status of our target list according to TV score is summarized in Figure 4.2. The target list was generated from literature searches and ongoing HD drug discovery research at partner laboratories. There are currently more than 225 targets at stages TV 1.0–2.5, 81 targets at TV 3.0, and a sharp decrease to only 12 at TV 3.5 and 8 at TV 4.0 (Figure 4.2a). The targets at or above TV 3.5 are identified in Table 4.2. Many of the targets at TV 1.0–2.0 were identified in the HD interactome derived from expression of full-length or partial fragments of Htt in yeast (Goehler et al., 2004; Kaltenbach et al., 2007; Tam et al., 2006; Wanker et al., 1997) or from coimmunoprecipitation studies (Kaltenbach et al., 2007; Kegel et al., 2005; Li et al., 2002; Sittler et al., 1998). Although in most cases it is unclear whether they play a critical role in the course of the disease, some of the proteins that interact with Htt have also been shown to be modifiers in a *Drosophila* model of HD (Kaltenbach et al., 2007). Figure 4.2a, a snapshot of the current state of the target list, reveals a bottleneck at TV 3.5—the first direct demonstration of *in vivo* efficacy

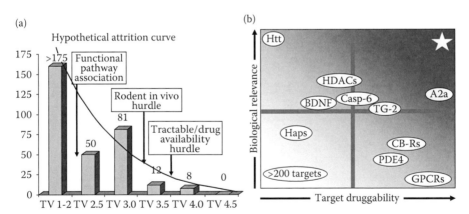

FIGURE 4.2 Number of targets binned by scoring criteria. (a) Current number of targets scored by the TV scale and binned by associative (TV 1–2), mechanistic (TV 2.5), causal *in vitro* (TV 3.0), causal *in vivo* (TV 3.5), and drug-like molecules (TV 4.0). A desirable attrition curve that we seek for an effective target validation "pipeline" is overlaid (curve). (b) Target validation for drug discovery progresses along two axes: biological relevance to disease and target tractability. Some sample targets are plotted on the graph with an extreme case exemplified by huntingtin (most relevant but nondruggable). Target profile residing in top right quadrant is desired (star).

TABLE 4.2
HD Targets Scored at TV 3.5–4.0

Targets at TV 4.0	Targets at TV 3.5
Htt	Transglutaminase 2
Histone deacetylase Class I and II	Caspase-6
Pyruvate dehydrogenase complex	Caspase-1
GRM5 (mGluR5)	Tumor protein p53
GRM2 (mGluR2)	Adenosine A2a receptor
SLC6A4 (SERT)	Nitric oxide synthase 1
GDNF	APOE
Fibroblast growth factor 2	BDNF
	CREB1
	STUB1 (CHIP)
	SIRT1 (sirtuin 1)
	PGC-1α peroxisome proliferator-activated receptor γ coactivator 1α

APOE, apolipoprotein E; GDNF, glial cell-derived neurotrophic factor; BDNF, brain-derived neurotrophic factor; CREB1, cAMP response element-binding protein 1; STUB1, STIP1 homology and U-box containing protein 1.

for a target. This bottleneck is expected to worsen as systematic target acquisition from whole genome screens in mammalian cell-based systems and model organisms yields large datasets. In the following section, we will address some of our efforts to relieve this bottleneck.

In Figure 4.2b, the biological relevance and chemical tractability of some of the targets at stages TV 3.5 and TV 4.0 are plotted. As shown, we have yet to identify an ideal target—one that is both highly relevant and chemically tractable, as indicated by the star in the upper right corner of Figure 4.2b, a point at which the full power of modern pharma drug development can be brought to bear. The current list of TV 4.0 and TV 3.5 targets is shown in Table 4.2. Because a very limited number of these targets is expected to advance in terms of small molecule development, it is critically important to steadily replenish this TV level. Also, the HDAC family of 11 isozymes at TV 4.0 is a prime candidate to be deconvoluted to identify the specific HDAC target of relevance to HD. Finally, some targets interrogated in POC studies using drug candidates already in clinical trials (see Chapter 12, this volume) showed no convincing efficacy for treating HD. Antidepressants in the serotonin reuptake inhibitor class are prescribed as symptomatic treatment of psychiatric symptoms in HD (Higgins, 2006; Slaughter et al., 2001), but the serotonin reuptake transporter SERT is scored only at TV 4.0, as its modulation has not yet been shown to delay onset or slow progression of the disease (Duan et al., 2004; Mattson et al., 2004; Wang et al., 2007). As an illustration of our target validation process, we selected several targets at TV 4.0 (fibroblast growth factor 2 and its receptor), TV 3.5 (brain-derived neurotrophic factor [BDNF] and its receptor), TV 3.0 (huntingtin-interacting protein 1), and TV 2.5 (phosphodiesterase 10a) and summarized in Table 4.3 the dataset and

TABLE 4.3

Representative Targets at TV 4.0–2.5 and Their Associated Evidence

TV	Criteria Definitions	FGF2 (TV 4.0) Assertion	BDNF (TV 3.5) Assertion	HIP14 (TV 3.0) Assertion	PDE10A (TV 2.5) Assertion
4a	Pharma intervention at target in higher species model effective	Jin (2005): FGF2 given to R6/2 increases basal neurogenesis in SVZ and improves lifespan and motor performance and decreases aggregates.			Hebb (2007): Treatment of wild-type and R6/1 and R6/2 transgenic HD mice with papaverine induces learning and memory deficits in the Morris water maze, motoric dysfunction in the rotarod test, and increased anxiety in the light-dark test (counterindication).
4b	**OR** direct gene therapy at target in higher species model effective				
3.5a	Genetic manipulation in higher species HD model effective		Alberch (2005): BDNF Val66Met polymorphism is associated with HD age of onset. Pineda (2005): R6/1 crossed to BDNF +/− mice exacerbates HD pathology.		Hebb (2007): PDE10A KO reduces locomotor behavior to novel stimuli and has deficit in spatial learning (Morris water maze) (counterindication).

continued

TABLE 4.3 (continued)
Representative Targets at TV 4.0–2.5 and Their Associated Evidence

TV	Criteria Definitions	FGF2 (TV 4.0)	BDNF (TV 3.5)	HIP14 (TV 3.0)	PDE10A (TV 2.5)
		Assertion	Assertion	Assertion	Assertion
3.5b	OR genetic manipulation exhibits HD-like phenotype		Cho (2007): ad-virus injection of BDNF/ noggin increases neurogenesis and improves motor function and lifespan in R6/2. Baquet (2004): Cortical BDNF KO mice exhibit HD-like phenotype.		
3.5c	AND conceivable path for pharma intervention at target in vivo				
3a	Genetic manipulation in cell, fly, and worm HD model effective			Yanai (2006): C214S N-term htt Q128 accelerates inclusion formation (cytoplasmic) and increases neuronal toxicity. KD of HIP14 in mouse neurons expressing fl-Htt (wild-type) increases nuclear inclusion and sensitivity to NMDA toxicity. Overexpression of HIP14 reduces inclusion by mutant Htt.	

3b	**OR** pharma manipulation in cell, fly, and worm HD models effective	Zala (2005): BDNF rescues NeuroD in Lenti-HTT transduced rat striatal primary neurons.	
2.5a	Target pathway aberrant in HD or a model system	Gines (2006): TrkB expression is down in HD and HD mice.	Yanai (2006): wild-type-Htt associates with HIP14 immediately downstream of C214, and enhances HIP14 activity. Htt regulation of HIP14 is unique; mutated Htt is less modified and poorly enhances HIP14 activity.	Hu (2004): mRNA decrease is caused by transcriptional suppression via mutated Htt. Three transcription initiation sites for PDE10A2 were differentially affected by mutant huntingtin transgene.
		Zuccato (2005): BDNF levels are reduced in R6/2 brain.	Fernández-Chacón (2004): CSP is a key substrate for HIP14, where the palmitoylated form is trafficked to presynapse, where it modulates Ca^{2+} channel activity and exocytosis.	

continued

TABLE 4.3 (continued)
Representative Targets at TV 4.0–2.5 and Their Associated Evidence

TV		FGF2 (TV 4.0)	BDNF (TV 3.5)	HIP14 (TV 3.0)	PDE10A (TV 2.5)
	Criteria Definitions	Assertion	Assertion	Assertion	Assertion
				Stowers (2007): *Drosophila* HIP14 loss of function produces photoreceptor cell neurodegeneration, leading to progressive blindness. Might deficits in CSP as a result of mHtt-HIP14 dysfunction cause the observed retinal degeneration and blindness in R6/2 mice?	
2.5b	**OR** target activity is altered in HD or a model system	Tooyama (1993): bFGF-positive astrocytes are slightly increased in HD brain.	Gauthier et al. (2004): mutant Htt impairs axonal transport of BDNF.	Ohyama (2007): *Drosophila* HIP14 mutants show exocytotic defects at low-frequency stimulation and a nearly complete loss of synaptic transmission at higher temperature. Interestingly, two exocytotic components known to be palmitoylated, cysteine string	Hebb (2004): PDE10A mRNA and protein decrease by 60% in the striatum of R6/1 and R6/2 HD mice before motor symptoms. PDE10A also decreased in the caudate-putamen of Grade 3 HD patients.

2a	Target binds mutant Htt	protein (CSP) and SNAP25, are severely mislocalized at HIP14 mutant synapses. HIP14 exocytotic defects can be suppressed by targeting CSP to synaptic vesicles using a chimeric protein approach.	Yanai (2006): Htt palmitoylated at C214 by HIP14 is essential for its trafficking. Mutated Htt associates poorly with HIP14 and is less palmitoylated.	Fujishige (1999) and Sasaki (2004) PDE7b1 and PDE10 expression is stimulated by D1 agonist.
2b	OR expression altered in HD or a model system		Zuccato (2003): Normal but not mHtt stimulates BDNF transcription via REST/NRSE; Zuccato (2001): BDNF levels are reduced in HD brain.	
2c	OR cellular distribution altered in HD or a model system			Gross-Langenhoff (2006): cGMP modulation by PDE10A; some controversy on activity regarding cAMP.

continued

TABLE 4.3 (continued)
Representative Targets at TV 4.0–2.5 and Their Associated Evidence

TV	Criteria Definitions	FGF2 (TV 4.0) Assertion	BDNF (TV 3.5) Assertion	HIP14 (TV 3.0) Assertion	PDE10A (TV 2.5) Assertion
1a	Target identified expressed in brain	Tooyama (1993): aFGF and bFGF are expressed in brain and striatum.		Huang (2004): HIP14 catalyzes SNAP25, PSD95, GAD65, synaptotagmin I, and htt and is expressed in the CNS and elsewhere. Overexpression enhances vesicular transport.	Hu (2004): PDE10A2 is the predominant isoform of the gene expressed in the striatum.
1b	Target identified expressed in striatum		Kells (2004): AAV-BDNF is protective in rat QA model.		
1c	Target mechanism implicated in HD	Maksimovic (2002): FGF is protective in QA rat model.	Ferrer (2000): BDNF is decreased in HD brain.		
0a	Target identified from proteomic screen			Singaraja (2002): HIP14 is isolated by Y2H to interact with Htt and targets Htt to clathrin vesicles and endosomes.	
0b	**OR** target identified from genomic screen				

FGF2, fibroblast growth factor (basic) 2; SVZ, subventricular zone; BDNF, brain-derived neurotrophic factor; PDE, phosphodiesterase; KO, knockout; HIP, huntingtin-interacting protein; NMDA, N-methyl-D-aspartic acid; Lenti-HTT, lentiviral huntingtin; CSP, cystein–string protein; REST/NRSE, RE1-silencing transcription factor; aFGF, fibroblast growth factor (acidic); SNAP25, synaptosomal associated protein 25; cAMP, cyclic adenosine monophosphate; GAD65, glutamate decarboxylase 65 kDa; CNS, central nervous system; AAV, adeno-associated virus; Y2H, yeast two-hybrid; QA, quinolinic acid.

rationale that qualified them for these TV scores. Although the current TV scale does not provide finer granularity within each TV category, data quality, richness, robustness, and reproducibility are of utmost importance in prioritizing targets for further evaluation. BDNF, for example, is supported by a number of independent genetic approaches showing effects on HD-relevant outcomes and would be given a higher priority than a target with a single positive genetic validation approach.

ADDRESSING BOTTLENECKS IN TARGET VALIDATION

PRIORITIZATION OF TV 3.0 TARGETS BY *IN VITRO* AND INVERTEBRATE ORGANISMAL MODELS IN EARLY TARGET VALIDATION

As is apparent in Figure 4.2a, there is a large and ever-growing list of targets at stages TV 1.0–3.0 that must be efficiently prioritized to face the TV 3.5–4.0 hurdles, which already constitute a major bottleneck. In this section, we discuss our effort to triage these targets with intermediate target validation assays that are efficient and have putative face validity for HD. Unfortunately, there is only a limited repertoire of *in vitro* and *ex vivo* assays and *in vivo* nonmammalian model organisms, and the value of these assays and models is uncertain, as none have yet been proven to be predictive of TV 3.5–4.0 *in vivo* rodent model testing. Thus, we currently use the entire existing panel of assays, and targets that show activity in a higher number of assays receive priority. Over time, however, we intend to improve and expand the current panel of assays, and eventually we will use only the most appropriate assays for the particular target under query, as well as mechanism-agnostic assays that are predictive of *in vivo* testing and therefore can serve a gating function for triaging targets.

Cell-Based Validation Assays

We currently use a series of cell-based screening assays as a validation panel to ascertain the biological consequences of manipulating a particular target, either by RNAi knockdown or by overexpression of its native or mutant forms. Because the relevance of these assays for any particular target is unknown, we use them as a panel and assign a higher priority to a target that shows positive effects in multiple assays. These assays include mutant Htt-induced cytotoxicity in the transformed cell line PC12 (Aiken et al., 2004), the striatal progenitor cell line ST14A (Bosch et al., 2004; Bottcher et al., 2003), primary striatal neurons, and conformational changes of mutant Htt in a cell-based assay (Arrasate et al., 2004; Brooks et al., 2004). There is clearly a heavy bias toward acute cytotoxicity of transformed cell lines induced by overexpression of mutant Htt, and we are striving to expand the repertoire of these assays by biasing toward neuronal cell types and functional readouts, including transcription (Cha, 2007; Cui et al., 2006; Sadri-Vakili and Cha, 2006; Sadri-Vakili et al., 2007; Zhai et al., 2005), clearance of mutant Htt (Bursch and Ellinger, 2005; Ravikumar et al., 2004; Rubinsztein, 2006; Shibata et al., 2006; Yamamoto et al., 2006), and neurite outgrowth (Lynch et al., 2007). We are investing in new assay development, including cocultures of cortical and striatal neurons as well as of neurons and glia (for interrogating targets such as kynurenine monooxygenase

[KMO]). We are also generating neuronal lines from stem cell or neural progenitor cells to develop a system with greater physiological relevance for central nervous system (CNS) neurons. However, even primary cocultures of neurons and glia will not model to any significant degree the *in vivo* circuitry and physiological context of HD. On the other hand, cell culture is one area in which the single clear-cut advantage of HD research—the monogenic, Q repeat length-dependent causation of disease—may be best leveraged to search for endpoints that clearly depend on this most salient of clinical features.

Ex Vivo Validation Assays

The major disadvantage of cell-based cytotoxicity assays is that they are acute and wholly out of the context of the *in vivo* CNS complexity and the slow, progressive nature of neurodegenerative diseases. To partly address these issues, we use a more complex tissue-based platform in which survival of biolistically transfected neurons in early postnatal rodent brain slices are monitored over the course of 4–7 days (Khoshnan et al., 2004; Lo et al., 1994; Riddle et al., 1997; Varma et al., 2007; Wang and Qin, 2006) (see also Chapter 5, this volume). The expression of mutant Htt in these CNS neurons *in situ* results in their demise. This platform can be multiplexed to coexpress the target gene product (overexpression or knockdown), Htt in its wild-type or mutant form, and a fluorescent protein marker, allowing the effects of target gene expression on mutant Htt-dependent toxicity to be assessed. However, because the level of overexpression or knockdown in this system cannot be easily examined on a routine basis, negative data will not be conclusive. Thus, the failure to observe an effect of target gene expression is not a kill criterion for a target. The accumulation of positive data in multiple types of assays will move a target forward in the queue. Pharmacological validation in this *ex vivo* brain slice platform is straightforward; compounds can be directly applied to the slices to assess their effects (Varma et al., 2007; Wang et al., 2006). As with the target gene expression assays, only a positive result is meaningful, as a negative result could be the result of insufficient compound penetration into the tissue; actual concentrations of compounds at their site of action inside the tissue are difficult to ascertain.

The limitations of the *ex vivo* platform are similar to those of the cell-based assays; it relies on overexpression of mutant Htt as the toxic insult and cytotoxicity as the endpoint. Brain slices also are harvested from early postnatal animals (days P7–10). Although an improvement over the reconstitution of embryonic neurons in culture, the cells in this system may be too immature to serve as an appropriate substrate for HD-relevant cell death. Possible adaptations of the platform to address some of these issues include the use of transgenic HD animals to bypass the need for acute overexpression of mutant Htt, as well as advances in automation, image capture, and quantization to measure alternative endpoints such as neurite outgrowth. However, the throughput of this platform will always be lower than any cell-based platform.

A version of the brain slice *ex vivo* assay that takes advantage of known electrophysiological changes in conventional brain slices is currently under development. In this assay, brain slices from adult transgenic HD animals are maintained in culture for a single day. This approach bypasses the issues of developmental immaturity, inappropriate levels of overexpression of mutant Htt, and exclusive focus on cytotoxicity. Unfortunately, this method only allows for pharmacological validation studies at present, as it has not yet been adapted to allow genetic manipulation of target gene expression. One possibility for future development is the overexpression or knockdown of target genes *in vivo* in transgenic HD animals by infection with recombinant viral expression vectors, followed by the preparation of brain slices for electrophysiological measurements. However, such protocols are labor-intensive and challenging to institute on a large scale and are therefore only appropriate for a limited list of targets.

Whole Organism-Based Validation Using Invertebrate Models

Invertebrate, whole organism models of human disease provide *in vivo* context in which to assay progressive, multimodal disease-linked phenotypes. The best-characterized models are *C. elegans* (Brignull et al., 2006; Faber et al., 1999; Link, 2006; Parker et al., 2001) and *Drosophila* (Link, 2001; Sipione and Cattaneo, 2001), both long-standing models that recapitulate the adult-onset, progressive nature of HD on transgenic expression of mutant Htt (see also Chapter 8, this volume). In these models, multiple endpoints can be monitored, including cell loss, behavioral deficits, and shortened lifespan (Faber et al., 1999; Marsh and Thompson, 2004; Marsh et al., 2003; Morley et al., 2002; Parker et al., 2001). Interestingly, the behavioral phenotypes in *Drosophila* have not yet been associated with specific neuropathology *in situ* and may be caused by subtle cellular and ensemble defects, as in transgenic mouse models. These model systems provide a valuable tool to assist in target validation by providing both high-throughput and complex phenotypes reflective of cell circuitry and cell–cell interactions that are impossible to replicate in any *in vitro* or *ex vivo* context. However, the nonmammalian biological context will produce fly-specific positives and miss genes not present in *Drosophila*. Whole genome screens with mutant Htt in these models have not yet been done, but there are two examples of pharmacological validation in which positive hits in *Drosophila* have been successfully translated to transgenic mouse models: the HDAC inhibitor suberoylanilide hydroxamic acid (SAHA) (Hockly et al., 2003) and the mutant Htt aggregation modulator C-8 (Chopra et al., 2007). These data suggest that the invertebrate models will be useful for genetic target validation. A major challenge will be the transitioning of positives from testing in these models into mammalian systems, as the rationale of the former is their ability to provide *in vivo* validation, whereas the latter are mostly *in vitro* cell-based assays. Targets that already have some link to HD at the TV 1.0–2.5 levels (e.g., localization in brain, association with Htt, alteration in HD models) can be qualified for TV 3.0. Targets that are novel and have no known association with HD will require a focused effort to provide additional supporting data. Obviously target hits from both model organism tests *and* cell-based tests will have higher priority than those from only a single type of test.

Deconvolution of Target Family

In certain cases, pharmacological probes that have shown efficacy in HD models are not sufficiently selective to clearly reveal their mechanisms of action and cognate targets. Nevertheless, the identification of such efficacious probes can lead to the deconvolution of the responsible target(s). An interesting example is that of the

HDAC Class I, II, and IV families, which consist of 11 distinct enzymes. In concert with histone acetyltransferases (HATs), these family members regulate transcription by modifying histone acetylation state and chromatin structure (Sadri-Vakili and Cha, 2006). These enzymes also regulate the activity of other proteins such as transcription factors and tubulin, mediate pleiotropic biological actions, and are prime targets for several disease indications, including cancer. It is believed that isozyme-selective HDAC inhibitors may be more effective and less toxic than the currently available broad-spectrum HDAC inhibitor (Langley et al., 2005). Early on, three lines of evidence suggested the relevance of HDACs in HD. First, weak pan-inhibitors of HDACs, such as SAHA, sodium butyrate, phenyl sodium butyrate, and valproate, showed moderate efficacy in some trials of transgenic HD mouse models (Fava, 1997; Grove et al., 2000; Hockly et al., 2003; Minamiyama et al., 2004). Second, genetic modification of *Drosophila* HDACs (four orthologues in Class I and II; Marsh et al., unpublished data), and potent pan-inhibitor trichostatin A showed efficacy in HD flies (Steffan et al., 2001). Third, mutant Htt appears to interact with HATs such as CREB-binding protein to modify transcription, and with axonal transport regulated by HDAC6, possibly as part of pathogenesis in HD (Dompierre et al., 2007; Iwata et al., 2005; Nucifora et al., 2001; Steffan et al., 2000). On this basis, HDACs were scored at TV 4.0, but it is not yet known whether a single isozyme or some combination of the 11 family members constitutes a therapeutic target for HD.

Parallel efforts currently underway include pharmaceutical development of isozyme-selective HDAC inhibitors, of which HDAC6 is the most notable success (Dokmanovic et al., 2007; Khan et al., 2008), and genetic validation studies in which HDAC KO mice are crossed into HD models to assess the effect of each HDAC isozyme on the HD phenotype. Because some of the HDAC isozyme KOs are embryonically lethal (Chang et al., 2006; Menegola et al., 2006; Vega et al., 2004), conditional KOs must be created. Meanwhile, HDAC inhibitors have shown no effects in most of the intermediate-stage *in vitro* and *ex vivo* assays, including mutant Htt-dependent cytotoxicity models, and limited information has been generated in the brain slice neuroprotection platform by using biolistics transfection to knock down expression of specific HDACs. An alternative brain slice assay may involve testing of HDAC inhibitors by electrophysiological methods, as the broad-spectrum inhibitors are known to increase long-term potentiation (Alarcon et al., 2004), a well-characterized cellular readout of synaptic plasticity that is affected in HD slices (Cummings et al., 2006, 2007; Gibson et al., 2005; Lynch et al., 2007; Picconi et al., 2006). It is hoped that these multiple parallel efforts will pinpoint the HD-relevant HDAC(s) on which to focus pharmaceutical development.

Pathway-Functional Network Development as a Tool for Additional TV Scoring

The large and continually increasing number of targets residing at TV 3.0 and lower reinforces the current and ongoing need to prioritize the list to focus our efforts on the best candidates. One method for prioritizing the list is to group these targets into functional networks and pathways, an exercise that may also lead to an improved understanding of HD pathophysiology. At the simplest level, this process will reveal whether there are multiple members of particular networks or pathways residing on the TV scoring list. Presumably, the members of such pathways should receive a higher priority. On a deeper level, such pathway analysis may provide a systems biology-oriented perspective on HD pathophysiology by providing valuable insights that would not be observed with a gene-by-gene approach. Although this approach is still in its beginning stages, the increased availability of genome-wide datasets and knowledge is expected to make this endeavor feasible in the short term. Our approach to the development of a gene network view of HD is described further at the end of this chapter.

OPTIMIZING THE PROCESS OF LATE-STAGE TARGET VALIDATION

Choice of Genetic HD Models for Target Validation

As summarized in previous chapters, several different mouse and rat models of HD have been described. Although each model displays features of HD, none is clearly superior to the others, and all have yet to be validated by the development of a clinically successful "gold standard" treatment. The older, acute neurotoxin-based models, such as quinolinic acid and 3-nitroprionic acid, which are based on acute cytotoxic insults that target striatal neurons *in vivo* have largely been superseded by the development of a variety of transgenic models after the identification and cloning of the *Htt* gene. A recent side-by-side comparison of transcriptional deficits of the R6/2, yeast artificial chromosome (YAC) 128, Q111 KI, and Hdh150 KI mouse models showed a remarkable similarity in gene sets that are up- or down-regulated (Kuhn et al., 2007). The main difference observed between the models is the age of onset of transcriptional changes that occurs earliest in the most aggressive R6/2 model (6 weeks) but is more delayed in the KI models (18 and 22 months, respectively, for Q111 and Hdh150) and most delayed in the YAC128 full-length model (onset of changes at 24 months). These data further support the hypothesis that fragment-based, KI, and full-length transgenic models may have common disease features and that the appropriateness of each model depends on the specific target and the particular set of phenotypic outcomes to be assayed. The short fragment R6/2 model develops HD phenotypes within a 3-month timeframe and is thus the most feasible choice for large-scale validation studies. Thus, purely for practical considerations, the R6/2 model is used for the primary testing of compounds. The full-length YAC128 and bacterial artificial chromosome 103 (BAC103) models develop their phenotypes within a 6- to 12-month timeframe, which poses logistical barriers for both genetic and pharmacological validation studies. In addition, these YAC and BAC models still overexpress full-length mutant Htt and show weight gain (Van Raamsdonk et al., 2006), features that are not manifest in HD patients. However, in cases in which the target under query requires a full-length HD model, one of these must suffice. In contrast, the KI mouse models, which include Q140, Q111, and Hdh150, are arguably the best genetic mimics of HD but do not develop robust behavioral deficits until advanced ages (~2 years). In some instances, however, the KI models may be suitable for monitoring specific outcome measures, such as inclusion bodies or transcriptional signatures, as the protracted timeframe allows for a thorough regional and temporal analysis. Finally, the HD transgenic rat (von Horsten et al., 2003), which expresses a rat mutant truncated Htt with 51 CAGs driven from the rat Htt promoter, has been shown to develop motoric deficits and neuropathology in the brain in a 1-year timeframe, as well as decreased survival (median of ~20 months). However, the transgenic rat model is only appropriate for TV 4.0 pharmacological testing because of the lack of transgenic rat models of most targets of interest. The rat model's main utility may be for testing gene- or cell-based therapies, as the model affords more accurate stereotaxic injections and availability of a variety of robust behavioral paradigms, particularly in terms of cognitive functions, that may be of interest in the context of HD.

Genetic Validation of Targets *In Vivo* (TV 3.5)

For TV 3.5, the first mammalian *in vivo* hurdle in the target validation pipeline, target gene expression must be altered genetically in the context of a rodent HD model to assess its effect on the progression of HD. In principle, altering expression of the target gene would mimic the actions of a chemical agonist or antagonist and thus predict the outcome of the pharmacologic tests required for TV 4.0. The genetically modified mice include transgenic overexpressers, dominant negatives, mutant KI mice, conventional or conditional KO mice, and transgenic RNAi mice, each of which modifies the target gene in a different way. The murine HD models used are typically the R6/2, YAC128, BAC103, or Q140 KI as described previously. Because no phenotypic outcome in these HD models has yet been shown to be predictive of clinical success in human HD trials, all possible phenotypic outcomes must be considered potentially relevant. Currently, the outcome measures associated with these models include motor, cognitive, histopathological, and biochemical readouts (e.g., aggregation, transcriptional changes) and survival.

Although the genetic validation studies of TV 3.5 precede the pharmacological studies at TV 4.0, the slow and expensive genetic approach may be bypassed in favor of a pharmacological demonstration of efficacy if a good pharmacological agent is available (Li et al., 2005). In such a case, pharmacological demonstration of efficacy can directly elevate the target to a *bona fide* "therapeutic target." (This is also true for gene therapy strategies such as viral expression of a protein or delivery of antisense RNA or RNAi.) Before initiating efficacy testing of an agent, certain critical information, including brain penetration, selectivity toward the presumed target, and tolerability in chronic dosing, should be obtained. Typically, even if pharmacological agents penetrate the brain in sufficient quantities to act on their putative targets, they are often not as selective as originally believed, making definitive TV score assessments problematic. It then becomes necessary to undertake efficacy testing of related compounds to generate a pharmacological profile, and these studies must be preceded by the same sequence of brain penetration and tolerability studies. It is also possible that a compound's efficacy is the result of action on multiple targets, which may lead to discordance between genetic and pharmacological validation approaches. Thus, pharmacological validation studies must be carried out with a fair amount of preparatory work, and the time saved compared with genetic validation may be less than expected. Nevertheless, we will undertake both genetic and pharmacological validation studies in parallel whenever possible, as we believe that the development of a complementary dataset using both approaches will make the most powerful argument in moving the target forward in drug development.

A major issue we must confront is the number of targets (currently 81) at stage TV 3.0. Genetic validation experiments, which can take at least 1–2 years to complete, form a significant bottleneck. To develop a more rapid, higher-throughput technology for advancing targets through the TV 3.5 genetic validation studies, we have adopted RNAi transgenic mouse technology from Taconic-Artemis (Cologne, Germany). In this system, small hairpin RNA (shRNA) sequences directed against the gene of interest are inserted into a tetracycline-inducible expression cassette at the mouse *rosa 26* genetic locus (Figure 4.3). This resident cassette allows rapid insertion of different shRNA sequences at the identical genomic site (the modified *rosa 26* locus), driving the tetracycline-regulated expression of each shRNA, and thus the knockdown of gene targets, in a temporally and spatially identical manner. We are currently evaluating shRNA transgenic mice engineered to knock down expression of the adenosine receptor 2a (A2a), KMO, and caspase-6 (C6) gene products, which will then be crossed into the appropriate murine HD models. The RNAi mice for A2a will be crossed with R6/2 mice because A2a mRNA and binding activities are altered in the R6/2 mouse (Tarditi et al., 2006). The RNAi mice for KMO will also be crossed with R6/2 mice because 3-HK, a toxic metabolite generated by the kynurenine pathway, is up-regulated in the brains of these mice (Guidetti et al., 2006), suggesting that KMO inhibition could be therapeutic. C6, which putatively

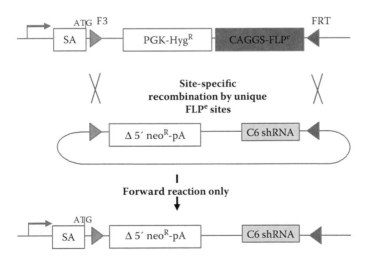

FIGURE 4.3 Scheme for generating targeted tetracycline-inducible shRNA transgenic mice. A caspase-6 (C6) shRNA is inserted in a landing pad cassette resident at the *rosa 26* locus of the mouse genome. The recombination is afforded by the action of FLP enzyme (expressed from CAGGS-FLP) acting on resident DNA recombination sites F3 and FRT that flank the DNA cassette. C6 shRNA-positive clones are identified by selection in G418 as a result of expression of the neomycin-resistance gene (neoR) carried in on the recombined insert. SA, splice acceptor; HygR, hygromycin resistance gene. (Slide courtesy of Artemis Pharmaceuticals GmbH.)

cleaves the huntingtin protein at amino acid 586 to generate a toxic proteolytic fragment (Graham et al., 2006), must be crossed with one of the full-length Htt mouse models; the R6/2 exon 1 transgene does not contain the relevant C6 cleavage site. If successful, this technology platform would greatly accelerate the genetic target validation process, helping to relieve the bottleneck by allowing faster development of the genetic models required for target validation.

An additional benefit of this approach is that RNAi knockdown mice may be better genetic mimics of drug treatments than conventional KOs, as their reduction of target gene expression is typically in the 70%–90% range *in vivo*. Similarly, pharmacological inhibition will reduce but not completely eliminate target protein activity. However, if target gene expression is not sufficiently reduced in the RNAi mice, a conventional KO approach will be required. In the case of C6, conventional KO mice are made in parallel to be crossed with YAC128 and BAC103 full-length Htt mice. Given the results of Graham et al. (2006) showing that mutation of the C6 cleavage site on mutant huntingtin at amino acid 586 (C586R) results in complete rescue of the HD phenotype in YAC128 mice, direct genetic manipulation of the C6 target is of particularly high priority. Although these experiments are seemingly sufficient to advance the TV score of C6 to 3.5, it is absolutely necessary to demonstrate that direct inhibition of the target in question, the C6 enzyme, is capable of altering phenotypic outcome in an HD mouse model. Such a demonstration is necessary to achieve the TV 3.5 level and thus justify a comprehensive search for C6 inhibitors suitable to allow TV 4.0 experiments and drug development efforts to go forward.

The experimental evidence required for a target to achieve TV scores of 3.5 and 4.0 is reasonably well defined, which allows for a systematic standardized approach and direct comparisons of data. At this stage of the TV process, the experiments have moved from the realm of exploratory biology to that of drug discovery. Therefore, standardized experimental practices are necessary to allow clear decision-making for drug development campaigns.

Pharmacologic Validation of Targets *In Vivo* (TV 4.0)

A number of pharmacologic agents (i.e., agonists or antagonists) have been tested in HD mouse models, primarily the R6/2 model and to a lesser extent the N171-82Q (Schilling et al., 1999) and YAC128 models (Slow et al., 2003). We have implemented standardized protocols with our partner, PsychoGenics Inc. (Tarrytown, NY), and have systematically evaluated and compared each of the murine HD models to determine each model's suitability for compound screening. The most significant differences between the models are whether they carry a mutant Htt fragment or the full-length mutant *Htt* gene and the expression level of the transgene. The genetic background of the mouse strain may be another important factor. The models showed a broad spectrum of severity of phenotypes, including motoric and cognitive deficits and neuropathological changes such as inclusion body formation. The R6/2 mouse, the most widely used model, is a transgenic C57BL6/CBA hybrid that expresses a mutant human *Htt* gene fragment encoding the first exon with a CAG repeat size that originated as 142 in the founder (Mangiarini et al., 1996, 1997). The advantages of the R6/2 mouse in drug testing are its robust and reproducible HD-like phenotype and the short duration of the studies (i.e., 3–4 months) compared with the other transgenic models. However, the mutant *Htt* expressed in this model contains only the first exon of a very large gene. As a result, this model is not suitable for testing compounds whose mechanisms of action depend on regions of the protein downstream of exon 1. Additionally, the rapid disease progression in the R6/2 mouse, thought to mimic juvenile HD and represent mid- to late-stage disease, creates a high standard for drugs to show benefit. Furthermore, the R6/2 mouse suffers CAG repeat instabilities over generational time. If not closely monitored, the CAG number can increase or decrease, which would affect the onset and severity of disease phenotypes (Stack et al., 2005) and confound drug studies. For these reasons, compounds will also be tested in the full-length models described previously. The BAC103 and YAC128 models offer the best compromise between rapidity and robustness of phenotypes and the practicality of large-scale drug testing.

Compound evaluation to attain a score of TV 4.0 follows a three-stage testing design to increase throughput (Figure 4.4). In Stage 1 testing, compounds are scored for tested motoric, general health, survival, and, if appropriate, cognitive endpoints. If results are statistically significant or if trends $(0.05 < P < 0.3)$ are observed in multiple outcomes measures, the compound will enter Stage 2 testing, which uses multiple drug doses and higher numbers of mice to extend and confirm the Stage 1 findings. Cognitive and neuropathology endpoints are also

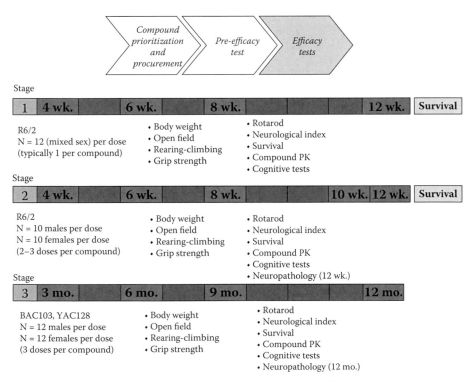

FIGURE 4.4   (See color insert following page 172.) Staged testing design for evaluation of compounds in HD mouse models.

evaluated more fully in Stage 2. Stage 1 allows for the testing of large numbers of compounds to determine which ones will most likely succeed in later stages of testing. Because this approach will lead to false-negative results in Stage 1, compounds that are negative in Stage 1 will be set aside for re-evaluation if additional supporting data become available. Compounds that succeed in the multiple outcome measures of Stage 2 will then be evaluated in Stage 3, which currently uses the BAC103 or YAC128 mice to evaluate motoric, general health, neuropathology, and cognitive endpoints. In certain instances, compounds of high priority and with a strong preclinical data package will bypass the first stage and go directly into evaluation using the paradigm shown for Stage 2 and/ or Stage 3.

Standardization of Testing at TV 4.0

We believe that compound evaluation for TV 4.0 assessment must be highly standardized to allow consistency and cross-comparisons of data over time and between different compounds. The testing of compounds in murine HD models typically follows the path shown in Figure 4.5. First, the best-in-class compound that hits the target of interest is identified. Then the optimal formulation for *in vivo* delivery is determined; the optimal route of administration to maximize plasma and brain concentrations, the maximum tolerated or optimal dose, and the optimal doses for efficacy studies are determined by additional testing.

The efficacy testing in the HD mice follows a protocol of "Best Practices" that have been implemented to provide consistency of testing at contract research organizations, including PsychoGenics Inc., as well as at academic laboratories. The protocol includes a well-controlled, balanced, and blinded study design, consistency in

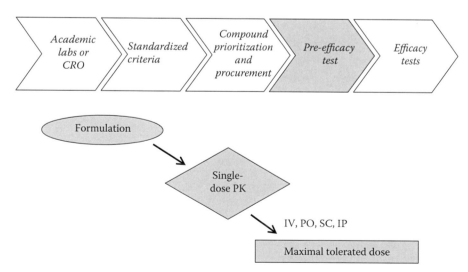

FIGURE 4.5 Preparation for testing of compounds in HD mice requires pre-efficacy work. PK, pharmacokinetic; IV, intravenous; PO, oral; SC, subcutaneous; IP, intraperitoneal. Each refers to the mode of dosing in the mouse.

the genetic background and CAG repeat size of the HD mice being used, and optimal husbandry to ensure care and overall general health of the mice being studied.

A critical component of the Best Practices is the husbandry of mice. All mice are housed in same sex groups in which HD and wild-type mice from different litters are mixed. Mixing R6/2 transgenic and wild-type mice has been shown to improve overall body weight and prevent early deaths of the mutant mice (Carter et al., 2000). The R6/2 mice show reduced social behaviors, such as nest building and huddling with cagemates, and inclusion of wild-type mice is postulated to enhance normal social interactions and to aid the R6/2 mice in maintaining better thermoregulation as a result of increased huddling (Carter et al., 2000). Fifty to seventy percent of the variation in the onset of HD is the result of CAG repeat size (Li et al., 2006), and the remaining 30%–50% of the variation is thought to be the result of genetic modifiers and environmental factors. A role for environmental factors is supported by studies of monozygotic twins that have shown differences in phenotypic presentation of disease (Anca et al., 2004). Therefore, we believe that testing compounds in mice housed under enrichment conditions, meant to simulate a more natural environment and to promote a basic level of social interaction and decreased stress, will more accurately predict therapeutic benefits. The enrichment consists of the addition of play tunnels, shredded paper, and plastic bones for all mice and has been shown to improve motor function and cortical brain atrophy in R6/2 mice (Hockly et al., 2002; van Dellen et al., 2000). R6/2 mice also have difficulty eating pelleted food from cage hoppers and reaching water bottle spouts (Carter et al., 2000). Therefore, to prevent dramatic weight loss caused by malnutrition, wet powdered food at bedding level and modified water spouts that reach further down into the cage are provided. These conditions ensure that R6/2 mice tested are more likely to succumb to effects of the HD mutation rather than to effects of malnutrition or dehydration, which would complicate the interpretation of results. In addition to the standardized husbandry conditions outlined above, the Best Practices also include the following considerations:

- Mice from different litters have been shown to have subtle phenotypic differences (Bates and Hockly, 2003). Therefore, transgenic and wild-type mice are randomized into testing groups to avoid "litter effects." Mice should also be weighed early after weaning, and each testing group should be counterbalanced according to mouse body weight.
- Mice should be housed in groups of four or five and separated by gender. In each cage, two wild-type mice of the same gender but different litters should be included in an attempt to provide normal social stimulation.
- Mice should be allowed to acclimate to the experimental room for at least 1 hour before the beginning of any experiment.
- Experimentation should be conducted in a blinded manner regarding genotype of the mouse and whether it has received drug or vehicle.
- Data analysis should not be performed until the end of the study to prevent bias in ongoing measurements.
- Tail samples should be taken at the end of the study for verification of genotypes and CAG sizes of individual mice.

- Validated protocols for phenotypic measures, which have been shown to detect significant differences between wild-type and the HD mouse, should be used. Such protocols should not be altered mid-experiment. Validated protocols generated by PsychoGenics Inc. are available to the research community.

In Vivo Target Validation at 4.5 and 5.0

Once a target has achieved validation at TV 4.0, the preclinical data package will be reviewed to determine its worthiness for the clinical studies required to achieve TV 4.5 and TV 5.0 scores. TV 4.5 requires positive outcomes in Phase 2 clinical trials in HD patients and/or positive outcomes in higher species such as nonhuman primates or even pig or sheep models of disease. Many of these nonhuman genetic models of HD are still in the process of being established, but if and when they become available, they may offer the opportunity to test therapeutic candidates in animal models that have brain neuroanatomy more comparable to that of humans. Testing of many therapeutic agents may not be necessary in nonhuman TV 4.5 models, as Phase 2 human clinical trials may be a more logical choice. However, in cases such as gene therapy, where site of brain delivery, degree of brain penetration, and extent of brain coverage are major issues, testing in higher mammalian models may be warranted. For example, delivery of adeno-associated virus (AAV)-small interfering RNA against Htt into the brains of transgenic HD mice shows beneficial outcomes (Harper et al., 2005). However, translation of these results to humans may require a demonstration of positive outcomes on the delivery of the AAV agent to a larger brain structure to justify the initiation of a human clinical trial. Progress has been made toward establishing genetic models of HD in sheep, pigs, and nonhuman primates. For the latter it is hoped that impairment of cognitive functions will better mirror the deficits observed in HD patients than those that can be observed in rodent HD models.

TV 5.0, the highest level of validation, is achieved when a pharmaceutical agent shows clinical efficacy in Phase 3 trials in HD patients. Our standard of clinical effectiveness is that a pharmaceutical agent must "delay onset or slow progression of HD" and cannot merely relieve symptoms. For that reason we do not currently consider Phase 3 clinical success of tetrabenazine (Huntington Study Group, 2006; Kenney and Jankovic, 2006), an inhibitor of the vesicle monoamine transporter 2 target, to be sufficient in ameliorating chorea symptoms to qualify this target for TV 5. Further investigation of the potential of tetrabenazine to affect HD onset or progression is warranted in preclinical HD models and in additional clinical trials. Furthermore, targets that reach TV 5.0 will be of great importance in the re-evaluation of preclinical cell-based and *in vivo* models of disease to determine which models and assays are the most predictive of success in human trials.

LONG-TERM CHALLENGES: DEVELOPMENT OF SYSTEMS BIOLOGY APPLICATIONS IN HD

As shown in the target scoring chart, it is difficult to translate the large number of potential targets at the TV 1.0–2.0 levels into pathway- or mechanism-relevant targets at TV 2.5. We attribute this observation to several factors: technological advances that enable frequent interrogation of genome-wide resources and produce a torrent of candidate targets; the complexity of HD biology and Htt function, which gives rise to a large dataset supportive of many functional roles for Htt; residence of nonstandardized datasets in various public and private locations, which leads to difficulty in correlation of these datasets; and publication pressure to further interrogate the most novel targets, which leaves a much larger list of targets with potentially greater value to drug discovery research unexamined. Thus, there is a critical need to address the handling of these early targets over the longer term.

One solution is to reduce the numerical complexity and clutter of these early data by assigning targets into gene networks (Zabel et al., 2006). Indeed, most gene products do not function alone but in pathways or networks in concerted action with other gene products. The abundant published literature regarding the cellular functions of many genes can then provide a basis for deducing the pathway-relatedness of the target genes. As a result of this approach, we may make indirect connections to key regulators and ultimately acquire better targets (see also Chapters 3 and 6, this volume). For a complex disease such as HD, this approach may also lead to the identification of the few key genes that regulate interlocking pathways involved in disease onset or progression. Ultimately, this approach may accelerate decision-making in target validation, although it may also demonstrate that a single gene approach may not be sufficient.

One way to assign targets to pathways and networks is to merge "wet-bench data" with *in silico* information curated at PubMed (MeSH categories), HD and neurodegenerative network overlaps at Kyoto Encyclopedia of Genes and Genomes (Limviphuvadh et al., 2007), and Gene Ontology. Data regarding tissue expression and distribution can also be overlaid with these datasets. HD-relevant information for the creation of HD gene networks should use the HD interactome (Giorgini and Muchowski, 2005; Goehler et al., 2004; Horn et al., 2006; Kaltenbach et al. 2007; Zabel et al., 2006), transcriptional changes observed in HD brain tissues (Hodges et al., 2006; Kuhn et al., 2007; Runne et al., 2007), and the list of potential genetic modifiers for HD onset identified from human and/or model organism studies (Chattopadhyay et al., 2003, 2005; Gusella and Macdonald, 2006; Li et al., 2006; MacDonald et al., 1999; Metzger et al., 2006a, 2006b; Wexler et al., 2004). In addition, modifiers of Htt expression in murine models can be included; this information is partly available at WebQTL (http://www.genenetwork.org). Several potential methods to assign targets into pathways or networks can then be developed for HD. Significant effort in annotating these groupings, with iterative hypothesis testing of newly realized pathways or networks, should improve the list of HD-relevant mechanisms summarized in Table 4.1. Because some pathways or mechanisms may be shared among other neurodegenerative disorders, we also seek to include knowledge from the study of these disorders. Open resources at Alzforum and the SWAN project (Clark and Kinoshita, 2007) are additional sources of information to further hone hypothetical strategies for building networks. The goal is to develop a systematic method to (1) prioritize targets within each TV level to be promoted for testing; (2) to diversify target prosecution and broaden coverage of pathways and mechanisms of HD progression; and (3) to select targets that function either early or late in affected pathways.

CONCLUSIONS

We view a stepwise process toward target validation as a key to building and filling the "target pipeline" with disease-relevant targets for drug discovery in HD. Some process hurdles remain, and significant scientific gaps must be filled. We propose short-term solutions to some of these issues, such as standardization of processes and scientific risk management, and also offer our long-term thoughts on the development of an integrated systems biology approach to gain new insight into HD and generate new hypotheses for target prosecution. Through these efforts, we will continue to identify and promote the most promising targets for drug discovery toward an effective therapy for HD.

ACKNOWLEDGMENTS

We thank Allan Tobin, Ethan Signer, Robert Pacifici, Michael Palazzolo, and Robi Blumenstein for their input and critical evaluation of the manuscript.

REFERENCES

Aiken CT, Tobin AJ, Schweitzer ES (2004) A cell-based screen for drugs to treat Huntington's disease. *Neurobiol Dis* 16(3): 546–555.

Alarcon JM, Malleret G, Touzani K, Vronskaya S, Ishii S, et al. (2004) Chromatin acetylation, memory, and LTP are impaired in CBP+/- mice: a model for the cognitive deficit in Rubinstein-Taybi syndrome and its amelioration. *Neuron* 42(6): 947–959.

Anca MH, Gazit E, Loewenthal R, Ostrovsky O, Frydman M, et al. (2004) Different phenotypic expression in monozygotic twins with Huntington disease. *Am J Med Genet A* 124(1): 89–91.

Anne SL, Saudou F, Humbert S (2007) Phosphorylation of huntingtin by cyclin-dependent kinase 5 is induced by DNA damage and regulates wild-type and mutant huntingtin toxicity in neurons. *J Neurosci* 27(27): 7318–7328.

Arrasate M, Mitra S, Schweitzer ES, Segal MR, Finkbeiner S (2004) Inclusion body formation reduces levels of mutant huntingtin and the risk of neuronal death. *Nature* 431(7010): 805–810.

Bailey CD, Johnson GV (2005) Tissue transglutaminase contributes to disease progression in the R6/2 Huntington's disease mouse model via aggregate-independent mechanisms. *J Neurochem* 92(1): 83–92.

Baquet ZC, Gorski JA, Jones KR (2004) Early striatal dendrite deficits followed by neuron loss with advanced age in the absence of anterograde cortical brain-derived neurotrophic factor. *J Neurosci* 24(17): 4250–4258.

Bates GP, Hockly E (2003) Experimental therapeutics in Huntington's disease: are models useful for therapeutic trials? *Curr Opin Neurol* 16(4): 465–470.

Bjugstad KB, Zawada WM, Goodman SI, Freed CR (2001) IGF-1 and bFGF reduce glutaric acid and 3-hydroxyglutaric acid toxicity in striatal cultures. *J Inherit Metab Dis* 24(6): 631–647.

Bonifati DM, Kishore U (2007) Role of complement in neurodegeneration and neuroinflammation. *Mol Immunol* 44(5): 999–1010.

Borrell-Pages M, Zala D, Humbert S, Saudou F (2006) Huntington's disease: from huntingtin function and dysfunction to therapeutic strategies. *Cell Mol Life Sci* 63(22): 2642–2660.

Bosch M, Pineda JR, Sunol C, Petriz J, Cattaneo E, et al. (2004) Induction of GABAergic phenotype in a neural stem cell line for transplantation in an excitotoxic model of Huntington's disease. *Exp Neurol* 190(1): 42–58.

Bottcher T, Mix E, Koczan D, Bauer P, Pahnke J, et al. (2003) Gene expression profiling of ciliary neurotrophic factor-overexpressing rat striatal progenitor cells (ST14A) indicates improved stress response during the early stage of differentiation. *J Neurosci Res* 73(1): 42–53.

Brignull HR, Morley JF, Garcia SM, Morimoto RI (2006) Modeling polyglutamine pathogenesis in C. elegans. *Methods Enzymol* 412: 256–282.

Brooks E, Arrasate M, Cheung K, Finkbeiner SM (2004) Using antibodies to analyze polyglutamine stretches. *Methods Mol Biol* 277: 103–128.

Bursch W, Ellinger A (2005) Autophagy—a basic mechanism and a potential role for neurodegeneration. *Folia Neuropathol* 43(4): 297–310.

Carter RJ, Hunt MJ, Morton AJ (2000) Environmental stimulation increases survival in mice transgenic for exon 1 of the Huntington's disease gene. *Mov Disord* 15(5): 925–937.

Cattaneo E (2003) Dysfunction of wild-type huntingtin in Huntington disease. *News Physiol Sci* 18: 34–37.

Cha JH (2007) Transcriptional signatures in Huntington's disease. *Prog Neurobiol* 83(4): 228–248.

Chang S, Young BD, Li S, Qi X, Richardson JA, et al. (2006) Histone deacetylase 7 maintains vascular integrity by repressing matrix metalloproteinase 10. *Cell* 126(2): 321–334.

Chattopadhyay B, Baksi K, Mukhopadhyay S, Bhattacharyya NP (2005) Modulation of age at onset of Huntington disease patients by variations in TP53 and human caspase activated DNase (hCAD) genes. *Neurosci Lett* 374(2): 81–86.

Chattopadhyay B, Ghosh S, Gangopadhyay PK, Das SK, Roy T, et al. (2003) Modulation of age at onset in Huntington's disease and spinocerebellar ataxia type 2 patients originated from eastern India. *Neurosci Lett* 345(2): 93–96.

Chien S, Reiter LT, Bier E, Gribskov M (2002) Homophila: human disease gene cognates in Drosophila. *Nucleic Acids Res* 30(1): 149–151.

Cho SR, Benraiss A, Chmielnicki E, Samdani A, Economides A, et al. (2007) Induction of neostriatal neurogenesis slows disease progression in a transgenic murine model of Huntington disease. *J Clin Invest* 117(10): 2889–2902.

Chopra V, Fox JH, Lieberman G, Dorsey K, Matson W, et al. (2007) A small-molecule therapeutic lead for Huntington's disease: preclinical pharmacology and efficacy of C2-8 in the R6/2 transgenic mouse. *Proc Natl Acad Sci U S A* 104: 16685–16689.

Clark T, Kinoshita J (2007) Alzforum and SWAN: the present and future of scientific web communities. *Brief Bioinform* 8(3): 163–171.

Cui L, Jeong H, Borovecki F, Parkhurst CN, Tanese N, et al. (2006) Transcriptional repression of PGC-1alpha by mutant huntingtin leads to mitochondrial dysfunction and neurodegeneration. *Cell* 127(1): 59–69.

Cummings DM, Milnerwood AJ, Dallerac GM, Vatsavayai SC, Hirst MC, et al. (2007) Abnormal cortical synaptic plasticity in a mouse model of Huntington's disease. *Brain Res Bull* 72(2–3): 103–107.

Cummings DM, Milnerwood AJ, Dallerac GM, Waights V, Brown JY, et al. (2006) Aberrant cortical synaptic plasticity and dopaminergic dysfunction in a mouse model of Huntington's disease. *Hum Mol Genet* 15(19): 2856–2868.

Curtis MA, Penney EB, Pearson AG, van Roon-Mom WM, Butterworth NJ, et al. (2003) Increased cell proliferation and neurogenesis in the adult human Huntington's disease brain. *Proc Natl Acad Sci U S A* 100(15): 9023–9027.

Di Maria E, Marasco A, Tartari M, Ciotti P, Abbruzzese G, et al. (2006) No evidence of association between BDNF gene variants and age-at-onset of Huntington's disease. *Neurobiol Dis* 24(2): 274–279.

Djousse L, Knowlton B, Hayden MR, Almqvist EW, Brinkman RR, et al. (2004) Evidence for a modifier of onset age in Huntington disease linked to the HD gene in 4p16. *Neurogenetics* 5(2): 109–114.

Dokmanovic M, Perez G, Xu W, Ngo L, Clarke C, et al. (2007) Histone deacetylase inhibitors selectively suppress expression of HDAC7. *Mol Cancer Ther* 6(9): 2525–2534.

Dompierre JP, Godin JD, Charrin BC, Cordelieres FP, King SJ, et al. (2007) Histone deacetylase 6 inhibition compensates for the transport deficit in Huntington's disease by increasing tubulin acetylation. *J Neurosci* 27(13): 3571–3583.

Dorsman JC, Smoor MA, Maat-Schieman ML, Bout M, Siesling S, et al. (1999) Analysis of the subcellular localization of huntingtin with a set of rabbit polyclonal antibodies in cultured mammalian cells of neuronal origin: comparison with the distribution of huntingtin in Huntington's disease autopsy brain. *Philos Trans R Soc Lond B Biol Sci* 354(1386): 1061–1067.

Duan W, Guo Z, Jiang H, Ladenheim B, Xu X, et al. (2004) Paroxetine retards disease onset and progression in Huntingtin mutant mice. *Ann Neurol* 55(4): 590–594.

Ducker CE, Stettler EM, French KJ, Upson JJ, Smith CD (2004) Huntingtin interacting protein 14 is an oncogenic human protein: palmitoyl acyltransferase. *Oncogene* 23(57): 9230–9237.

Ekshyyan O, Aw TY (2004) Apoptosis: a key in neurodegenerative disorders. *Curr Neurovasc Res* 1(4): 355–371.

Ernfors P, Lee KF, Jaenisch R (1994) Mice lacking brain-derived neurotrophic factor develop with sensory deficits. *Nature* 368(6467): 147–150.

Faber PW, Alter JR, MacDonald ME, Hart AC (1999) Polyglutamine-mediated dysfunction and apoptotic death of a Caenorhabditis elegans sensory neuron. *Proc Natl Acad Sci U S A* 96(1): 179–184.

Fava M (1997) Psychopharmacologic treatment of pathologic aggression. *Psychiatr Clin North Am* 20(2): 427–451.

Gauthier LR, Charrin BC, Borrell-Pages M, Dompierre JP, Rangone H, et al. (2004) Huntingtin controls neurotrophic support and survival of neurons by enhancing BDNF vesicular transport along microtubules. *Cell* 118(1): 127–138.

Gibson HE, Reim K, Brose N, Morton AJ, Jones S (2005) A similar impairment in CA3 mossy fibre LTP in the R6/2 mouse model of Huntington's disease and in the complexin II knockout mouse. *Eur J Neurosci* 22(7): 1701–1712.

Gil JM, Mohapel P, Araujo IM, Popovic N, Li JY, et al. (2005) Reduced hippocampal neurogenesis in R6/2 transgenic Huntington's disease mice. Neurobiol Dis 20(3): 744–751.

Giorgini F, Muchowski PJ (2005) Connecting the dots in Huntington's disease with protein interaction networks. *Genome Biol* 6(3): 210.

Goehler H, Lalowski M, Stelzl U, Waelter S, Stroedicke M, et al. (2004) A protein interaction network links GIT1, an enhancer of huntingtin aggregation, to Huntington's disease. *Mol Cell* 15(6): 853–865.

Gorski JA, Balogh SA, Wehner JM, Jones KR (2003) Learning deficits in forebrain-restricted brain-derived neurotrophic factor mutant mice. *Neuroscience* 121(2): 341–354.

Graham RK, Deng Y, Slow EJ, Haigh B, Bissada N, et al. (2006) Cleavage at the caspase-6 site is required for neuronal dysfunction and degeneration due to mutant huntingtin. *Cell* 125(6): 1179–1191.

Grove VE Jr, Quintanilla J, DeVaney GT (2000) Improvement of Huntington's disease with olanzapine and valproate. *N Engl J Med* 343(13): 973–974.

Groves MR, Hanlon N, Turowski P, Hemmings BA, Barford D (1999) The structure of the protein phosphatase 2A PR65/A subunit reveals the conformation of its 15 tandemly repeated HEAT motifs. *Cell* 96(1): 99–110.

Guidetti P, Bates GP, Graham RK, Hayden MR, Leavitt BR, et al. (2006) Elevated brain 3-hydroxykynurenine and quinolinate levels in Huntington disease mice. *Neurobiol Dis* 23(1): 190–197.

Guidetti P, Schwarcz R (2003) 3-Hydroxykynurenine and quinolinate: pathogenic synergism in early grade Huntington's disease? *Adv Exp Med Biol* 527: 137–145.

Gusella JF, Macdonald ME (2006) Huntington's disease: seeing the pathogenic process through a genetic lens. *Trends Biochem Sci* 31(9): 533–540.

Hackam AS, Hodgson JG, Singaraja R, Zhang T, Gan L, et al. (1999) Evidence for both the nucleus and cytoplasm as subcellular sites of pathogenesis in Huntington's disease in cell culture and in transgenic mice expressing mutant huntingtin. *Philos Trans R Soc Lond B Biol Sci* 354(1386): 1047–1055.

Hackam AS, Singaraja R, Wellington CL, Metzler M, McCutcheon K, et al. (1998) The influence of huntingtin protein size on nuclear localization and cellular toxicity. *J Cell Biol* 141(5): 1097–1105.

Hannan AJ (2005) Novel therapeutic targets for Huntington's disease. *Expert Opin Ther Targets* 9(4): 639–650.

Hansson O, Nylandsted J, Castilho RF, Leist M, Jaattela M, et al. (2003) Overexpression of heat shock protein 70 in R6/2 Huntington's disease mice has only modest effects on disease progression. *Brain Res* 970(1–2): 47–57.

Harper SQ, Staber PD, He X, Eliason SL, Martins IH, et al. (2005) RNA interference improves motor and neuropathological abnormalities in a Huntington's disease mouse model. *Proc Natl Acad Sci U S A* 102(16): 5820–5825.

Hebb AL, Robertson HA, Denovan-Wright EM (2004) Striatal phosphodiesterase mRNA and protein levels are reduced in Huntington's disease transgenic mice prior to the onset of motor symptoms. *Neuroscience* 123(4): 967–981.

Higgins DS Jr. (2006) Huntington's disease. Curr Treat Options Neurol 8(3): 236–244.

Hockly E, Cordery PM, Woodman B, Mahal A, van Dellen A, et al. (2002) Environmental enrichment slows disease progression in R6/2 Huntington's disease mice. *Ann Neurol* 51(2): 235–242.

Hockly E, Richon VM, Woodman B, Smith DL, Zhou X, et al. (2003) Suberoylanilide hydroxamic acid, a histone deacetylase inhibitor, ameliorates motor deficits in a mouse model of Huntington's disease. *Proc Natl Acad Sci U S A* 100(4): 2041–2046.

Hodges A, Strand AD, Aragaki AK, Kuhn A, Sengstag T, et al. (2006) Regional and cellular gene expression changes in human Huntington's disease brain. *Hum Mol Genet* 15(6): 965–977.

Horn SC, Lalowski M, Goehler H, Droge A, Wanker EE, et al. (2006) Huntingtin interacts with the receptor sorting family protein GASP2. *J Neural Transm* 113(8): 1081–1090.

Hu H, McCaw EA, Hebb AL, Gomez GT, Denovan-Wright EM (2004) Mutant huntingtin affects the rate of transcription of striatum-specific isoforms of phosphodiesterase 10A. *Eur J Neurosci* 20(12): 3351–3363.

Huang K, Yanai A, Kang R, Arstikaitis P, Singaraja RR, et al. (2004) Huntingtin-interacting protein HIP14 is a palmitoyl transferase involved in palmitoylation and trafficking of multiple neuronal proteins. *Neuron* 44(6): 977–986.

Huntington Study Group (2006) Tetrabenazine as antichorea therapy in Huntington disease: a randomized controlled trial. *Neurology* 66(3): 366–372.

Iwata A, Riley BE, Johnston JA, Kopito RR (2005) HDAC6 and microtubules are required for autophagic degradation of aggregated huntingtin. *J Biol Chem* 280(48): 40282–40292.

Jin K, LaFevre-Bernt M, Sun Y, Chen S, Gafni J, et al. (2005) FGF-2 promotes neurogenesis and neuroprotection and prolongs survival in a transgenic mouse model of Huntington's disease. *Proc Natl Acad Sci U S A* 102(50): 18189–18194.

Kaltenbach LS, Romero E, Becklin RR, Chettier R, Bell R, et al. (2007) Huntingtin interacting proteins are genetic modifiers of neurodegeneration. *PLoS Genet* 3(5): e82.

Kegel KB, Sapp E, Yoder J, Cuiffo B, Sobin L, et al. (2005) Huntingtin associates with acidic phospholipids at the plasma membrane. *J Biol Chem* 280(43): 36464–36473.

Kenney C, Jankovic J (2006) Tetrabenazine in the treatment of hyperkinetic movement disorders. *Expert Rev Neurother* 6(1): 7–17.

Khan N, Jeffers M, Kumar S, Hackett C, Boldog F, et al. (2008) Determination of the class and isoform selectivity of small molecule HDAC inhibitors. *Biochem J* 409(2): 581–589.

Khoshnan A, Ko J, Watkin EE, Paige LA, Reinhart PH, et al. (2004) Activation of the IkappaB kinase complex and nuclear factor-kappaB contributes to mutant huntingtin neurotoxicity. *J Neurosci* 24(37): 7999–8008.

Kishikawa S, Li JL, Gillis T, Hakky MM, Warby S, et al. (2006) Brain-derived neurotrophic factor does not influence age at neurologic onset of Huntington's disease. *Neurobiol Dis* 24(2): 280–285.

Kuhn A, Goldstein DR, Hodges A, Strand AD, Sengstag T, et al. (2007) Mutant huntingtin's effects on striatal gene expression in mice recapitulate changes observed in human Huntington's disease brain and do not differ with mutant huntingtin length or wild-type huntingtin dosage. *Hum Mol Genet* 16(15): 1845–1861.

La Spada AR (2005) Huntington's disease and neurogenesis: FGF-2 to the rescue? *Proc Natl Acad Sci U S A* 102(50): 17889–17890.

Langley B, Gensert JM, Beal MF, Ratan RR (2005) Remodeling chromatin and stress resistance in the central nervous system: histone deacetylase inhibitors as novel and broadly effective neuroprotective agents. *Curr Drug Targets* 4(1): 41–50.

Lazic SE, Grote H, Armstrong RJ, Blakemore C, Hannan AJ, et al. (2004) Decreased hippocampal cell proliferation in R6/1 Huntington's mice. *Neuroreport* 15(5): 811–813.

Leegwater-Kim J, Cha JH (2004) The paradigm of Huntington's disease: therapeutic opportunities in neurodegeneration. *NeuroRx* 1(1): 128–138.

Li JL, Hayden MR, Warby SC, Durr A, Morrison PJ, et al. (2006) Genome-wide significance for a modifier of age at neurological onset in Huntington's disease at 6q23-24: the HD MAPS study. *BMC Med Genet* 7: 71.

Li JY, Popovic N, Brundin P (2005) The use of the R6 transgenic mouse models of Huntington's disease in attempts to develop novel therapeutic strategies. *NeuroRx* 2(3): 447–464.

Li SH, Cheng AL, Zhou H, Lam S, Rao M, et al. (2002) Interaction of Huntington disease protein with transcriptional activator Sp1. *Mol Cell Biol* 22(5): 1277–1287.

Li SH, Li XJ (2004) Huntingtin and its role in neuronal degeneration. *Neuroscientist* 10(5): 467–475.

Limviphuvadh V, Tanaka S, Goto S, Ueda K, Kanehisa M (2007) The commonality of protein interaction networks determined in neurodegenerative disorders (NDDs). *Bioinformatics* 23(16): 2129–2138.

Link CD (2001) Transgenic invertebrate models of age-associated neurodegenerative diseases. *Mech Ageing Dev* 122(14): 1639–1649.

Link CD (2006) C. elegans models of age-associated neurodegenerative diseases: lessons from transgenic worm models of Alzheimer's disease. *Exp Gerontol* 41(10): 1007–1013.

Lo DC, McAllister AK, Katz LC (1994) Neuronal transfection in brain slices using particle-mediated gene transfer. *Neuron* 13(6): 1263–1268.

Lynch G, Kramar EA, Rex CS, Jia Y, Chappas D, et al. (2007) Brain-derived neurotrophic factor restores synaptic plasticity in a knock-in mouse model of Huntington's disease. *J Neurosci* 27(16): 4424–4434.

MacDonald ME, Vonsattel JP, Shrinidhi J, Couropmitree NN, Cupples LA, et al. (1999) Evidence for the GluR6 gene associated with younger onset age of Huntington's disease. *Neurology* 53(6): 1330–1332.

Mangiarini L, Sathasivam K, Mahal A, Mott R, Seller M, et al. (1997) Instability of highly expanded CAG repeats in mice transgenic for the Huntington's disease mutation. *Nat Genet* 15(2):197–200.

Mangiarini L, Sathasivam K, Seller M, Cozens B, Harper A, et al. (1996) Exon 1 of the HD gene with an expanded CAG repeat is sufficient to cause a progressive neurological phenotype in transgenic mice. *Cell* 87(3): 493–506.

Maragakis NJ, Rothstein JD (2006) Mechanisms of disease: astrocytes in neurodegenerative disease. *Nat Clin Pract Neurol* 2(12): 679–689.

Marsh JL, Pallos J, Thompson LM (2003) Fly models of Huntington's disease. *Hum Mol Genet* 12 Spec No 2: R187–193.

Marsh JL, Thompson LM (2004) Can flies help humans treat neurodegenerative diseases? *Bioessays* 26(5): 485–496.

Mattson MP, Maudsley S, Martin B (2004) BDNF and 5-HT: a dynamic duo in age-related neuronal plasticity and neurodegenerative disorders. *Trends Neurosci* 27(10): 589–594.

McBride JL, Ramaswamy S, Gasmi M, Bartus RT, Herzog CD, et al. (2006) Viral delivery of glial cell line-derived neurotrophic factor improves behavior and protects striatal neurons in a mouse model of Huntington's disease. *Proc Natl Acad Sci U S A* 103(24): 9345–9350.

Menalled LB, Sison JD, Wu Y, Olivieri M, Li XJ, et al. (2002) Early motor dysfunction and striosomal distribution of huntingtin microaggregates in Huntington's disease knock-in mice. *J Neurosci* 22(18): 8266–8276.

Menegola E, Di Renzo F, Broccia ML, Giavini E (2006) Inhibition of histone deacetylase as a new mechanism of teratogenesis. *Birth Defects Res C Embryo Today* 78(4): 345–353.

Metzger S, Bauer P, Tomiuk J, Laccone F, Didonato S, et al. (2006a) The S18Y polymorphism in the UCHL1 gene is a genetic modifier in Huntington's disease. *Neurogenetics* 7(1): 27–30.

Metzger S, Bauer P, Tomiuk J, Laccone F, Didonato S, et al. (2006b) Genetic analysis of candidate genes modifying the age-at-onset in Huntington's disease. *Hum Genet* 120(2): 285–292.

Minamiyama M, Katsuno M, Adachi H, Waza M, Sang C, et al. (2004) Sodium butyrate ameliorates phenotypic expression in a transgenic mouse model of spinal and bulbar muscular atrophy. *Hum Mol Genet* 13(11): 1183–1192.

Morley JF, Brignull HR, Weyers JJ, Morimoto RI (2002) The threshold for polyglutamine-expansion protein aggregation and cellular toxicity is dynamic and influenced by aging in Caenorhabditis elegans. *Proc Natl Acad Sci U S A* 99(16): 10417–10422.

Nemeth H, Toldi J, Vecsei L (2006) Kynurenines, Parkinson's disease and other neurodegenerative disorders: preclinical and clinical studies. *J Neural Transm* (70): 285–304.

Nucifora FC Jr, Sasaki M, Peters MF, Huang H, Cooper JK, et al. (2001) Interference by huntingtin and atrophin-1 with cbp-mediated transcription leading to cellular toxicity. *Science* 291(5512): 2423–2428.

Oliveira JM, Chen S, Almeida S, Riley R, Goncalves J, et al. (2006) Mitochondrial-dependent Ca2+ handling in Huntington's disease striatal cells: effect of histone deacetylase inhibitors. *J Neurosci* 26(43): 11174–11186.

Ona VO, Li M, Vonsattel JP, Andrews LJ, Khan SQ, et al. (1999) Inhibition of caspase-1 slows disease progression in a mouse model of Huntington's disease. *Nature* 399(6733): 263–267.

Palmer TD, Ray J, Gage FH (1995) FGF-2-responsive neuronal progenitors reside in proliferative and quiescent regions of the adult rodent brain. *Mol Cell Neurosci* 6(5): 474–486.

Parker JA, Connolly JB, Wellington C, Hayden M, Dausset J, et al. (2001) Expanded polyglutamines in Caenorhabditis elegans cause axonal abnormalities and severe dysfunction of PLM mechanosensory neurons without cell death. *Proc Natl Acad Sci U S A* 98(23): 13318–13323.

Pattison LR, Kotter MR, Fraga D, Bonelli RM (2006) Apoptotic cascades as possible targets for inhibiting cell death in Huntington's disease. *J Neurol* 253(9): 1137–1142.

Perez-De La Cruz V, Santamaria A (2006) Integrative hypothesis for Huntington's disease: a brief review on experimental evidence. *Physiol Res* 56(5): 513–526.

Petersen A, Mani K, Brundin P (1999) Recent advances on the pathogenesis of Huntington's disease. *Exp Neurol* 157(1): 1–18.

Petrasch-Parwez E, Nguyen HP, Lobbecke-Schumacher M, Habbes HW, Wieczorek S, et al. (2007) Cellular and subcellular localization of Huntingtin [corrected] aggregates in the brain of a rat transgenic for Huntington disease. *J Comp Neurol* 501(5): 716–730.

Phillips W, Morton AJ, Barker RA (2005) Abnormalities of neurogenesis in the R6/2 mouse model of Huntington's disease are attributable to the in vivo microenvironment. *J Neurosci* 25(50): 11564–11576.

Picconi B, Passino E, Sgobio C, Bonsi P, Barone I, et al. (2006) Plastic and behavioral abnormalities in experimental Huntington's disease: a crucial role for cholinergic interneurons. *Neurobiol Dis* 22(1): 143–152.

Pineda JR, Canals JM, Bosch M, Adell A, Mengod G, et al. (2005) Brain-derived neurotrophic factor modulates dopaminergic deficits in a transgenic mouse model of Huntington's disease. *J Neurochem* 93(5): 1057–1068.

Ramaswamy S, Shannon KM, Kordower JH (2007) Huntington's disease: pathological mechanisms and therapeutic strategies. *Cell Transplant* 16(3): 301–312.

Ravikumar B, Vacher C, Berger Z, Davies JE, Luo S, et al. (2004) Inhibition of mTOR induces autophagy and reduces toxicity of polyglutamine expansions in fly and mouse models of Huntington disease. *Nat Genet* 36(6): 585–595.

Reiter LT, Potocki L, Chien S, Gribskov M, Bier E (2001) A systematic analysis of human disease-associated gene sequences in Drosophila melanogaster. *Genome Res* 11(6): 1114–1125.

Riddle DR, Katz LC, Lo DC (1997) Focal delivery of neurotrophins into the central nervous system using fluorescent latex microspheres. *Biotechniques* 23(5): 928–934, 936–927.

Ross CA, Poirier MA (2004) Protein aggregation and neurodegenerative disease. *Nat Med* 10(Suppl): S10–17.

Rubinsztein DC (2006) The roles of intracellular protein-degradation pathways in neurodegeneration. *Nature* 443(7113): 780–786.

Runne H, Kuhn A, Wild EJ, Pratyaksha W, Kristiansen M, et al. (2007) Analysis of potential transcriptomic biomarkers for Huntington's disease in peripheral blood. *Proc Natl Acad Sci U S A* 104(36):14424–14429.

Sadri-Vakili G, Bouzou B, Benn CL, Kim MO, Chawla P, et al. (2007) Histones associated with downregulated genes are hypo-acetylated in Huntington's disease models. *Hum Mol Genet* 16(11): 1293–1306.

Sadri-Vakili G, Cha JH (2006) Histone deacetylase inhibitors: a novel therapeutic approach to Huntington's disease (complex mechanism of neuronal death). *Curr Alzheimer Res* 3(4): 403–408.

Schilling G, Becher MW, Sharp AH, Jinnah HA, Duan K, et al. (1999) Intranuclear inclusions and neuritic aggregates in transgenic mice expressing a mutant N-terminal fragment of huntingtin. *Hum Mol Genet* 8(3): 397–407.

Shibata M, Lu T, Furuya T, Degterev A, Mizushima N, et al. (2006) Regulation of intracellular accumulation of mutant Huntingtin by Beclin 1. *J Biol Chem* 281(20): 14474–14485.

Singaraja RR, Hadano S, Metzler M, Givan S, Wellington CL, et al. (2002) HIP14, a novel ankyrin domain-containing protein, links huntingtin to intracellular trafficking and endocytosis. *Hum Mol Genet* 11(23): 2815–2828.

Sipione S, Cattaneo E (2001) Modeling Huntington's disease in cells, flies, and mice. *Mol Neurobiol* 23(1): 21–51.

Sittler A, Walter S, Wedemeyer N, Hasenbank R, Scherzinger E, et al. (1998) SH3GL3 associates with the Huntingtin exon 1 protein and promotes the formation of polygln-containing protein aggregates. *Mol Cell* 2(4): 427–436.

Slaughter JR, Martens MP, Slaughter KA (2001) Depression and Huntington's disease: prevalence, clinical manifestations, etiology, and treatment. *CNS Spectr* 6(4): 306–326.

Slow EJ, van Raamsdonk J, Rogers D, Coleman SH, Graham RK, et al. (2003) Selective striatal neuronal loss in a YAC128 mouse model of Huntington disease. *Hum Mol Genet* 12(13): 1555–1567.

Stack EC, Kubilus JK, Smith K, Cormier K, Del Signore SJ, et al. (2005) Chronology of behavioral symptoms and neuropathological sequela in R6/2 Huntington's disease transgenic mice. *J Comp Neurol* 490(4): 354–370.

Steffan JS, Bodai L, Pallos J, Poelman M, McCampbell A, et al. (2001) Histone deacetylase inhibitors arrest polyglutamine-dependent neurodegeneration in Drosophila. *Nature* 413(6857): 739–743.

Steffan JS, Kazantsev A, Spasic-Boskovic O, Greenwald M, Zhu YZ, et al. (2000) The Huntington's disease protein interacts with p53 and CREB-binding protein and represses transcription. *Proc Natl Acad Sci U S A* 97(12): 6763–6768.

Subba Rao K (2007) Mechanisms of disease: DNA repair defects and neurological disease. *Nat Clin Pract* 3(3): 162–172.

Takano H, Gusella JF (2002) The predominantly HEAT-like motif structure of huntingtin and its association and coincident nuclear entry with dorsal, an NF-kB/Rel/dorsal family transcription factor. *BMC Neurosci* 3: 15.

Tam S, Geller R, Spiess C, Frydman J (2006) The chaperonin TRiC controls polyglutamine aggregation and toxicity through subunit-specific interactions. *Nat Cell Biol* 8(10): 1155–1162.

Tarditi A, Camurri A, Varani K, Borea PA, Woodman B, et al. (2006) Early and transient alteration of adenosine A2A receptor signaling in a mouse model of Huntington disease. *Neurobiol Dis* 23(1): 44–53.

Tarlac V, Storey E (2003) Role of proteolysis in polyglutamine disorders. *J Neurosci Res* 74(3): 406–416.

Tooyama I, Kremer HP, Hayden MR, Kimura H, McGeer EG, et al. (1993) Acidic and basic fibroblast growth factor-like immunoreactivity in the striatum and midbrain in Huntington's disease. *Brain Res* 610(1): 1–7.

Truant R, Atwal R, Burtnik A (2006) Hypothesis: Huntingtin may function in membrane association and vesicular trafficking. *Biochem Cell Biol* 84(6): 912–917.

Trzesniewska K, Brzyska M, Elbaum D (2004) Neurodegenerative aspects of protein aggregation. *Acta Neurobiol Exp (Wars)* 64(1): 41–52.

van Dellen A, Blakemore C, Deacon R, York D, Hannan AJ (2000) Delaying the onset of Huntington's in mice. *Nature* 404(6779): 721–722.

Van Raamsdonk JM, Gibson WT, Pearson J, Murphy Z, Lu G, et al. (2006) Body weight is modulated by levels of full-length huntingtin. *Hum Mol Genet* 15(9): 1513–1523.

Van Raamsdonk JM, Pearson J, Slow EJ, Hossain SM, Leavitt BR, et al. (2005) Cognitive dysfunction precedes neuropathology and motor abnormalities in the YAC128 mouse model of Huntington's disease. *J Neurosci* 25(16): 4169–4180.

Varma H, Voisine C, DeMarco CT, Cattaneo E, Lo DC, et al. (2007) Selective inhibitors of death in mutant huntingtin cells. *Nat Chem Biol* 3(2): 99–100.

Vega RB, Matsuda K, Oh J, Barbosa AC, Yang X, et al. (2004) Histone deacetylase 4 controls chondrocyte hypertrophy during skeletogenesis. *Cell* 119(4): 555–566.

von Horsten S, Schmitt I, Nguyen HP, Holzmann C, Schmidt T, et al. (2003) Transgenic rat model of Huntington's disease. *Hum Mol Genet* 12(6): 617–624.

Wang H, Guan Y, Wang X, Smith K, Cormier K, et al. (2007) Nortriptyline delays disease onset in models of chronic neurodegeneration. *Eur J Neurosci* 26(3): 633–641.

Wang JK, Portbury S, Thomas MB, Barney S, Ricca DJ, et al. (2006) Cardiac glycosides provide neuroprotection against ischemic stroke: discovery by a brain slice-based compound screening platform. *Proc Natl Acad Sci U S A* 103(27):10461–10466.

Wang LH, Qin ZH (2006) Animal models of Huntington's disease: implications in uncovering pathogenic mechanisms and developing therapies. *Acta Pharmacol Sin* 27(10): 1287–1302.

Wanker EE (2000) Protein aggregation in Huntington's and Parkinson's disease: implications for therapy. *Mol Med Today* 6(10): 387–391.

Wanker EE, Rovira C, Scherzinger E, Hasenbank R, Walter S, et al. (1997) HIP-I: a huntingtin interacting protein isolated by the yeast two-hybrid system. *Hum Mol Genet* 6(3): 487–495.

Wexler NS, Lorimer J, Porter J, Gomez F, Moskowitz C, et al. (2004) Venezuelan kindreds reveal that genetic and environmental factors modulate Huntington's disease age of onset. *Proc Natl Acad Sci U S A* 101(10): 3498–3503.

Xia J, Lee DH, Taylor J, Vandelft M, Truant R (2003) Huntingtin contains a highly conserved nuclear export signal. *Hum Mol Genet* 12(12): 1393–1403.

Yamamoto A, Cremona ML, Rothman JE (2006) Autophagy-mediated clearance of huntingtin aggregates triggered by the insulin-signaling pathway. *J Cell Biol* 172(5): 719–731.

Yanai A, Huang K, Kang R, Singaraja RR, Arstikaitis P, et al. (2006) Palmitoylation of huntingtin by HIP14 is essential for its trafficking and function. *Nat Neurosci* 9(6): 824–831.

Zabel C, Sagi D, Kaindl AM, Steireif N, Klare Y, et al. (2006) Comparative proteomics in neurodegenerative and non-neurodegenerative diseases suggest nodal point proteins in regulatory networking. *J Proteome Res* 5(8): 1948–1958.

Zala D, Benchoua A, Brouillet E, Perrin V, Gaillard MC, et al. (2005) Progressive and selective striatal degeneration in primary neuronal cultures using lentiviral vector coding for a mutant huntingtin fragment. *Neurobiol Dis* 20(3): 785–798.

Zhai W, Jeong H, Cui L, Krainc D, Tjian R (2005) In vitro analysis of huntingtin-mediated transcriptional repression reveals multiple transcription factor targets. *Cell* 123(7): 1241–1253.

Zhang Y, Ona VO, Li M, Drozda M, Dubois-Dauphin M, et al. (2003) Sequential activation of individual caspases, and of alterations in Bcl-2 proapoptotic signals in a mouse model of Huntington's disease. *J Neurochem* 87(5): 1184–1192.

Zourlidou A, Gidalevitz T, Kristiansen M, Landles C, Woodman B, et al. (2007) Hsp27 overexpression in the R6/2 mouse model of Huntington's disease: chronic neurodegeneration does not induce Hsp27 activation. *Hum Mol Genet* 16(9): 1078–1090.

Zuccato C, Belyaev N, Conforti P, Ooi L, Tartari M, et al. (2007) Widespread disruption of repressor element-1 silencing transcription factor/neuron-restrictive silencer factor occupancy at its target genes in Huntington's disease. *J Neurosci* 27(26): 6972–6983.

Zuccato C, Cattaneo E (2007) Role of brain-derived neurotrophic factor in Huntington's disease. *Prog Neurobiol* 81(5–6): 294–330.

Zuccato C, Ciammola A, Rigamonti D, Leavitt BR, Goffredo D, et al. (2001) Loss of huntingtin- mediated BDNF gene transcription in Huntington's disease. *Science* 293(5529): 493–498.

Zuccato C, Tartari M, Crotti A, Goffredo D, Valenza M, et al. (2003) Huntingtin interacts with REST/NRSF to modulate the transcription of NRSE-controlled neuronal genes. *Nat Genet* 35(1): 76–83.

5 High-Throughput and High-Content Screening for Huntington's Disease Therapeutics

Hemant Varma, Donald C. Lo, and
Brent R. Stockwell

CONTENTS

INTRODUCTION

Since the isolation of the gene and mutation that cause Huntington's disease (HD) more than a decade ago, there has been optimism that this knowledge would lead to the rapid discovery of therapeutic agents for this fatal and incurable disease [1]. In fact, considerable effort has been invested in HD drug discovery, predominantly in academic environments but also in the biopharmaceutical industry to some certain extent. Historically, however, interest in HD research and development in large pharmaceutical firms has been limited by the relatively small size of the HD patient population (~30,000 affected persons in United States), which has led to the perception that the market size for HD might be too small to justify the investment of substantial resources necessary to bring a new drug to clinical trials. However, there is a growing appreciation for the actual market size of "first in class" drugs for otherwise unaddressable diseases, as well as for the idea that HD may prove to be a paradigmatic disease for other, much more prevalent neurodegenerative diseases, such as Alzheimer's disease and Parkinson's disease [2].

One hurdle in drug discovery for most neurodegenerative disorders is their incompletely understood, multigenic, and multifactorial etiology. Thus, animal models of these diseases may often recapitulate only some aspects of such diseases and rarely reproduce the full pathophysiology of the human diseases [3, 4]. In contrast, HD is caused by a well-defined mutation in a single autosomal gene that has a dominant and fully penetrant phenotype. This simplicity in the genetics of HD has allowed the creation of a variety of cells cultures and transgenic animal models for HD in which there can be greater confidence that these models do capture important aspects of human disease initiation and progression [5].

This chapter will focus on the development and implementation of *high-throughput* and *high-content in vitro* assays for the discovery of HD therapeutics. High-throughput screening (HTS) of small molecules allows the rapid interrogation of the effects of thousands to hundreds of thousands of small molecules in a variety of *in vitro* and cell-based assays, whereas high-content screening (HCS) approaches may sacrifice some of these high-throughput capabilities in return for great biological and phenotypic complexity in the assay endpoints used. Why do we need such high-throughput and high-content methods? Simply because we do not currently have sufficient knowledge of the molecular targets and pathways that may be therapeutic for HD nor do we know how to design *a priori* custom small molecule compounds that will be guaranteed to have the desired effect on such biological targets. Hence rapid testing of many tens of thousands or more drug molecule candidates in HD models offers the potential for systematized serendipity, that we will encounter effective compounds in such discovery campaigns that will prove to be clinically relevant, using controllable and predictable *in vitro* screening processes.

Active compounds emerging from HTS and HCS screens—termed "hits"—are the templates on which eventually drug "leads" are developed through further combinatorial and medicinal chemistry efforts (Figure 5.1). Such efforts in the field have identified a number of hits that are being pursued as drug leads (see Chapters 8 and 12, this volume). In the following sections we will describe different strategies and approaches in the design and implementation of HTS and HCS screens for HD, and will also discuss the development of prioritization schemes for potential drug leads identified, future screening approaches, and the use of these compounds for gaining new insights into mechanisms underlying HD.

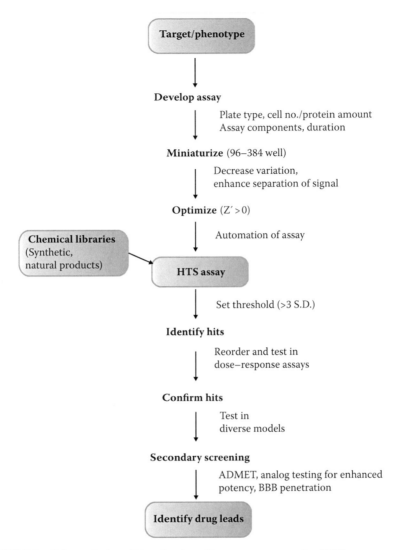

FIGURE 5.1   Schematic describing the drug discovery process using HTS.

HIGH-THROUGHPUT SCREENING

HIGH-THROUGHPUT AND HIGH-CONTENT SCREENING

HTS is a technology that has evolved and matured considerably during the past 10–15 years and lies at the heart of modern drug discovery in the biopharmaceutical sector and now increasingly in academia through university efforts and through the molecular screening initiatives that are part of the new National Institutes of Health (NIH) Roadmap [6, 7]. HTS is a method of testing thousands of compounds rapidly in parallel for their activity in one or more biological assays. The basic components of HTS are a miniaturized biological assay, automated transfers and liquid handling steps, and automated quantitative readout of the assays. The development of this field has been fueled by advances in robotics and engineering that allow automation of assays, as well as by miniaturization of cellular and biochemical assays. The technology needed to detect more complex cellular phenotypes in a high-throughput format is also rapidly evolving, allowing screens that yield more biologically complex information. This can be especially useful for disease states such as HD, in which particular cellular phenotypes relating to disease are known but in which the specific disease pathways leading to these phenotypic changes are not known with certainty. Many of these more complex screens are image based and can assay subcellular biological processes. They have depended on dramatic improvements in recent years in automated microscopy and image analysis. This emerging field is appropriately named *high-content screening* because the information content in each test is increased relative to conventional *in vitro* assays, such as protein enzyme assays.

For any high-throughput or high-content screen, the initial and the most critical step is the choice of assay endpoints. At this stage, one seeks to choose an assay that most closely reflects the disease process and yet is amenable to miniaturization and compatible with automation. The choice of assay can be complicated by a lack of understanding of the exact disease process to target or by the ability to generate a model that recapitulates all aspects of a complex disease. Thus, there is always a tradeoff between complexity and screenability. This is particularly true for HD, for which a wide variety of pathological mechanisms have been proposed to be relevant for disease but none has been biologically, much less clinically, validated. To compensate for such uncertainty, a broad range of phenotypes in different HD model assays has been brought to bear in HTS screening campaigns with the hope that consensus targets and pathways will emerge across several screening assays. Although this leads to a multiplicity of assays and accompanying time and costs, it holds the promise of identifying leads that can affect several pathological processes important in HD pathogenesis.

Not surprisingly, the choice of compounds that are to be screened can have a great impact on the quality of hits retrieved and their ultimate developmental potential as drug lead candidates. Here we discuss compound library assembly before we move on to assay development.

CHEMICAL COMPOUND LIBRARIES

A wide spectrum of screening compounds is available from a range of commercial vendors. These compounds range from natural products to purely synthetic compounds. Natural products are obtained from plants, animals, or microorganisms and have been a mainstay for new drug discovery since the beginning of modern pharmaceutical development [8]. Although natural product compounds are frequently superior in terms of biological activity and chemical complexity, they have the disadvantages of higher costs, limited availability, and difficulty of chemical synthesis [9]. Purely synthetic compounds are usually made from a few simple chemical building blocks, and synthetic libraries are often assembled by altering the starting chemical structure of the building blocks and then joining them together in a large number of different combinations. This methodology, known as combinatorial chemistry, allows dense coverage of a particular "chemical space" and the creation of novel chemical structures not available in nature [10], as well as an enormous variety of structural variations, limited only by the synthetic tools available. Synthetic compounds have the advantage of straightforward and inexpensive resynthesis, as well as the ability of making numerous analogs for improving potency and efficacy, pharmacodynamic properties, and reducing unwanted side effects such as cellular toxicity and other off-target actions. Although the costs of these compounds are lower than natural products, a number of these synthetic compounds suffer from poor solubility in aqueous solutions and lack the structural complexity of natural products, largely owing to the difficulty of synthesizing complex natural product-like structures. Because of the pros and cons of each compound class, the authors have typically used a mix of natural and synthetic compounds in their screening efforts.

Compounds can also be divided into two categories, based on functionality: compounds with known biological activity and those with unknown activity. Biologically active compounds include Food and Drug Administration (FDA)-approved drugs. This subset of compounds is especially useful for screening because much biological and pharmacological information is available for these drugs. Furthermore, any identified hits may rapidly progress to clinical trials. Several such collections have been used for screening and include the National Institute of Neurodegenerative Disorders and Stroke (NINDS) custom collection of 1,040 biologically active compounds (Microsource Inc., Gaylordsville, CT), the LOPAC collection of 1,280 compounds (Sigma-Aldrich, St. Louis, MO), the Prestwick Collection of 1,120 compounds, and 1,990 compounds assembled by the National Cancer Institute. These are commercially available to researchers. Similar collections have been assembled by a few academic groups [11, 12]. One drawback of these biologically active compound collections is the limited range of targets they affect. Estimates suggest that such known drugs target fewer than 500 of the potentially thousands of cellular targets [13]. Thus, assuming that effective treatment of neurodegenerative disorders such as HD may require modulating the function of novel molecular targets, known biologically active compounds that target a limited subset of cellular targets may not yield effective hits. However, many of these known drugs are promiscuous and target proteins other than the ones against which they were originally developed and may yet lead to the development of novel and efficacious lead candidates.

COMPOUND FILTERING AND LIBRARY ASSEMBLY

Commercial vendors maintain compound collections that range from a few thousand to more than 1 million compounds. Most of these compounds are novel, and there is limited or no biological information about them. Therefore, carefully sifting through these compound collections before acquisition is recommended and can save much time and expense in the screening process, which ultimately will lead to hits of improved quality. Several *in silico* filters have been developed for this purpose. These filters are based on empirical rules that predict "drug-likeness," a concept based on consensus chemical properties of known effective drugs that are FDA approved or have reached clinical trials [14]. The most commonly used criteria are the Lipinski's and Veber's rules for predicting oral bioavailability of non-natural products [15, 16].

These rules are based on polarity, size of the molecules, and molecular rigidity. In general, large molecules (molecular mass > 500 Da) or very hydrophobic or hydrophilic compounds are unlikely to be orally bioavailable because of poor solubility or membrane permeability. Although some vendors apply these criteria to their compound collections, this practice is not universal, and it may be desirable to filter compounds based on these criteria. In addition, it is valuable to eliminate compounds with toxic, reactive, or unstable groups. In the authors' experience, these filters can eliminate a large fraction (up to 50%) of commercial libraries.

Another important consideration in selecting compounds to screen in neurological disease assays is blood–brain barrier (BBB) penetrability because a compound with poor brain penetration is unlikely to be useful as therapy for neurological diseases. Although *in silico* filters of BBB penetrability are not as predictive as Lipinski's rules, they can enhance the quality of a library [17]. Most of these calculations can be performed using commercially available software, such as Molecular Operating Environment (MOE) developed by the Chemical Computing Group (Montreal, QC, Canada).

Finally, there are practical considerations including the ease of reordering compounds from vendors; this can be a major bottleneck in reconfirming hits if the vendor is slow in resupply or has limited availability of library compounds. In addition, quality and purity of compounds can vary considerably across vendors. This variability can arise from differences in the quality of the original syntheses or from differences in storage conditions and ongoing quality assessment.

ASSEMBLY OF COMPOUND LIBRARIES

Once a decision is made regarding the composition of the compounds in the libraries to be screened, the process of assembling a library for screening involves identifying commercial vendors, obtaining compounds in 96-well or 384-well format, preparing multiple copies of libraries in solvent (usually dimethyl sulfoxide [DMSO]), providing adequate storage (compounds are kept frozen at −80°C or −20°C), and maintaining record-keeping and tracking for the compounds obtained. The decision on the number, quantity, and type of compounds will depend on the budget and scope of the screening project. Most large companies maintain compound collections of 1 million or more compounds. However, with a hit rate of 0.1%, which is usual in most HTS efforts, the authors have typically found that screening a collection of ~50,000 compounds with compounds from different classes (natural and synthetic compounds) is likely to obtain a sufficient number of hits such that a few will survive the attrition rate in progression from hit to lead development. However, larger screens can enhance the likelihood of identifying more potent and active starting scaffolds.

ASSAY DEVELOPMENT

HTS assays can be *in vitro*, cell-based assays and whole organism based. Compounds are added directly in these assay formats, and activity, binding, or another endpoint is usually measured using some automated readout that involves fluorescence, luminescence, or absorbance measurement. After selecting an appropriate assay endpoint, whether one that has a molecular or cellular basis, a key step in establishing a high-throughput assay is adaptation of the assay into miniaturized format. This requires optimization of multiple assay parameters, such as types of plates (clear, opaque), well density, coating of wells, manufacturer, concentrations of components (cells, assay reagents), liquid handling and washing steps, duration of assay, and the choice of assay readout. A widely used metric of the quality and usability of an assay for HTS is the Z'-factor [18], a statistical index of the signal variability of an assay in relation to the separation in signal between its positive and negative controls, such that:

$$Z' = 1 - (3*sigma_p + 3*sigma_n)/(signal_p - signal_n)$$

where $sigma_p$ and $sigma_n$ are the standard deviations of the positive and negative controls, and $signal_p$ and $signal_n$ are the mean values of the positive and negative controls, respectively.

Most robust HTS assays have Z' values greater than 0.5, although complex assays with multiple steps are still usable with lower Z' values (0–0.5). An assay with a Z' value less than zero is not considered useful for HTS because of poor signal to noise relative to the variance of the measured values.

Because most compound libraries use DMSO as solvent, the effect of DMSO solvent at a concentration that will be used in screening needs to be assessed during optimization; a typically tolerated final concentration for DMSO is 0.1%. Most assays use 96-well or 384-well plates; higher-density assay formats are less suitable for most assays because of issues with optimization of highly miniaturized (1,536 or higher) assays and the cost of plate readers and liquid handlers required for such formats. Once optimized, multiple independent runs of the assay are performed to assess the reproducibility of the assay. At this stage, most reagents used in the optimized assay should be obtained in sufficient quantity for the whole screening program; this prevents variation in different batches of these components that can affect assay quality.

Most workflow in HTS is automated and uses robotic liquid handling, plate transfers, and plate reading. The workflow is carefully annotated at each stage by bar-coding plates. Large-scale HTS screens are generally done at a single compound concentration in multiple replicates (typically 10 µM for libraries in the tens of thousands and 2 µM for libraries in the hundreds of thousands), although it can be a good strategy to test smaller, focused compound collections, such as FDA-approved drugs, at multiple concentrations. Because of the greater clinical potential of such focused libraries, the increase in the cost of screening at multiple concentrations is more than offset by the rapid progress to clinical trials for any such identified drug.

Data from screens can be stored and analyzed using information management systems [19] or, more laboriously, in Excel spreadsheets (Microsoft, Redmond, WA). The data are analyzed to identify "hits," data points that cross a certain threshold defined to identify a positive outcome. These thresholds are somewhat arbitrary, but a value of 3 standard deviations from the mean signal of DMSO-treated wells is a reasonable and common cutoff, as it provides a manageable, statistical false-positive hit rate (0.15%). Alternatively, one can simply select the maximum number of hits that can be processed via "cherry picking," usually several hundred compounds. If the screening is performed in triplicate, the median rather than the mean for a particular compound should be used to determine hits; this guards against the excessive impact of large outlier effects, as is common in screening.

Once hit compounds are identified, their activity is reconfirmed in full dose-response studies. Thereafter, compounds are reordered from or resynthesized by the vendors, and activity and potency are tested in dose–response curve experiments in replicate runs. Finally, it is critical to confirm the identity and purity of novel hits by high-resolution mass spectrometry and ^{1}H and ^{13}C NMR; it is not uncommon to have activity from contaminants in both natural product and synthetic compound samples. These confirmed hits provide a starting point for testing in other models, structure–activity relationship, mechanism of action studies, and ultimately in lead optimization efforts.

HTS ASSAYS IN HD AND RELATED POLYGLUTAMINE DISORDERS

The vast majority of HTS assays are cell-based and *in vitro* assays. The latter consist of assays in which interaction or activity of a purified protein, RNA or DNA, or a simple mixture of a few molecules is assessed and reflects a known or suspected intermolecular interaction in the disease process. For example, this could be a ligand interaction with a receptor implicated in the disease process, an enzymatic process (e.g., a specific kinase), or a proteinopathic process such as huntingtin (htt) aggregation. In such *in vitro* assays, compounds are added directly to the reaction mix, and activity or binding is typically measured using an optical query such as fluorescence, luminescence, or absorbance. These assays have the advantage that any hits identified are by definition against a known target. However, in such cases even if the target is known, the molecular mechanism will then need to be determined, which can often be laborious. Furthermore, even when such mechanistic information can be obtained, it can be difficult to translate the activity of such compounds into the more complex cellular milieu because of a variety of factors, including cellular permeability and metabolism, toxicity, and selectivity in relation to other possible targets of the compound's action that may be present in the cell. Finally, the clinical potential of an *in vitro* hit will vary in proportion to the degree that the hypothesized target had originally been validated in the disease process.

In contrast, cell-based assays have the advantage of identifying compounds that affect a phenotype in the complexity of a cellular environment but suffer from the lack of knowledge about the target and mechanism of action. Indeed, there is no guarantee that the molecular target of a hit from a cell-based assay can be found within a short time. Furthermore, such assays are usually more costly and difficult to adapt to a miniature HTS assay.

In the following sections, we will describe several HTS/HCS assays that have been developed for HD and related polyglutamine disorders, encompassing both *in vitro* and cell-based assays (Table 5.1).

TABLE 5.1
HTS Assays for HD and Hits Identified in the Various Assays

Screen	Htt Construct	Assay	Compounds Tested	Main Hits	2° Screen	HD Mouse	Clinical Use/FDA Approval	Reference
Aggregation in vitro	Exon 1, Q51	Filter retardation	184,880	Benzothiazoles PGL-135	Cell-based aggregation	R6/2	None	[21, 23]
Aggregation in vitro	N1-171, Q58	Filter retardation	1,040	Celastrol, juglone	Mutant htt localization	NT	Yes	[24]
Aggregation in yeast	Exon 1, Q103, EGFP	Yeast growth	16,000	C2-8	PC12, R6/2 brain slice, *Drosophila*	NT	None	[28]
Aggregation in cells	*N1-127, AR, Q65	FRET	4,000	Y-27632, fosfosal, nadolol, gefitinib	PC12, *Drosophila*	NT	Yes	[30, 32]
Cell death	AR1-109, Q112	Caspase-3	1,040	Cardiac glycosides	Motor neuron cells	NT	Yes	[41]
Cell death	N63, Q148	PC12 viability	1,040	Cannabinoids	None	NT	Yes	[45]
Cell death	N63, Q103	PC12 viability	1,040	Acivicin, isoproterenol, propafenone	None	NT	Yes	[46]
Cell death	N548, Q128	ST14A viability	43,685	R1, R2, R4	PC12, yeast, C. *elegans*, brain slice	NT	None	[48]

AGGREGATION-BASED ASSAYS

In Vitro Aggregation Assays

Since the discovery of intracellular aggregates of mutant proteins in polyglutamine diseases including HD [20], it has been conjectured that aggregates are toxic and that inhibiting aggregation is a reasonable therapeutic goal. The first HTS screens in HD were devised to identify compounds that would disrupt aggregates *in vitro* [21]. Purified mutant htt (exon 1) on incubation at 37°C for 16 hours forms detergent-insoluble aggregates, and a simple and robust screen was devised to identify compounds that could prevent the formation of sodium dodecyl sulfate-insoluble aggregates. Mutant htt was incubated with individual compounds for 16 hours, and the protein mix was filtered through a cellulose acetate membrane. The membrane differentially retained aggregated proteins in this "filter retardation assay." Mutant htt aggregates that were retained on the membrane were quantified by Western blotting using an antibody that detects htt. Compounds known to interact with and thereby inhibit aggregates (thioflavine S and Congo red) were used as positive controls.

The authors screened 184,880 compounds for aggregate inhibition and identified 300 hits that had modest potencies (1–11 μM) and included 25 compounds in the class of benzothiazoles [21]. Riluzole, a closely related benzothiazole, has previously shown therapeutic benefit by slowing disease progression in patients with amyotrophic lateral sclerosis [22]. The authors chose to focus on this structural class of molecules for further development. These benzothiazole compounds and related analogs were tested for inhibition of aggregation in a cell-based aggregation assay. Unfortunately, all the compounds identified in the *in vitro* assay were toxic to cells. However, two analogs, designated PGL-135 and PGL-137, decreased aggregation in a cell culture model of mutant htt and at modest potencies (EC_{50}, 40–100 μM). PGL-135 was subsequently tested in an HD mouse model but was unfortunately found to be metabolically unstable and therefore ineffective [23].

Thus, a simple, robust, and economical assay yielded a substantial number of hits, of which a few were active in a cell-based aggregation assay. However, this screening effort also highlighted some major risks in the *in vitro* drug discovery process, notably the inherent uncertainty in translating *in vitro* results to cell culture, where a majority of the most potent hits were found to be toxic, and thence to animal models.

An *in vitro* aggregation assay similar to the one described above was performed by Wang et al. [24], in which a longer htt fragment (1–171 amino acids) was used. This assay identified 19 compounds that inhibited aggregation by more than 50%; of these, 10 compounds had an EC_{50} less than 15 μM. In a secondary assay, the authors found that two of the active compounds, juglone and Celastrol, reversed a nuclear localization phenotype of mutant htt protein in a cell culture model of HD. One possible explanation for this result is that direct binding compounds to mutant htt may alter the pathological conformation of htt and prevent its localization to the nucleus. Further work on the underlying mechanistic basis of this htt localization phenotype may reveal insights into the activity of these hits. Of the remaining compounds, Celastrol and meclocycline were interesting from a therapeutic viewpoint. Celastrol has anti-inflammatory and antioxidant properties and has been considered for treating Alzheimer's disease [25]. Meclocycline is from the class of tetracycline antibiotics, which includes minocycline, a compound that is efficacious in a mouse HD model and is currently undergoing clinical trials for HD [26, 27]. Although none of the identified compounds are approved for human use, the assay did show potential for finding compounds that could be useful for HD therapy.

CELL-BASED AGGREGATION ASSAYS

Although *in vitro* aggregation screens have the advantage of a simple assay format, they are unable to target other cellular mechanisms that are involved in aggregation. Zhang et al. [28] conducted a cell-based aggregation screen in yeast. In their model, inducible expression of a green fluorescent protein (GFP)-tagged mutant htt (Q103) fragment results in aggregate formation and growth suppression of yeast. This was the basis of a simple assay in which yeast growth acted as a surrogate for aggregation and mutant toxicity. Yeast growth was assayed by measuring absorbance (optical density, 600 nm), and GFP fluorescence was used to assay expression levels of mutant protein and aggregate formation; this latter parameter ensured the exclusion of compounds that suppressed expression of mutant htt. Using this assay, the authors screened a library of 16,000 compounds at 10 μM and identified nine compounds that enhanced yeast growth and GFP fluorescence by 25% or more. Of these compounds, four, which were designated C1–C4, also inhibited aggregation in a mammalian cell-based (PC12) model of HD.

The authors then tested a number of analogs of the four active compounds for aggregation inhibition in the PC12 model. Three potent analogs (IC_{50}, ~0.1 μM) representing three classes of compounds inhibited aggregation in PC12. Surprisingly, all active compounds, with one exception (C2-8), failed to inhibit aggregation in an *in vitro* aggregation assay. The C2-8 compound was further tested and found to suppress aggregation in an HD mouse brain slice aggregation assay and also suppressed neurodegeneration in a *Drosophila* model of HD.

Thus, this work described the use of a high-throughput aggregation-based assay to identify potent aggregation inhibitors that were subsequently found to be effective in a series of mechanistic and phenotypic screens of increasing complexity, including an *in vivo* HD model. Their work also raised interesting questions about the process of aggregation *in vitro* compared with that in cells, suggesting that processes by which aggregation is regulated in cells are more complex and likely to differ substantially from aggregation *in vitro*, and thereby may provide greater opportunities to inhibit aggregation *in vivo*.

FLUORESCENCE RESONANCE ENERGY TRANSFER-BASED AGGREGATION ASSAYS

Another approach to directly assay aggregate formation in cells uses fluorescence resonance energy transfer (FRET). This assay is based on the transfer of energy between two fluorophores that are in close spatial proximity. The key criteria for efficient FRET are a donor fluorophore whose emission spectrum overlaps the wavelength range that can excite an acceptor fluorophore, close spatial proximity between the acceptor and donor molecules (<100 Å), and that the emission of the acceptor should be above the emission of the donor to separate the signals.

Using a FRET-based approach, Pollitt et al. [32] developed a cellular aggregation assay for HTS. They used two constructs, a donor cyan fluorescent protein (CFP) and an acceptor yellow fluorescent protein (YFP), each fused to a mutant N-terminal fragment of androgen receptor (AR) protein expressing unexpanded (Q25) or expanded (Q65) glutamine repeats. The polyglutamine expansion in AR causes the neurodegenerative disorder X-linked spinobulbar muscular atrophy [29]. These constructs were cotransfected into cells along with a mutant form of the glucocorticoid receptor (ΔGR) that enhances aggregation in this model. The soluble proteins (Q25 containing AR) did not cause FRET, whereas the expanded polyglutamine containing AR forms aggregates, thereby bringing the CFP and YFP fusion proteins in close proximity, resulting in FRET. Using this FRET-based cellular aggregation assay, a robust high-throughput assay was devised (Z′ > 0.6). A total of 2,800 biologically active small molecules were tested in this assay, and numerous active compounds that caused a decrease in FRET, and hence aggregation, were identified. An inhibitor of Rho-activated serine/threonine kinase (Y-27632) was found to inhibit mutant AR aggregation, and this was confirmed biochemically. In addition, this compound also inhibited aggregation of an exon 1 htt fragment and alleviated degeneration in a *Drosophila* model of HD. Thus, as for the PC12-based HTS assay described in the preceding section, a series of secondary assays of increasing biological and clinical relevance provided validation of this assay system in terms of its ability to generate hits with high likelihood for efficacy in more complex systems.

Using this assay, the group further screened a set of biologically active compounds, including a collection of 1,040 compounds of the NINDS collection and a library of 300 kinase inhibitors [30]. The authors identified 10 compounds that inhibited aggregation by 20%–30% in a reproducible and dose-dependent manner. Further, 9 of the 10 compounds also inhibited mutant htt aggregation in a neuronal cell line. Surprisingly, only 2 of the 10 compounds inhibited aggregation of a pure polyglutamine stretch. This result suggested that the mechanisms for aggregation are modulated by flanking protein context and that these mechanisms are likely conserved for htt and AR protein; these compounds could be useful chemical tools to understand these mechanisms. Finally, five of these compounds were effective in alleviating neurodegeneration in the *Drosophila* HD model, and of these, three are FDA-approved drugs (Table 5.1). This result suggests a very high predictive value for this cell-based aggregation assay for activity *in vivo*. Although none of these three FDA-approved compounds (nadolol, fosfosal, and levonordefrin) are appropriate for long-term therapy because of toxicity, it raises the possibility that the development of less toxic analogs could result in potential HD drugs. Furthermore, it will be important to determine whether these compounds inhibit aggregate formation through their known mechanisms of action rather than off-target effects, that is, to identify and validate the targets of action of these compounds in the context of HD to provide alternative, target-based paths to new drug development.

In summary, a variety of *in vitro* [21, 24, 31] and cell-based aggregation assays [28, 30, 32] have led to the identification of a number of small molecule inhibitors of this process (see Table 5.1). In fact, because some of these aggregation inhibitors were also found to be active *in vivo* (*Drosophila* model), this result suggests that the *in vitro* processes targeted by these molecules are also operant within the cellular and whole organism context. However, limitations of this approach were also apparent. First, *in vitro* aggregation inhibitors are limited by their low probability in being effective in cell culture, both because of toxicity and the likelihood of different mechanisms being involved in aggregation in cells [21]. Second, the aggregation process appears to strongly depend on the regions flanking the polyglutamine tract, which may also limit the relevance of some hits emerging from *in vivo* aggregation screens using model polyglutamine protein constructs [30]. Finally, in light of recent data that some forms or stages of htt aggregation may be a protective cellular response [33], it is evident that there can be substantial risk in basing HTS screens on mechanisms and targets that lack clinical and/or biological validation.

CELL DEATH-RELATED ASSAYS

More complex biological assays can decrease such risks by focusing on phenotypic endpoints that are clearly implicated in the disease process. Neuronal loss, for example, occurs ubiquitously in HD patients and is reproduced in some HD animal models [5, 34]. Although the exact role of cell death in disease pathogenesis is debated, a role for caspases and cell death is likely contributory to disease progression [35]. Caspases are key players in the cell death cascade, and inhibition of caspases genetically or by using small molecules is beneficial in cell culture and mouse models of HD [36, 37]. Furthermore, caspases are likely involved in processing mutant htt to generate htt protein fragments that are toxic in their own right [38, 39]. This provides a rationale for screening for inhibitors of caspases and cell death induced by mutant htt as potential drug leads.

CASPASE INHIBITION-BASED ASSAYS

Caspases are a family of cysteine proteases that cleave protein substrates adjacent to aspartate residues with the flanking sequences being key to the specificity of the different caspases [36]. Both caspase-3 and caspase-7 are terminal "executioner" caspases in a cascade of caspase activation [36]. *In vitro* assays have been developed that can detect cellular caspase activity based on cleavage of specific peptide substrates that release a fluorogenic substrate [40].

Piccioni et al. [41] modified a cell viability assay where expression of mutant AR (Q112) induces caspase-3 activation and cell death in human embryonic kidney cells for use as a medium-throughput assay. In this assay, polyglutamine cytotoxicity was assessed by measuring caspase-3 activity in the transfected cells at 72 hours after transfecting mutant AR. The assay was sufficiently robust ($Z' > 0.3$) and had low variability over time. Using this assay, the authors screened the NINDS compound collection (1,040) and identified 15 compounds that reproducibly inhibited caspase-3 activity by more than 70%. Eight compounds were excluded because of metabolic cytotoxicity; four of the remaining seven compounds were confirmed to enhance cell viability.

Three of these hit compounds belonged to the class of cardiac glycosides, whereas one was a calcium channel blocker (suloctidil); subsequent testing of other cardiac glycosides found those to be active as well. The cardiac glycosides also prevented cell death in a motor neuron-derived cell line, where expression of mutant AR caused cell death. The mechanism of caspase-3 inhibition in these cases appeared to be different from the known cardiotonic effects of cardiac glycosides (via inhibition of the Na^+, K^+-ATPase) but remains to be uncovered. Nonetheless, from a therapeutic standpoint, these compounds are promising drug leads because there are two cardiac glycosides that are FDA approved (digoxin and digitoxin) and are still in active clinical use. It will be of great interest to test these compounds in mouse HD models.

A surprising outcome of this screen was that although the assay endpoint was a downstream point in the apoptotic cascade (caspase-3 inhibition), none of the active compounds transpired to inhibit caspases directly. One explanation for this result is that there are many more targets in the upstream cascade than at the level of caspase-3 itself. Alternatively, inhibiting caspase-3 alone may be difficult to achieve via a chemical inhibitor or may not be sufficient to block cell death. In fact, other studies have suggested that even if one component of the apoptotic cascade (e.g., caspase-3) is genetically inactivated, cells appear to continue to undergo apoptosis, albeit at slightly decreased rates [42, 43]. These results underscore the ability of more complex biological assays to identify unexpected targets that may be upstream of the actual molecular process that is the focus of the assay endpoint itself.

PC12 CELL-BASED VIABILITY ASSAYS

Whereas the caspase-activation screen described above focused on one specific molecular mechanism leading to cell death, adapting a cell-based assay to have a phenotypic endpoint rather than a molecular endpoint can further expand the possibility of using the screen to identify multiple intervention points that could not be predicted *a priori*. In this context, two groups have developed cell culture HTS models for HD in which cell death is induced by expression of mutant htt in PC12 cells, and this cell death is used directly as the assay endpoint. PC12 cells are derived from a rat pheochromocytoma and have been extensively characterized [44]. These cells have some neuronal features and can be induced to differentiate into neuron-like cells on treatment with nerve-derived growth factor (NGF) [44]. Although the mechanisms by which mutant htt induces cell death are not clear, this approach offers a multitude of potential mechanistic interventions without biasing to any particular mechanisms of cell death. Aiken et al. [45] developed a model of inducible mutant htt (exon 1 of htt with 103 glutamines) in which induction of GFP-mutant htt fusion protein expression caused rapid formation of aggregates and cell death within 48–72 hours. The authors confirmed decreased cell viability by multiple viability assays and decided to use a simple lactate dehydrogenase release assay for monitoring viability. They found that broad-spectrum caspase inhibitors such as BOC-D-fmk and Z-VAD-fmk were effective in preventing cell death induced by mutant htt. Importantly, this rescue by caspase inhibitors did not prevent the formation of visible aggregates in PC12 cells, indicating that formation of aggregates *per se* may not lead to obligatory cell death or, alternatively, that caspase inhibitors act downstream of the aggregate formation.

A robust HTS assay was developed around this model ($Z' = 0.5$) and used to screen the 1,040-compound NINDS library. The authors identified 18 compounds that completely rescued cell death in this model. Of these compounds, they focused on a group of cannabinoids that were effective in providing protection against cell death. However, their relatively high EC_{50}s (EC_{50}, 50–100 µM) and the reported lack of cannabinoid receptors in PC12 made the authors conclude that the cannabinoids likely act by a different mechanism that may be related to their reported antioxidant effects.

In another PC12-based viability assay, Wang et al. [46] used NGF-differentiated PC12 cells in which inducible expression of a c-*myc* tagged exon 1 htt with 148 glutamines caused aggregate formation within 2 days and cell death within 4 days. They screened the NINDS library and identified five compounds, in addition to the caspase inhibitor Z-VAD-fmk, that could prevent cell death. Interestingly, three of the protective compounds (acivicin, nipecotic acid, and mycophenolic acid) also decreased aggregate formation in this model.

It is notable that despite using the same parental cell line, a similar htt construct (exon 1 of htt), and screening the identical compound library, the two PC12 cell-based viability screens failed to identify any compounds in common except the broad-spectrum caspase inhibitors. The identifiable differences in these two assays were a seemingly minor difference in the length of polyglutamine repeats (103 vs. 148) and the use of undifferentiated versus differentiated PC12 cells. These results suggest that even small changes in assay parameters can lead to a widely different result in HTS, especially in cell-based screens in which cell context can dramatically alter the output functions of given biochemical targets and pathways [47].

ST14A Cell Viability Assays

Given the critical role of cell context in normal and disease-related biochemical processes, it can be advantageous to conduct HTS screens in cell lines that are as close as possible to the cell types that are known to be affected in the disease state. Thus, Varma et al. [48] have used an engineered stable striatal neuronal cell line expressing the N-terminal 548 amino acids of htt. These cells are susceptible to cell death on serum deprivation at 39°C. Using this cell model, Varma et al. [48] developed a high-throughput cell viability assay in a 384-well format. Cell viability was detected based on a fluorescent viability dye (calcein acetomethoxy). They screened 43,685 novel compounds from diverse sources and identified 29 novel compounds that rescued cell death. Importantly, these compounds selectively inhibited cell death in mutant but not in parental striatal cells that do not express mutant htt. This is an important point because selectivity indicates that these hits are more likely to target pathways that are specifically activated by mutant htt and thus are likely targets for therapy. For example, four active compounds inhibited caspase-3 activation in mutant cells but not in parental cells. These results, together with reports of increased caspase-3 processing in N548 mutant cells and expression in a transgenic mouse HD model, support the notion that caspase-3 activation by mutant htt is involved in HD. Further work will be needed to determine the mechanism of selective rescue by these compounds and is likely to generate insight into the connection between mutant htt and caspase activation.

To prioritize these active compounds and identify potential drug leads, Varma et al. [48] also tested the 29 hits in three additional HD models—PC12 cells, yeast, and *Caenorhabditis elegans*—and identified four candidates that were active in multiple models. One set of related compounds designated R1 and R2 was also active in a rat brain slice HD assay (see below for further discussion). Thus, although the mechanism of action of these compounds remains to be identified, it appears that these compounds target mechanisms of mutant htt toxicity that are shared across a number of diverse cell-, tissue-, and organism-based models for HD.

HIGH-CONTENT SCREENING APPROACHES TOWARD HD

On the generation of reproducible hits from an HTS screen, the greatest challenge is the translation of these hits into successful drug lead candidates that make it through preclinical development into human clinical trials. Although many of these issues relate to drug availability, stability, safety, and so on, and can be at least partially addressed in the design of the input chemical libraries, very often the first major hurdle is whether hits are still efficacious in whole animal models for the disease state.

To help reduce the high dropout rate that is typical at this juncture of the drug development process, there has been much effort in recent years to develop higher-throughput versions of assays that retain much more biological context—and hopefully clinical relevance—relating to the disease state. Thus, the goal is to provide "high-content" screens that may be more predictive of *in vivo* efficacy than more reductionist assay models, and to use such HCS approaches to serve as secondary screens to help bridge the gap between *in vitro* and *in vivo* drug development studies, or even as primary screens if relatively small focused libraries are available for screening.

Such "high-content" screens are all cell- or tissue-based assays rather than *in vitro* assays; indeed, it can be argued that any cell-based screen is intrinsically of higher biological content. Thus, a natural extension of the ST14A screens, described above and which were based on a transformed striatal neuronal cell line, would be the use of primary striatal neuronal cultures as the basis for medium-throughput drug discovery screens. The biotechnology company Trophos SA (Marseille, France) has used this approach to create an HCS screen for HD in which an amino-terminal (480-amino acid) fragment of Htt containing a 68-polyglutamine repeat expansion is transfected into primary striatal neurons via electroporation; this induces progressive cell death in transfected neurons over the course of 6 days [49]. The endpoint is image based and automated, and the numbers of surviving cells are assessed using automated image analysis programs.

The other "axis" of HCS is the nature of the endpoint itself, and there is an increasing move toward using more subtle phenotypic endpoints than overt cell death, such as changes in neurite extension, cell shape, and the translocation of intracellular markers, including the Htt protein itself. Such assays have heavily depended on the development of turnkey automated imaging and analysis hardware and software systems that can be used to screen up to thousands of wells per day using image analysis algorithms. These systems typically include automated plate handling, automated focus and image acquisition, and automated image analysis based on user-preset threshold parameters. Although decidedly not as high throughput as the faster *in vitro* assays, such HCS offers the potential of identifying hits with much higher probability of efficacy in animal models as a result of the biological sophistication of the assay itself. A lentiviral-based RNA interference (RNAi) library of 104,000 targeting vectors against human and mouse genes, for example, was screened using such an approach via high-content imaging for mitotic progression in human cancer cells [50].

Another rapidly evolving area is the use of tissue explants in HCS screens for neurological disorders [51]. Initial efforts in this area arose in the stroke field, in which all clinical candidates developed out of HTS screens have so far failed in clinical trials (>100 clinical trials to date). Thus, it appeared that there were important pathophysiological processes occurring in human stroke that were not adequately represented in even primary cultures of cortical or hippocampal neurons; therefore, compounds that were efficacious in cell culture would not work as predicted in the whole animal. Brain slice assays for stroke have generally been based on organotypic brain slice cultures (250–400 μm thick) taken from hippocampus or cortex, and have used direct cell death endpoints such as propidium iodide staining for dying cells [52].

Wang et al. [53] have recently described a brain slice-based HCS for stroke, based on using biolistic transfection of GFP to generate a "sentinel" population of cortical pyramidal neurons within chronic brain slice explants. The authors then assessed the numbers of healthy cortical neurons, as assessed morphologically, as a function of time after simulating stroke-like conditions in these explants using oxygen and glucose deprivation (OGD). By 3 days after OGD, most neurons in the brain slice explants had died, and this could be tracked by the GFP-expressing sentinel population of neurons, generating a robust Z'-factor for medium-throughput compound screening.

In a screen of ~5,000 synthetic and natural product compounds, including all FDA-approved drugs, this group identified ~74 primary and reproducible hits. Interestingly, one of the two most efficacious hits to emerge from this screen was also a member of the cardiac glycoside family, neriifolin, and was subsequently shown to be neuroprotective in two independent whole animal models of focal ischemia. This suggests that there may be some common mechanisms of action between the neuropathological sequelae of stroke and the neurodegenerative processes in polyglutamine disorders.

Similar brain slice-based approaches are also being undertaken in HD [48, 54–59]. As discussed above, cell-based versus *in vitro* htt aggregation has often yielded contrasting results; by extension, it may be important to determine potential drug effects on aggregation not only in cells but also in intact neural tissue. Accordingly, the Bates laboratory has developed a brain tissue-based aggregation screen based on organotypic slice explants from the R6/2 transgenic mouse model for HD. In this assay, measurements of aggregate numbers, intensity, and size are made via confocal microscopy, and drug effects on these endpoints are assessed over the course of 3–4 weeks in culture [54, 55, 59].

Several laboratories are now using another approach in which biolistic transfection is used to generate a sentinel population of medium spiny neurons (MSNs) within coronal brain slice explants containing striatum; these MSNs are cotransfected to express mutant htt constructs and a vital visual marker such as YFP, which is used to assess the numbers of healthy MSNs remaining as a function of time after htt transfection. Using this approach, Murphy and Messer [56] and Southwell et al. [58] have shown that scFv intrabodies against htt (see Chapter 10, this volume) can be protective against biolistically transfected htt DNA, and Khoshnan et al. [57] have shown the involvement of the nuclear factor κB pathway in mutant htt toxicity, all within the context of chronic brain slice explants.

Biolistics-based brain slice assays are also used in small molecule compound screening campaigns to help identify the most promising candidates for further preclinical development [48]. In an HCS screening mode, conditions were developed such that mutant htt-transfected neurons undergo gradual degeneration of dendrites and eventual cell death beginning on day 3 after transfection; by day 5 there is a 2- to 3-fold decrease in numbers of healthy MSNs in corticostriatal brain slices transfected with mutant htt. The expectation is that this brain slice assay should show good overlap in terms of the range of compounds that have been demonstrated to provide benefit in transgenic mouse models of HD; in fact, in a "cross-validation" survey of such compounds, several compound classes, such as histone deacetylase (HDAC) inhibitors, were found to also be efficacious in the brain slice model (D. Lo, unpublished data).

Because of the complexity of this assay, as well as the intrinsic variability from brain slice to brain slice, the Z'-factor of this assay is low and requires that larger numbers be used (6–12 brain slices per data point). This limits the throughput of this assay to ~1,000–10,000 compounds per year for a typical academic laboratory, depending on the number of concentrations to be tested for each compound; therefore, the principal role for this type of assay to date has been as a secondary screen to help prioritize hits from HTS screens for further preclinical development. As described previously, the authors have used a brain slice-based HD to help validate the hits emerging from an ST14A HTS screen for compounds protective against Htt-induced cell death [48].

Thus, cell- and tissue-based HCS screens are increasingly applied in earlier stages of the drug discovery process to provide a "preview" of potential drug action of compounds identified in primary screening assays with higher throughput but perhaps less direct physiological relation to the *in vivo* disease state. Such prioritization of hits from primary screens is a critical step in the drug development process, especially in cases where large-scale HTS screens have generated numerous drug lead candidates of different chemical classes, as the throughput of animal models for complex diseases such as HD is often restricted to only a few or a few tens of compounds per year (see Chapters 7 and 8, this volume).

HTS ACHIEVEMENTS

The future of drug discovery in HD seems to hold much promise. The HTS approach has identified small molecules that can alleviate toxic phenotypes of mutant HD *in vitro*, cell culture models, and some animal models, suggesting that new therapies may be found for HD. Numerous high-throughput assays have been developed successfully. Although limited phenotypes have been assayed in these screens (aggregation, caspase activity, and cell death), a number of active compounds have been identified. A few of these hold promise as drug leads. For example, FDA-approved drugs have been identified using the FRET-based aggregation assay and caspase inhibition assay [30, 41]. The PC12 cell viability assay has identified cannabinoids that have a history of both recreational and medicinal use [60]. Furthermore, these HTS assays can now be used for larger-scale screening efforts. The cell-based assays are also ripe for conducting genome-wide RNAi-based screens [61], which may provide drug targets for modulating aggregation and cell death induced by mutant htt.

In addition to the HTS assays used for the primary screens, numerous different medium- to low-throughput secondary assays have been developed for prioritizing hits identified in the primary screens. These include cell-based, brain slice-based, and whole animal (*Drosophila* and *C. elegans*) assays [46, 48, 62, 63]. A number of these assays have been somewhat validated because a few compounds, such as HDAC inhibitors and cystamine, that are active in mouse HD models were also active in these models [64–66]. These models should qualify as rapid and economic predictors of success in mouse HD models.

CHALLENGES IN PRIORITIZING HIT COMPOUNDS

It has been a very productive and informative beginning for the HTS approach to HD. However, most compounds identified in the screens have a long way to go before reaching clinical trials. Challenges remain in the development of a rational prioritization scheme for the identified compounds such that the "best" hits can be tested in mouse models of HD. The list of secondary assays that have been used are diverse and include cell-based (PC12) assays, more involved brain slice cultures that provide a more *in vivo*-like environment for drug testing, and finally whole organisms, such as *Drosophila* and *C. elegans* models. These are useful for prioritization, although it is still not clear what mechanisms are targeted in these diverse models. Furthermore, because of a lack of consistent use of these secondary assays, it is difficult to compare the effects of compounds identified in different assays.

Another obstacle in developing these compounds is assessing their metabolic stability, toxicity, and blood–brain barrier (BBB) penetration. Priority could be given to FDA-approved drugs that are active in at least two distinct HD models and cross the BBB. Novel compounds would need to pass a more rigorous set of filters, including high potency ($EC_{50} < 1$ μM), low toxicity, metabolic stability, and evidence of BBB penetration. We propose a rational prioritization scheme that encompasses a few critical criteria for progressing the compounds discovered in HD assays (Figure 5.2).

If the compounds meet these criteria, the choice of mouse HD models to test these hits is the next prioritization. Currently, mouse models that express the N-terminal fragment of htt (R6/2) are preferred for drug testing because of the rapid development of HD phenotype [5]. However, other mouse models that express a longer

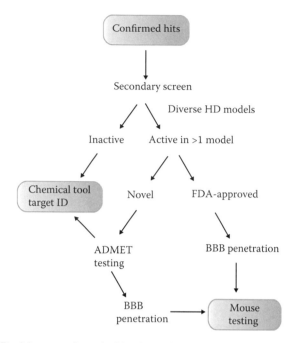

FIGURE 5.2 Decision tree for prioritization scheme hits identified in primary screens. Testing in multiple distinct HD models could identify hits active in at least more than one model; such hits likely target a conserved HD mechanism. These compounds could be prioritized for testing in mouse models of HD. Compounds that pass absorption, distribution, metabolism, and excretion and BBB criteria would be candidates for testing in mouse models, as well as tools for understanding HD mechanisms and target identification. Any compounds that fail these criteria could still be useful as chemical tools.

fragment of mutant htt (N1-171 or longer) are available and should be considered in certain cases. For example, compounds that interfere with htt processing, an important aspect of mutant toxicity, are not expected to be effective in the R6/2 model that expresses the small exon 1 htt fragment that does not appear to require processing for toxicity. In addition, neuronal loss is observed in the YAC128 and Q89CMV mice with full-length htt but not in the R6/2 mice with only exon 1 of htt [5]. Thus, the HD models expressing full-length or larger htt fragments may be more suitable for testing cell death inhibitors.

A consistent result of the screening efforts has been that few common hits were identified despite screening identical compound libraries in various HD assays. There are varied reasons for this result, including the initial screen being conducted at one compound concentration, interspecies differences in targets, and htt constructs of different length used in different assays. However, they raise two important questions. How should we best prioritize the various hits identified in the diverse models, and what should be done with hits that are not active in the multiple but limited HD assays? One strategy that has been applied in a few screening efforts has been to identify compounds that show activity in multiple distinct HD models. This seems to be a reasonable strategy, as it is based on the assumption that these compounds affect a conserved mechanism of mutant htt toxicity rather than some specific effect in one model. However, because the relevant mechanisms of HD are elusive and the aspects of HD toxicity represented in each model are not clear, this strategy could arbitrarily disregard promising compounds.

To give these compounds a more thorough appraisal, they and their closely related analogs could be assembled into a compound library and made available to investigators by a commercial vendor or an NIH-based resource for testing in other neurodegenerative diseases. Testing these compounds could reveal common mechanisms involved in neurodegeneration and help identify additional compounds for further study and drug development. The results for all the hits identified in HD and related polyglutamine disease screens (currently >100) could be stored in a comprehensive database and updated with their mechanism of action and activity in other neurodegenerative disease models. This database could be a valuable resource and a starting point for drug testing for the researchers in HD and other neurodegenerative disorders.

SMALL MOLECULES AS CHEMICAL TOOLS FOR UNDERSTANDING HD MECHANISMS

The value of compounds discovered in these assays may lie not only in their potential as drug leads but also in their use to enhance the understanding of HD mechanisms. Often, compounds with interesting activities are neglected if they are deemed unsuitable for drug development. However, such compounds may enhance understanding of disease mechanisms and in some cases may be used to identify targets for HD therapy. These small molecules can also be used to control different aspects of mutant htt toxicity and test hypotheses about mutant toxicity. For example, because aggregates are typically formed in HD, their role is always confounded. Hypotheses on the role of aggregates in HD vary from those suggesting they are harmful to those saying they are innocuous or even protective [33]. However, by controlling the aggregation process with small molecules, one can begin to address these hypotheses and to dissect the mechanisms involved in mutant htt aggregation. In this context, it is notable that a recent screen identified a small molecule that enhances aggregation and is protective in cell culture models of HD and Parkinson's disease [67]. Thus, experiments using small molecules can lead to hypothesis generation and testing. These compounds could also be developed as chemical tools to identify protein targets for therapy, thus complementing the growing armory of genetic and biochemical tools to address the mechanisms of HD.

FUTURE DIRECTIONS

Most current assays target aggregation or cell viability in HD models. However, as more insight into mechanisms of HD pathology is obtained, more complex assays that target specific cellular phenotypes need to be developed (Figure 5.3). In fact, assays that enhance htt clearance have already been devised, and hits have been identified [68]. The assay for mutant htt clearance is notable because it has identified a compound that selectively enhanced the degradation of mutant htt. Because decreasing

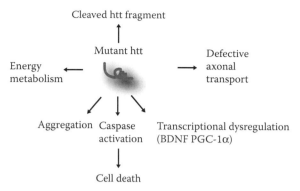

FIGURE 5.3 **(See color insert following page 172.)** Multiple mechanisms implicated in HD pathology. The ones highlighted in red text are mechanisms that have been targeted in HTS assays, whereas the others are potential mechanisms for HTS assays.

or knocking out the expression of mutant htt has been shown to reverse the effects of HD in a mouse model [69], it would appear that such an approach is ideal for targeting the source of toxicity. It remains to be seen whether this compound will be active in other relevant HD models. Recent work has shown a role for htt cleavage [39] and transcriptional dysregulation of a master regulator of mitochondrial metabolism, peroxisome proliferator-activated receptor γ coactivator 1α (PGC-1α), in HD pathology [70]. Therefore, assays involving htt cleavage and transcriptional up-regulation of PGC-1α would appear to be the next logical set of assays to develop [71]. A role for defective microtubule-based transport has also been suggested in HD, and this defect is another complex phenotype for HCS [72, 73]. In the future, the challenge for the HTS community and HD researchers will be to create assays that target specific molecular phenotypes of HD and thereby develop targeted therapies for HD.

REFERENCES

1. Huntington's Disease Collaborative Research Group. (1993) *Cell* 72, 971–983.
2. Roberson, E. D. & Mucke, L. (2006) *Science* 314, 781–784.
3. Newman, M., Musgrave, F. I., & Lardelli, M. (2007) *Biochimica et Biophysica Acta* 1772, 285–297.
4. Fleming, S. M., Fernagut, P. O., & Chesselet, M. F. (2005) *NeuroRx: Journal of the American Society for Experimental NeuroTherapeutics* 2, 495–503.
5. Menalled, L. B. & Chesselet, M. F. (2002) *Trends Pharmacol Sci* 23, 32–39.
6. Fox, S., Farr-Jones, S., Sopchak, L., Boggs, A., Nicely, H. W., Khoury, R., & Biros, M. (2006) *Journal of Biomolecular Screening: the Official Journal of the Society for Biomolecular Screening* 11, 864–869.
7. Geerts, H., Trojanowski, J. Q., & Lee, V. M. (2005) *Science of Aging Knowledge Environment: SAGE KE* 2005, pe4.
8. Newman, D. J. & Cragg, G. M. (2007) *Journal of Natural Products* 70, 461–477.
9. Koehn, F. E. & Carter, G. T. (2005) *Nature Reviews. Drug Discovery* 4, 206–220.
10. Schreiber, S. L. (2000) *Science* 287, 1964–1969.
11. Root, D. E., Flaherty, S. P., Kelley, B. P., & Stockwell, B. R. (2003) *Chemistry & Biology* 10, 881–892.

12. Chong, C. R., Chen, X., Shi, L., Liu, J. O., & Sullivan, D. J., Jr. (2006) *Nature Chemical Biology* 2, 415–416.
13. Hopkins, A. L. & Groom, C. R. (2002) *Nature Reviews. Drug Discovery* 1, 727–730.
14. Walters, W. P. & Namchuk, M. (2003) *Nature Reviews. Drug Discovery* 2, 259–266.
15. Lipinski, C. A., Lombardo, F., Dominy, B. W., & Feeney, P. J. (2001) *Advanced Drug Delivery Reviews* 46, 3–26.
16. Veber, D. F., Johnson, S. R., Cheng, H. Y., Smith, B. R., Ward, K. W., & Kopple, K. D. (2002) *Journal of Medicinal Chemistry* 45, 2615–2623.
17. Clark, D. E. (2001) *Combinatorial Chemistry & High-Throughput Screening* 4, 477–496.
18. Zhang, J., Chung, T. D., & Oldenburg, K. R. (1999) *Journal of Biomolecular Screening* 4, 67–83.
19. Kelley, B. P., Lunn, M. R., Root, D. E., Flaherty, S. P., Martino, A. M., & Stockwell, B. R. (2004) *Chemistry & Biology* 11, 1495–1503.
20. Scherzinger, E., Lurz, R., Turmaine, M., Mangiarini, L., Hollenbach, B., Hasenbank, R., Bates, G. P., Davies, S. W., Lehrach, H., & Wanker, E. E. (1997) *Cell* 90, 549–558.
21. Heiser, V., Engemann, S., Brocker, W., Dunkel, I., Boeddrich, A., Waelter, S., Nordhoff, E., Lurz, R., Schugardt, N., Rautenberg, S., et al. (2002) *Proceedings of the National Academy of Sciences of the United States of America* 99 Suppl 4, 16400–16406.
22. Lacomblez, L., Bensimon, G., Leigh, P. N., Guillet, P., & Meininger, V. (1996) *Lancet* 347, 1425–1431.
23. Hockly, E., Tse, J., Barker, A. L., Moolman, D. L., Beunard, J. L., Revington, A. P., Holt, K., Sunshine, S., Moffitt, H., Sathasivam, K., et al. (2006) *Neurobiology of Disease* 21, 228–236.
24. Wang, J., Gines, S., MacDonald, M. E., & Gusella, J. F. (2005) *BMC Neuroscience* 6, 1.
25. Allison, A. C., Cacabelos, R., Lombardi, V. R., Alvarez, X. A., & Vigo, C. (2001) *Progress in Neuro-Psychopharmacology & Biological Psychiatry* 25, 1341–1357.
26. Hersch, S., Fink, K., Vonsattel, J. P., & Friedlander, R. M. (2003) *Annals of Neurology* 54, 841; author reply 842–843.
27. Bonelli, R. M., Hèodl, A. K., Hofmann, P., & Kapfhammer, H. P. (2004) *International Clinical Psychopharmacology* 19, 337–342.
28. Zhang, X., Smith, D. L., Meriin, A. B., Engemann, S., Russel, D. E., Roark, M., Washington, S. L., Maxwell, M. M., Marsh, J. L., Thompson, L. M., et al. (2005) *Proceedings of the National Academy of Sciences of the United States of America* 102, 892–897.
29. Poletti, A., Negri-Cesi, P., & Martini, L. (2005) *Endocrine* 28, 243–262.
30. Desai, U. A., Pallos, J., Ma, A. A., Stockwell, B. R., Thompson, L. M., Marsh, J. L., & Diamond, M. I. (2006) *Human Molecular Genetics* 15, 2114–2124.
31. Berthelier, V. & Wetzel, R. (2006) *Methods in Enzymology* 413, 313–325.
32. Pollitt, S. K., Pallos, J., Shao, J., Desai, U. A., Ma, A. A., Thompson, L. M., Marsh, J. L., & Diamond, M. I. (2003) *Neuron* 40, 685–694.
33. Lansbury, P. T. & Lashuel, H. A. (2006) *Nature* 443, 774–779.
34. Hickey, M. A. & Chesselet, M. F. (2003) *Progress in Neuro-Psychopharmacology & Biological Psychiatry* 27, 255–265.
35. Tobin, A. J. & Signer, E. R. (2000) *Trends in Cell Biology* 10, 531–536.
36. Thornberry, N. A. & Lazebnik, Y. (1998) *Science* 281, 1312–1316.
37. Sanchez Mejia, R. O. & Friedlander, R. M. (2001) *Neuroscientist* 7, 480–489.
38. Wellington, C. L., Ellerby, L. M., Gutekunst, C. A., Rogers, D., Warby, S., Graham, R. K., Loubser, O., van Raamsdonk, J., Singaraja, R., Yang, Y. Z., et al. (2002) *Journal of Neuroscience* 22, 7862–7872.
39. Graham, R. K., Deng, Y., Slow, E. J., Haigh, B., Bissada, N., Lu, G., Pearson, J., Shehadeh, J., Bertram, L., Murphy, Z., et al. (2006) *Cell* 125, 1179–1191.

40. Thornberry, N. A., Rano, T. A., Peterson, E. P., Rasper, D. M., Timkey, T., Garcia-Calvo, M., Houtzager, V. M., Nordstrom, P. A., Roy, S., Vaillancourt, J. P., et al. (1997) *Journal of Biological Chemistry* 272, 17907–17911.

41. Piccioni, F., Roman, B. R., Fischbeck, K. H., & Taylor, J. P. (2004) *Human Molecular Genetics* 13, 437–446.

42. D'Mello, S. R., Kuan, C. Y., Flavell, R. A., & Rakic, P. (2000) *Journal of Neuroscience Research* 59, 24–31.

43. Oppenheim, R. W., Flavell, R. A., Vinsant, S., Prevette, D., Kuan, C. Y., & Rakic, P. (2001) *Journal of Neuroscience: the Official Journal of the Society for Neuroscience* 21, 4752–4760.

44. Tischler, A. S., Powers, J. F., & Alroy, J. (2004) *Histology and Histopathology* 19, 883–895.

45. Aiken, C. T., Tobin, A. J., & Schweitzer, E. S. (2004) *Neurobiology of Disease* 16, 546–555.

46. Wang, W., Duan, W., Igarashi, S., Morita, H., Nakamura, M., & Ross, C. A. (2005) *Neurobiology of Disease* 20, 500–508.

47. Sills, M. A., Weiss, D., Pham, Q., Schweitzer, R., Wu, X., & Wu, J. J. (2002) *Journal of Biomolecular Screening* 7, 191–214.

48. Varma, H., Voisine, C., DeMarco, C. T., Cattaneo, E., Lo, D. C., Hart, A. C., & Stockwell, B. R. (2007) *Nature Chemical Biology* 3, 99–100.

49. Valenza, M., Rigamonti, D., Goffredo, D., Zuccato, C., Fenu, S., Jamot, L., Strand, A., Tarditi, A., Woodman, B., Racchi, M., et al. (2005) *Journal of Neuroscience: the Official Journal of the Society for Neuroscience* 25, 9932–9939.

50. Moffat, J., Grueneberg, D. A., Yang, X., Kim, S. Y., Kloepfer, A. M., Hinkle, G., Piqani, B., Eisenhaure, T. M., Luo, B., Grenier, J. K., et al. (2006) *Cell* 124, 1283–1298.

51. Cho, S., Wood, A., & Bowlby, M. R. (2007) *Current Neuropharmacology* 5, 19–33.

52. Rytter, A., Cronberg, T., Asztely, F., Nemali, S., Wieloch, T., Rytter, A., Cronberg, T., Asztely, F., Nemali, S., & Wieloch, T. (2003) *Journal of Cerebral Blood Flow & Metabolism* 23, 23–33.

53. Wang, J. K., Portbury, S., Thomas, M. B., Barney, S., Ricca, D. J., Morris, D. L., Warner, D. S., & Lo, D. C. (2006) *Proceedings of the National Academy of Sciences of the United States of America* 103, 10461–10466.

54. Zhang, X., Smith, D. L., Meriin, A. B., Engemann, S., Russel, D. E., Roark, M., Washington, S. L., Maxwell, M. M., Marsh, J. L., Thompson, L. M., et al. (2005) *Proceedings of the National Academy of Sciences of the United States of America* 102, 892–897.

55. Hay, D. G., Sathasivam, K., Tobaben, S., Stahl, B., Marber, M., Mestril, R., Mahal, A., Smith, D. L., Woodman, B., Bates, G. P., et al. (2004) *Human Molecular Genetics* 13, 1389–1405.

56. Murphy, R. C. & Messer, A. (2004) *Brain Research Molecular Brain Research* 121, 141–145.

57. Khoshnan, A., Ko, J., Watkin, E. E., Paige, L. A., Reinhart, P. H., & Patterson, P. H. (2004) *Journal of Neuroscience* 24, 7999–8008.

58. Southwell, A. L., Khoshnan, A., Dunn, D. E., Bugg, C. W., Lo, D. C., & Patterson, P. H. (2008) *Journal of Neuroscience* 28, 9013–9020.

59. Smith, D. L., Portier, R., Woodman, B., Hockly, E., Mahal, A., Klunk, W. E., Li, X. J., Wanker, E., Murray, K. D., Bates, G. P., et al. (2001) *Neurobiology of Disease* 8, 1017–1026.

60. Mackie, K. (2006) *Annual Review of Pharmacology and Toxicology* 46, 101–122.

61. Rondinone, C. M. (2006) *Endocrinology* 147, 2650–2656.

62. Faber, P. W., Voisine, C., King, D. C., Bates, E. A., & Hart, A. C. (2002) *Proceedings of the National Academy of Sciences of the United States of America* 99, 17131–17136.

63. Steffan, J. S., Bodai, L., Pallos, J., Poelman, M., McCampbell, A., Apostol, B. L., Kazantsev, A., Schmidt, E., Zhu, Y. Z., Greenwald, M., et al. (2001) *Nature* 413, 739–743.
64. Dedeoglu, A., Kubilus, J. K., Jeitner, T. M., Matson, S. A., Bogdanov, M., Kowall, N. W., Matson, W. R., Cooper, A. J., Ratan, R. R., Beal, M. F., et al. (2002) *Journal of Neuroscience: the Official Journal of the Society for Neuroscience* 22, 8942–8950.
65. Apostol, B. L., Kazantsev, A., Raffioni, S., Illes, K., Pallos, J., Bodai, L., Slepko, N., Bear, J. E., Gertler, F. B., Hersch, S., et al. (2003) *Proceedings of the National Academy of Sciences of the United States of America* 100, 5950–5955.
66. Gardian, G., Browne, S. E., Choi, D. K., Klivenyi, P., Gregorio, J., Kubilus, J. K., Ryu, H., Langley, B., Ratan, R. R., Ferrante, R. J., et al. (2005) *Journal of Biological Chemistry* 280, 556–563.
67. Bodner, R. A., Outeiro, T. F., Altmann, S., Maxwell, M. M., Cho, S. H., Hyman, B. T., McLean, P. J., Young, A. B., Housman, D. E., & Kazantsev, A. G. (2006) *Proceedings of the National Academy of Sciences of the United States of America* 103, 4246–4251.
68. Coufal, M., Maxwell, M. M., Russel, D. E., Amore, A. M., Altmann, S. M., Hollingsworth, Z. R., Young, A. B., Housman, D. E., & Kazantsev, A. G. (2007) *Journal of Biomolecular Screening: the Official Journal of the Society for Biomolecular Screening* 12, 351–360.
69. Yamamoto, A., Lucas, J. J., & Hen, R. (2000) *Cell* 101, 57–66.
70. Cui, L., Jeong, H., Borovecki, F., Parkhurst, C. N., Tanese, N., & Krainc, D. (2006) *Cell* 127, 59–69.
71. McGill, J. K. & Beal, M. F. (2006) *Cell* 127, 465–468.
72. Gunawardena, S., Her, L. S., Brusch, R. G., Laymon, R. A., Niesman, I. R., Gordesky-Gold, B., Sintasath, L., Bonini, N. M., & Goldstein, L. S. (2003) *Neuron* 40, 25–40.
73. Gauthier, L. R., Charrin, B. C., Borrell-Pagáes, M., Dompierre, J. P., Rangone, H., Cordeliáeres, F. P., De Mey, J., MacDonald, M. E., Lessmann, V., Humbert, S., et al. (2004) *Cell* 118, 127–138.

6 Value of Invertebrate Genetics and Biology to Develop Neuroprotective and Preventive Medicine in Huntington's Disease

Christian Neri

CONTENTS

INTRODUCTION

During the past decade, *Drosophila* and *Caenorhabditis elegans* genetics have emerged as powerful approaches for the study of cellular responses to neurodegenerative disease proteins. Such studies have provided new strategies and rationales for the development of neuroprotective drugs, such as the pharmacological manipulation of longevity modulator networks. This chapter will describe how and why these model systems may be used as efficient translational research tools for Huntington's disease (HD) in the discovery and development of neuroprotective drugs.

Neurodegenerative diseases, including HD, constitute a large and clinically heterogeneous group of brain illnesses for which there are currently no neuroprotective drugs available. Although important methodologies have been developed by academic and industrial researchers to find therapies for these diseases, notably chemical screening tools, the large majority of the molecules so far evaluated in clinical trials have failed to show significant efficacy. In rodent models, the administration of neurotoxins may reproduce some features of the human diseases (von Bohlen Und Halbach, 2005), and this approach has been used extensively to search for new treatments with some success, such as the use of levodopa for symptomatic treatment of Parkinson's disease, although with significant CAG side effects (Tse, 2006).

However, neuroprotective treatments able to benefit large numbers of patients with minimal side effects have not yet been developed. Thus, to date research in the field of neurodegenerative disease pathogenesis has given limited benefit to patients, and drug discovery and development remain major challenges for industrial and preindustrial research and need to be improved. Drug development is an expensive and time-consuming process that works as a pipeline, with the proof of success being conclusive clinical trials.

Although there is a need to improve clinical trial design to evaluate the effects of neuroprotective drugs, improvement may also be needed at the entry points of the pipeline and at critical points along the preclinical discovery process (Hung and Schwarzschild, 2007). With the advent of physiological genomics and network biology, the concepts and paradigms used to study neurodegenerative diseases are rapidly evolving (Feany, 2000; Lim et al., 2006b). Implementing these concepts and paradigms early into the drug development process may strongly enhance the translational infrastructure used to tackle neurodegenerative diseases.

The aim of this chapter is to emphasize how and why invertebrate biology may allow new rationale(s) for drug discovery and development to be exploited for the development of new drugs for HD, notably in view of developing neuroprotective and preventive medicines. HD is a dominantly inherited disease caused by expanded polyglutamines (polyQs) in the huntingtin (htt) protein (Figure 6.1) and is clinically characterized by cortical and striatal degeneration accompanied by motor,

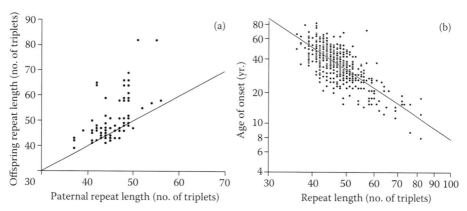

FIGURE 6.1 Inverse relationship between CAG repeat length in the *htt* gene and age at onset of the disease. Data are from the Baltimore HD Center. (a) Tendency of repeat length to increase with paternal transmission of HD. Points *above* the *diagonal line* represent cases in which the repeat length increased during transmission from father to child (*n* = 84 pairs; mean [SD] increase of repeat length, 4.2 [0.8] triplets). (b) Correlation of repeat length with age at onset in HD. As repeat length increases, age at onset of disease decreases (*n* = 480; *r²* = 0.57). (From Margolis, R. L., et al., *Arch Gen Psychiatry*, 56, 1019, 1999. With permission of R. Margolis and American Medical Association Press.)

cognitive, and neuropsychiatric symptoms (Walker, 2007). The toxic effects of polyQ-expanded htt forms and genetic modifiers of cytotoxicity are being studied in several cell systems and in model organisms, including yeast, nematodes, flies, and rodents (Levine et al., 2004; Rubinsztein, 2002; Sipione and Cattaneo, 2001). Thus, a large amount of knowledge is being accumulated on the roles of normal htt and the effects of polyQ-expanded htt at the neuronal cell level. There are also great efforts being made to collect detailed clinical data in a normalized manner (e.g., see http://www.euro-hd.net) and to characterize large cohorts of HD patients and presymptomatic individuals. Thus, HD has become a "model disease" to define the best ways to fight neuronal dysfunction and neurodegeneration in the brain and to improve drug discovery, which may foster the identification of a cure for this disease and perhaps other degenerative diseases.

INVERTEBRATE BIOLOGY AS A PATH TOWARD PREVENTIVE MEDICINE

Neurodegenerative diseases like HD are devastating. There is an urgent need to find drugs that protect diseased neurons, and quantitative and easy-to-use biomarkers arc nccded to monitor the evolution of the disease and the effects of experimental therapies in clinical trials (Ward, 2007). In the quest for a cure for HD, one may consider that we may not need to thoroughly understand the mechanisms that underlie the decline of diseased neurons to find ways to slow disease progression. Using the "best models and detection techniques available" may be sufficient to find drug leads (Varma et al., 2007) and biomarkers (Hye et al., 2006). On the other hand, a clear understanding of the molecular logic underlying neuronal decline/survival may favor the identification of neuroprotective drugs and instructive biomarkers based on strong biological rationales. In practice, the two approaches are run in parallel because they are complementary.

However, HD is a multistep, progressive process that contains several points for a potential therapeutic intervention, and hence the rationale models and methods for drug discovery may considerably differ depending on the specific point of intervention that is being pursued. If the aim is to identify any drugs that may provide some benefit (symptomatic or neuroprotective) to individuals showing the major disease symptoms, for example, then screening without a known biological mechanism could well be successful (Natesan et al., 2006). However, if the aim is to develop neuroprotective drugs that could be given to a large number of individuals before they show advanced symptoms based on the use of biomarkers that allow disease progression and drug effects to be monitored (preventive medicine)—in other words, if the idea is to consistently slow down disease progression by chemical means as early as possible and over an extended period—then it becomes essential to understand the neuronal cell decision processes under pathological conditions, and to target the biology of the disease early in the drug discovery process (Griffioen, 2007).

A comprehensive analysis of such cell decision processes requires well-defined simple model systems that allow a large number of modifiers (i.e., genes, pathways, active compounds) to be evaluated for activity. This analysis also needs to be performed under pathophysiological model conditions that recapitulate the early phases (e.g., neuronal dysfunction before cell death) or specific components of proteotoxicity (e.g., soluble forms and/or oligomers of disease proteins) in neurodegenerative diseases. Drug discovery focused on the early biological components of degenerative disease processes will also benefit from new applications of model systems; one example may be to move from drug discovery aimed at blocking cell death cascades (Toulmond et al., 2004) to drug discovery aimed at promoting intrinsic neuronal cell defenses as recently emphasized by studies in nematodes (Anekonda, 2006; Tang and Chua, 2008).

COMPLEXITY IN HUNTINGTON'S DISEASE: A NETWORK OF DISEASE MODIFIERS?

The causes for neurodegenerative diseases are likely a mix of genetic, epigenetic, and environmental factors that together with the individual's life history translate into clinical heterogeneity. Clinical heterogeneity is an obstacle in developing effective treatments for neurodegenerative diseases as there are no biomarkers known at present to allow disease variability and progression to be precisely assessed. The early detection of manifestations of the disease in presymptomatic individuals is difficult for the same reason, the lack of appropriate biomarkers, which is an additional obstacle to the development of neuroprevention.

Clinical heterogeneity holds true even for autosomal dominant forms of neurodegenerative disorders like HD. The disease shows anticipation (earlier age at onset with successive generations, a phenomenon linked to increasing CAG repeat lengths in the *htt* gene), and age of onset is greatly variable for a given extent of polyQ expansion (Figure 6.1). Strikingly, the most frequent mutant *htt* alleles (containing 40–50 glutamines) account for only 50% of the variability (20 years) of the age at onset based on motor symptoms.

In addition to environmental factors, genetic modifiers likely play important roles in HD clinical heterogeneity (Andresen et al., 2007; Li et al., 2003; Shelbourne et al., 2007; Wexler et al., 2004). Consistent with the notion that several genes may modify HD pathogenesis, there are numerous htt partner proteins that have been described (Goehler et al., 2004; Kaltenbach et al., 2007) and associated mechanisms proposed, including transcriptional dysregulation, abnormal protein cleavage, misfolding and degradation, abnormal trafficking, oxidative stress/mitochondrial dysfunction, and synaptic dysfunction (Harjes and Wanker, 2003; Landles and Bates, 2004).

It is important to note that these abnormalities may result from either a loss of normal htt function or a gain of new functional properties by mutant htt, or both. Such altered cellular processes may be mediated by a large number of htt partner proteins in the nucleus and cytoplasm (Goehler et al., 2004; Kaltenbach et al., 2007), and several of these abnormalities, such as oxidative stress and transcriptional dysregulation, also are implicated in the pathogenesis of other neurodegenerative diseases. Therefore, an understanding of the molecular bases of HD variability may also allow the pathogenesis and variability of other degenerative disorders to be understood as well.

A major trend in HD research is to search for genetic/biological modifiers of polyQ-expanded htt cytotoxicity in model systems. This information may help phenotypic variability in HD to be better understood because the modifiers of polyQ-expanded htt cytotoxicity might also modify the disease (Arango et al., 2006; Holbert et al., 2001). Additionally, genetic and biological modifiers may constitute targets of high interest to develop neuroprotective strategies and provide a useful source of candidate biomarkers.

NETWORK BIOLOGY AND NEUROPROTECTIVE DRUG DISCOVERY

The use of "network biology" or "systems biology" has emerged as a powerful approach to study complex biological processes based on the analysis of signaling and regulatory networks (Curtis et al., 2006; Klamt et al., 2006; Liu et al., 2006; Rast, 2003). A disease process may be seen as dynamic disturbances of networks, notably of signaling networks within the cell and between cells and tissues, and may be of signaling networks that link the inner and outer body. Although a network-oriented view of disease processes may be reductionist, it has the advantage of providing a unified concept for the study and modification of degenerative disorders. Furthermore, the use of network analysis greatly favors data integration and candidate gene and drug prioritization in disease research. Signaling and regulatory networks within the cell are classically found to be scale-free, small-world graphs with a power-law degree distribution of interactions on nodes (Almaas, 2007), meaning that different parts of the network are highly interconnected, forming "functional modules."

In this rich phenomenological context, a pressing issue is to identify the key functional modules that modulate the survival of diseased neurons because this information may be invaluable in prioritizing candidate targets, markers, and drugs for neurodegenerative diseases. Because the cell survival machinery is highly complex as indicated by the genetic analyses of stress response and longevity modulators (Carter and Brunet, 2007), one needs a model system that allows for the manipulation of neuronal cell survival at the genome-scale level to emphasize the "biological principles" underlying the protection or rescue of diseased neurons. One also needs a convenient system in which to identify the survival pathways that may be used by neuroprotective compounds to document the "chemical genetics" of neuronal cell survival. Here comes invertebrate genetics.

THE INVERTEBRATE TOOL BOX

Several recent reviews have emphasized the power of invertebrates, notably the fruit fly *Drosophila melanogaster* and the nematode worm *Caenorhabditis elegans*, to study and manipulate the toxic effects of neurodegenerative disease proteins in transgenic animals (Bonini, 2001; Cauchi and van den Heuvel, 2006; Kaletta and Hengartner, 2006; Segalat, 2007; Tickoo and Russell, 2002; Zoghbi and Botas, 2002). *C. elegans* is a transparent animal that has been widely studied at the anatomical, behavioral, and genetic levels, and therefore constitutes a very well-defined model system for the study of diverse biological phenomena (Sulston, 1983). *C. elegans* has about 300 neurons out of 1,000 cells and is amenable to fast molecular manipulation (transgenesis, mutagenesis, classical genetics, RNA interference [RNAi] experiments, and microarray analysis using whole animal RNA or purified cell RNA), as well as drug screening *in vivo* (Parker et al., 2004). *C. elegans* cells are generated by an invariant lineage in normal animals; thus, one may directly monitor the effects of pharmacological and genetic manipulations on exactly the same neuron in all animals of a given population by using well-characterized behaviors. Additionally, synchronized populations of nematodes can be grown in liquid cultures in 96-well plates, and cellular phenotypes may be assessed directly by *in vivo* imaging of animals engineered to express fluorescently labeled proteins (Sieburth et al., 2005), a process that has great drug screening potential. Several reports have emphasized how genome-wide RNAi screens can be instructive to study the modulators of complex biological processes, including transposon silencing (Vastenhouw et al., 2003), body fat storage (Ashrafi et al., 2003), regulation of lifespan (Lee et al., 2003), synapse structure and function (Sieburth et al., 2005), and regulation of microRNA pathways (Parry et al., 2007). Interestingly, these datasets may now be analyzed using the connectivity and modularity of a single and high-coverage *C. elegans* gene network (Lee et al., 2008).

Thus, *C. elegans* is a genetically and pharmacologically tractable model system, in which one may easily and precisely study how genes and drugs may modulate cytotoxicity in transgenic animals expressing a neurodegenerative disease protein (Kraemer et al., 2003; Parker et al., 2001). Additionally, the use of *C. elegans* is cost- and time-effective compared with vertebrate model organisms, and can enable the identification and study of key modulators of conserved biological processes well in advance of their discovery and evaluation in more complex models such as the mouse. These advantages apply to developmental genetics, cell survival genetics, and—critically for drug discovery—chemical genetics.

Appropriate Use of Invertebrates in Drug Discovery

What exactly do we mean by modeling neurodegenerative diseases in invertebrates? Many people working in the field of neurodegenerative disease research believe that modeling illnesses such as Parkinson's disease or HD in distantly related organisms such as invertebrates is not relevant to the disease because the human brain is far more complex than the fly brain or the *C. elegans* neuronal circuitry. This might be true if the primary aims were to recapitulate disease complexity at the whole brain level.

However, this is often not the rationale for using invertebrates: here the rationale is primarily to study events at the cellular level by testing large numbers of genes/pathways and small molecules (Nollen et al., 2004; Perrimon et al., 2007). Thus, speaking of "invertebrate models of degenerative diseases" is probably a misnomer, at least as it is often used by many researchers. Models of disease typically refer to rodent and primate models that show neurological endpoints and neuropathological features as they may present in disease patients; among these features are behavioral and metabolic phenotypes, protein aggregates, and neurodegeneration.

A key issue is whether using endpoint phenotypes will allow effective neuroprotective treatments to be developed. What if the most effective targets for drug intervention are actually at the earliest stages of proteotoxicity, namely, cell dysfunction produced by soluble or poorly aggregated disease proteins? What if large aggregates may participate in cell protection rather than cell death? These provocative notions are now well documented (Arrasate et al., 2004; Cheng et al., 2007; Saudou et al., 1998), and a major effort in the field is to target the early phases of neurodegenerative disease processes.

In this context, invertebrates are good model systems to study *in vivo* how postmitotic cells may respond to the stress produced by disease proteins during developmental and adult life. Like developmental mechanisms, stress response and cell survival mechanisms are genetically regulated, and they are selected for conservation in the course of evolution (Lithgow, 2006). In this respect, it should thus be accurate to speak of "invertebrate models of cellular response" as cellular responses, and perhaps variations of cell–cell communications are the components of disease pathogenesis most likely to be well recapitulated in flies and nematodes. This essential aspect of comparative biology has been extensively illustrated by the prominent value of *Drosophila* and *C. elegans* in the study of longevity and aging (Lim et al., 2006a; Olsen et al., 2006). However, nematodes and flies are distantly related to humans, and despite a high level of genetic conservation for numerous essential pathways involved in development and/or cell maintenance, invertebrate physiology and human physiology may obviously be greatly different. Thus, whatever may be found using invertebrates in terms of disease targets and drug candidates has to be evaluated in mammalian models before one may really believe in therapeutic potential.

The promise for therapeutic relevance of nematodes and flies relies on the fact that several signaling pathways known to influence neurodegeneration are represented in invertebrates, including chaperone proteins, heat shock response pathway components, and protein degradation pathways (proteasome-mediated degradation and autophagy) (Ravikumar et al., 2004). Longevity/cell maintenance and metabolism pathways may also be important players in degenerative diseases, and conveniently these pathways are a subject of intense study in invertebrates (Curtis et al., 2006; Hwangbo et al., 2004; Murphy et al., 2003). Thus, the conservation and simplification of signaling pathways make invertebrates attractive model systems for the identification of conserved signaling networks critical for neuronal cell protection or rescue. However, some limitations may apply. Although a small molecule compound may be active in both invertebrate and mammalian models of proteotoxicity, one has to consider the possibility that the primary target of the compound may be different because of amino acid variations in orthologous proteins. However, this issue can be easily addressed by testing for drug binding to the putative targets in invertebrates and their mammalian counterparts.

Therefore, invertebrates can be seen as model systems, allowing a large number of biological processes to be probed genetically and a large number of drugs to be evaluated to establish strong rationales for protecting or rescuing diseased neurons. However, studying the gene networks involved in such cell response processes is almost impossible in mammals; thus, there is a need for more simple models in which genetic studies can be performed relatively quickly, that is, within months rather than years. The mechanisms elucidated in invertebrate models that underlie the activity of genes and drugs of high interest may then be characterized further in mammalian models. Thus, invertebrate-based translational research may work as an efficient "filter" to start from a large number of potentially interesting genetic and chemical modifiers (i.e., drug target candidates) to a manageable group of promising markers and drugs for further development in mammalian models and humans. In this respect, nematodes and flies may be used early in the target and drug discovery pipeline, and they represent an important step to prioritize genes and compounds for evaluation in cell-based assays or to gate cell-based assays with experiments in mice.

TARGET AND DRUG DISCOVERY IN *DROSOPHILA*

Several laboratories have generated transgenic flies expressing polyglutamine proteins, most often in photoreceptor neurons in the compound eye (Fernandez-Funez et al., 2000; Jackson et al., 1998; Marsh et al., 2000; Warrick et al., 1998). These animals have allowed several genetic modifiers of expanded polyQ cytotoxicity to be identified (Bilen and Bonini, 2007; Fernandez-Funez et al., 2000; Warrick et al., 1999) and rationales for neuroprotection to be defined, including the use of suppressor polypeptides (Kazantsev et al., 2002), inhibition of mammalian target of rapamycin (Ravikumar et al., 2004), and inhibition of sumoylation (Steffan et al., 2004). Interestingly, studies with *Drosophila* might inform combinatorial therapies, namely, the potential effectiveness of drug combinations used at low threshold concentrations (Agrawal et al., 2005). Additionally, studies with polyglutamine *Drosophila* models have emphasized targets that may be either common or specific to different polyglutamine diseases, such as spinocerebellar ataxia 1 and HD (Branco et al., 2008).

Thus, *Drosophila* animals and cells have great potential for target identification and small molecule compound screens (Perrimon et al., 2007; Stilwell et al., 2006). For polyglutamine disease genes, *Drosophila* has been used to screen the National Institute of Neurodegenerative Disorders and Stroke (NINDS) Custom Collection (1,040 Food and Drug Administration-approved drugs) within the framework of the Drug Screening Consortium Initiative run by NINDS (Heemskerk et al., 2002a, 2002b). The rationale was to search for promising drugs and drug targets based on activity found in multiple simple models (biochemical assays, nematodes, flies, mammalian cells) recapitulating either protein aggregation or cytotoxicity, or both. Although no drug consistently showed activity across these disparate screens, individual assays allowed interesting hits to be identified in cell-free and cell-based assays (Aiken et al., 2004; Desai et al., 2006; Stavrovskaya et al., 2004) and in flies (Juan Botas, personal communication), with overlap observed between screens run in *Drosophila* and the other model systems (Desai et al., 2006). This also applied to nematodes (see below). Thus, one outcome of the NINDS initiative was to demonstrate the predictive value of invertebrate model systems for drug screening, which has encouraged several laboratories to further develop and improve this type of approach.

TARGET AND DRUG DISCOVERY IN *C. ELEGANS*

Target validation is a normalized process, and appropriate scientific language should be used when talking about "targets." There is a consensus for the classification of targets according to their "validation score." Notably, a difference is made between a "disease target" (a target emphasized in simple model systems such as cells or invertebrates) and a "therapeutic target" (a target validated in rodent/primate models of the disease and/or associated to the human disease).

As for the *Drosophila* models discussed above, several laboratories have generated and characterized *C. elegans* transgenics expressing disease-relevant polyglutamine proteins (Caldwell et al., 2003; Faber et al., 1999; Khan et al., 2006; Parker et al., 2001; Satyal et al., 2000). These animals have allowed several conserved modifiers of polyglutamine cytotoxicity to be identified, including insulin signaling (Morley et al., 2002); autophagy (Jia et al., 2007), which was consistent with findings in *Drosophila* (Berger et al., 2006); torsins (Caldwell et al., 2003); sirtuins (Parker et al., 2005); and synaptic proteins (Parker et al., 2007), which was consistent again with findings in *Drosophila* (Kaltenbach et al., 2007). Some of these modifiers like sirtuins are druggable and are likely to elicit cell protection through several effectors, including Forkhead box (FOXO) transcription factors and their transcriptional targets (Parker et al., 2005). Thus, they may constitute disease targets of high interest to develop neuroprevention in HD (Antebi, 2004; Giannakou and Partridge, 2004; Guarente, 2006; Westphal et al., 2007) and other neurodegenerative diseases, including Alzheimer's disease and Parkinson's disease (Chen et al., 2005; Outeiro et al., 2007; Tang and Chua, 2008). However, genetic and pharmacological experiments in mouse models of the disease are needed to confirm the therapeutic potential initially emphasized by these studies of polyglutamines and sirtuins in nematodes.

Similarly to *Drosophila*, *C. elegans* may also inform combinatorial therapies (Voisine et al., 2007). Besides the evaluation of individual compounds for rescue of diseased cells, functional screens have been successfully undertaken in *C. elegans* transgenics expressing N-terminal htt in a small number of sensory (touch receptor) neurons (Heemskerk et al., 2002a, 2002b; Parker et al., 2004). Because these screens use behavior (restoration of response to light touch) as a readout for activity, the throughput is much lower compared with high-throughput screening (HTS) based on cell-free assays. Nonetheless, the value of drug screening in invertebrates is not to saturate chemical diversity but rather to test medium-sized compound libraries and collections (typically 1,000–20,000 molecules) in a setting that best reflects *in vivo* physiology, thus, the importance of functional/behavioral readouts. One important recent example is the identification of a new calcium channel antagonist from a screen of 14,100 small molecules in a nematode functional screen based on egg-laying behavior (Kwok et al., 2006).

Importantly, nematode screens can be performed in 96-well plates because the animals can be grown in liquid cultures. Typically, synchronized L1 larvae can be incubated with food and compound in 50-µL liquid cultures until they reach adulthood (Parker et al., 2004, 2005). In contrast to pharmacological assays in Petri dishes, this much higher throughput allows low-dose incubations (in the 0.1–100 µM

range) to be included in primary screening. Screening at such low doses is important because activities detected at high doses only may not be specific or may have
limited drug development potential. In liquid cultures, compounds may be ingested
together with bacteria or, more likely, they may diffuse across the cuticle into the
tissues of the animal. If the main route for small molecules in liquid culture is across
the cuticle, then well-accessible cells like sensory and cord neurons are more likely
to respond to compounds.

Such use of transgenic *C. elegans* expressing expanded polyglutamines in touch
receptor neurons has been successful for the identification of structurally related
compounds from the NINDS Custom Collection (Parker et al., 2004), including
antipsychotic drugs. Importantly, some of these compounds were also found to be
active in mammalian cell models of polyglutamine-expanded htt toxicity. Many
researchers think that false-negative results (as a result of compounds being unable
to get into the animal) may be problematic; however, one must remember that the
occurrence of false-negative findings is common to any screening assay. Here, as for
most drug screening assays, compound solubility is important; poorly soluble compounds or promiscuous aggregators such as protoporphyrins may not show activity
or may show strong and hard-to-believe effects, such as very low EC_{50} (Figure 6.2),

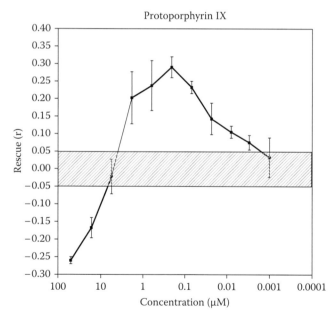

FIGURE 6.2   Dose–response curve for rescue (r) of neuronal dysfunction (loss of response
to light touch at the tail, indicating dysfunction of two touch receptor neurons, the PLM cells)
by protoporphyrin IX in transgenic nematodes expressing polyQ-expanded exon 1 htt in
touch receptor neurons (Parker et al., 2001). Compound concentration in the animals is likely
to be 100× less than the one in the culture medium. Most hits found in the screen showed
EC_{50} in the 10–0.1 μM range, and the very low EC_{50} (<0.1 μM) shown by protoporphyrin IX
is likely caused by promiscuous aggregation of the molecule. (From E. Lambert, personal
communication.)

which is unlikely to reflect a true activity but rather the fact that compound concentration in the animal is higher than in the culture medium because of the accumulation of the molecule in tissues.

Another powerful approach in *C. elegans* is true chemical genetics, not just the use of many compounds with known/putative drug targets to probe the "genetic" sensitivity of a disease model to drug intervention. With the increasingly large panel of viable and homozygotic mutants available, it is possible to generate a genetic profile of drug activity *in vivo* (Parker et al., 2005; Voisine et al., 2007), which, in the context of network biology, may provide invaluable information on the targets affected by a given compound. Here, targets does not necessarily mean the target directly bound by the compound (which can be determined by other means) but rather the effectors (proteins, pathways, biological processes) used by the compound to mediate the desired therapeutic outcome. The aim, as well as the method used to identify these follow-on targets, is to determine whether compound activity is lost in transgenic animals null for specific genes—the critical advantage of the nematode is that a significant portion of the genome could be practically screened.

As for drug screening, it is important to use low concentrations of compounds (typically in the micromolar range) in liquid cultures because using high concentrations may be misleading should the compound have secondary targets for activity, which is usually the case with many drugs. Should no mutant be available for the target of interest, combining compound incubation and RNAi feeding is feasible. However, this approach can be laborious if there is an unwanted interaction between the compound and the RNAi clone or if combined drug treatment with neuronal RNAi is necessary (Simmer et al., 2003), which requires F1 screens and food (hence compound) refeeding. Another potential problem is that the kinetics of compound activity and RNAi effect may be greatly different. Finally, negative results with RNAi are difficult to interpret because they may be caused by an insufficient level of target suppression.

Alternatively, the molecular profiling of drug activity using microarrays and mRNA from fluorescence-activated cell sorting (FACS)-purified, GFP-positive cells is also possible; however, it depends on the good maintenance of primary cell cultures. Additionally, this approach is not really "competitive" compared with microarray profiling of compound activity in models closer to the situation in humans, namely, mammalian cells. Thus, the best rationale for microarray profiling in cultured nematode cells may be to complement chemical genetics studies rather than use it as a primary discovery tool. Finally, it is important to note that medicinal chemistry and drug optimization in *C. elegans* or *Drosophila* is unlikely to be useful except for specific occasions where the binding site(s) of the compound is truly conserved between the invertebrate target and its mammalian counterpart.

Thus, together with the strong value of *C. elegans* for disease target discovery, the amenability of *C. elegans* to chemical screening and chemical genetics *in vivo* makes this model organism a powerful tool that may be used at different stages of the drug discovery and development process (Figure 6.3), from its initial phases (target identification, discovery of active compounds) through the biological characterization of therapeutic targets and drug lead activities.

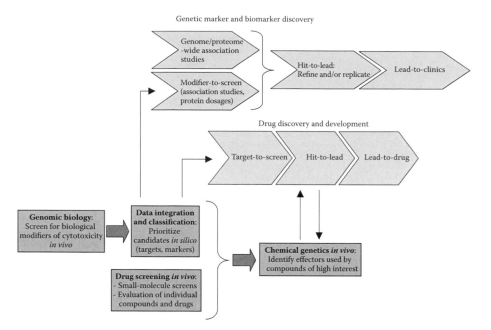

FIGURE 6.3   (See color insert following page 172.) Implementation of *in vivo* biology and systems biology approaches into the drug and marker discovery process. Ways to combine invertebrate genetics and systems biology with the traditional components of drug and marker discovery are shown.

CONCLUSIONS

Although drug discovery relies strongly on chemistry and HTS in cell-free or cell-based systems, there is now compelling evidence that invertebrate genetics is a useful component of translational research on neurodegenerative diseases (Driscoll and Gerstbrein, 2003). Research on HD and other polyglutamine diseases indicates that invertebrates may provide unique insights into the biological processes that are altered in the disease and into the biological processes that modulate neuronal cell survival. As for RNAi screens (Nollen et al., 2004; Sieburth et al., 2005), the future value of drug discovery with invertebrates will probably be mostly in the development of "physiological" drug screening in which the effects of compounds on specific biological processes (e.g., developmental defects, protein translocation, protein interactions, vesicle trafficking, morphology of subcellular compartments) are monitored in live animals using fluorescent reporters. Another trend is the combination of drug screening with genetics and genomics to identify drug effectors, which may provide strong guidance to neuropreventive medicine and pharmacogenomics.

It is important to note, however, that because many stress response and cell survival processes are finely tuned, the effects of proteotoxicity modifiers (genes, drugs) *in vivo* may highly depend on: (1) the severity of the stress (e.g., type of disease protein, level of transgenic protein expression, susceptibility of target neurons to injury); and (2) gene and drug dosages. Thus, one should be cautious when comparing the outcome of genetic and pharmacological manipulations between invertebrate models of cell response (such as cell death endpoints) versus early phenotypes (such as cell dysfunction), which may lead to different conclusions as to the way a target should be manipulated (inhibition vs. activation), as well as the primary mode of action of compounds.

Given that the biological modifiers of proteotoxicity in the neuronal cell may also modify HD pathogenesis, and furthermore may be the targets directly or indirectly used by many neuroprotective compounds, we may be just at the beginning of what will be a way paved by invertebrate genetics ("scan *in vivo* for biological rationales"), network biology ("prioritize candidates *in silico*"), and mammalian biology and human genetics ("select pharmaceutical strategies") toward developing neuroprevention for HD and other degenerative diseases.

REFERENCES

Agrawal, N., Pallos, J., Slepko, N., Apostol, B.L., Bodai, L., Chang, L.W., Chiang, A.S., Thompson, L.M. and Marsh, J.L. (2005) Identification of combinatorial drug regimens for treatment of Huntington's disease using Drosophila. *Proc Natl Acad Sci U S A*, **102**, 3777–3781.

Aiken, C.T., Tobin, A.J. and Schweitzer, E.S. (2004) A cell-based screen for drugs to treat Huntington's disease. *Neurobiol Dis*, **16**, 546–555.

Almaas, E. (2007) Biological impacts and context of network theory. *J Exp Biol*, **210**, 1548–1558.

Andresen, J.M., Gayan, J., Cherny, S.S., Brocklebank, D., Alkorta-Aranburu, G., Addis, E.A., Cardon, L.R., Housman, D.E. and Wexler, N.S. (2007) Replication of twelve association studies for Huntington's disease residual age of onset in large Venezuelan kindreds. *J Med Genet*, **44**, 44–50.

Anekonda, T.S. (2006) Resveratrol—a boon for treating Alzheimer's disease? *Brain Res Rev*, **52**, 316–326.

Antebi, A. (2004) Tipping the balance toward longevity. *Dev Cell*, **6**, 315–316.

Arango, M., Holbert, S., Zala, D., Brouillet, E., Pearson, J., Regulier, E., Thakur, A.K., Aebischer, P., Wetzel, R., Deglon, N., et al. (2006) CA150 expression delays striatal cell death in overexpression and knock-in conditions for mutant huntingtin neurotoxicity. *J Neurosci*, **26**, 4649–4659.

Arrasate, M., Mitra, S., Schweitzer, E.S., Segal, M.R. and Finkbeiner, S. (2004) Inclusion body formation reduces levels of mutant huntingtin and the risk of neuronal death. *Nature*, **431**, 805–810.

Ashrafi, K., Chang, F.Y., Watts, J.L., Fraser, A.G., Kamath, R.S., Ahringer, J. and Ruvkun, G. (2003) Genome-wide RNAi analysis of Caenorhabditis elegans fat regulatory genes. *Nature*, **421**, 268–272.

Berger, Z., Ravikumar, B., Menzies, F.M., Oroz, L.G., Underwood, B.R., Pangalos, M.N., Schmitt, I., Wullner, U., Evert, B.O., O'Kane, C.J., et al. (2006) Rapamycin alleviates toxicity of different aggregate-prone proteins. *Hum Mol Genet*, **15**, 433–442.

Bilen, J. and Bonini, N.M. (2007) Genome-wide screen for modifiers of ataxin-3 neurodegeneration in Drosophila. *PLoS Genet*, **3**, 1950–1964.

Bonini, N.M. (2001) Drosophila as a genetic approach to human neurodegenerative disease. *Parkinsonism Relat Disord*, **7**, 171–175.

Branco, J., Al-Ramahi, I., Ukani, L., Perez, A.M., Fernandez-Funez, P., Rincon-Limas, D. and Botas, A.J. (2008) Comparative analysis of genetic modifiers in Drosophila points to common and distinct mechanisms of pathogenesis among polyglutamine diseases. *Hum Mol Genet*, **17**, 376–390.

Caldwell, G.A., Cao, S., Sexton, E.G., Gelwix, C.C., Bevel, J.P. and Caldwell, K.A. (2003) Suppression of polyglutamine-induced protein aggregation in Caenorhabditis elegans by torsin proteins. *Hum Mol Genet*, **12**, 307–319.

Carter, M.E. and Brunet, A. (2007) FOXO transcription factors. *Curr Biol*, **17**, R113–114.

Cauchi, R.J. and van den Heuvel, M. (2006) The fly as a model for neurodegenerative diseases: is it worth the jump? *Neurodegener Dis*, **3**, 338–356.

Chen, J., Zhou, Y., Mueller-Steiner, S., Chen, L.F., Kwon, H., Yi, S., Mucke, L. and Gan, L. (2005) SIRT1 protects against microglia-dependent beta amyloid toxicity through inhibiting NF-kappa B signaling. *J Biol Chem*, **280**, 40364–40367.

Cheng, I.H., Scearce-Levie, K., Legleiter, J., Palop, J.J., Gerstein, H., Bien-Ly, N., Puolivali, J., Lesne, S., Ashe, K.H., Muchowski, P.J., et al. (2007) Accelerating amyloid-beta fibrillization reduces oligomer levels and functional deficits in Alzheimer disease mouse models. *J Biol Chem*, **282**, 23818–23828.

Curtis, R., O'Connor, G. and DiStefano, P.S. (2006) Aging networks in Caenorhabditis elegans: AMP-activated protein kinase (aak-2) links multiple aging and metabolism pathways. *Aging Cell*, **5**, 119–126.

Desai, U.A., Pallos, J., Ma, A.A., Stockwell, B.R., Thompson, L.M., Marsh, J.L. and Diamond, M.I. (2006) Biologically active molecules that reduce polyglutamine aggregation and toxicity. *Hum Mol Genet*, **15**, 2114–2124.

Driscoll, M. and Gerstbrein, B. (2003) Dying for a cause: invertebrate genetics takes on human neurodegeneration. *Nat Rev Genet*, **4**, 181–194.

Faber, P.W., Alter, J.R., MacDonald, M.E. and Hart, A.C. (1999) Polyglutamine-mediated dysfunction and apoptotic death of a Caenorhabditis elegans sensory neuron. *Proc Natl Acad Sci U S A*, **96**, 179–184.

Feany, M.B. (2000) Studying human neurodegenerative diseases in flies and worms. *J Neuropathol Exp Neurol*, **59**, 847–856.

Fernandez-Funez, P., Nino-Rosales, M.L., de Gouyon, B., She, W.C., Luchak, J.M., Martinez, P., Turiegano, E., Benito, J., Capovilla, M., Skinner, P.J., et al. (2000) Identification of genes that modify ataxin-1-induced neurodegeneration. *Nature*, **408**, 101–106.

Giannakou, M.E. and Partridge, L. (2004) The interaction between FOXO and SIRT1: tipping the balance towards survival. *Trends Cell Biol*, **14**, 408–412.

Goehler, H., Lalowski, M., Stelzl, U., Waelter, S., Stroedicke, M., Worm, U., Droege, A., Lindenberg, K.S., Knoblich, M., Haenig, C., et al. (2004) A protein interaction network links GIT1, an enhancer of huntingtin aggregation, to Huntington's disease. *Mol Cell*, **15**, 853–865.

Griffioen, G. (2007) Targeting disease, not disease targets: innovative approaches in tackling neurodegenerative disorders. *IDrugs*, **10**, 259–263.

Guarente, L. (2006) Sirtuins as potential targets for metabolic syndrome. *Nature*, **444**, 868–874.

Harjes, P. and Wanker, E.E. (2003) The hunt for huntingtin function: interaction partners tell many different stories. *Trends Biochem Sci*, **28**, 425–433.

Heemskerk, J., Tobin, A.J. and Bain, L.J. (2002a) Teaching old drugs new tricks. Meeting of the Neurodegeneration Drug Screening Consortium, 7–8 April 2002, Washington, DC, USA. *Trends Neurosci*, **25**, 494–496.

Heemskerk, J., Tobin, A.J. and Ravina, B. (2002b) From chemical to drug: neurodegeneration drug screening and the ethics of clinical trials. *Nat Neurosci*, **5 Suppl 1**, 1027–1029.

Holbert, S., Denghien, I., Kiechle, T., Rosenblatt, A., Wellington, C., Hayden, M.R., Margolis, R.L., Ross, C.A., Dausset, J., Ferrante, R.J., et al. (2001) The Gln-Ala repeat transcriptional activator CA150 interacts with huntingtin: neuropathologic and genetic evidence for a role in Huntington's disease pathogenesis. *Proc Natl Acad Sci U S A*, **98**, 1811–1816.

Hung, A.Y. and Schwarzschild, M.A. (2007) Clinical trials for neuroprotection in Parkinson's disease: overcoming angst and futility? *Curr Opin Neurol*, **20**, 477–483.

Hwangbo, D.S., Gersham, B., Tu, M.P., Palmer, M. and Tatar, M. (2004) Drosophila dFOXO controls lifespan and regulates insulin signalling in brain and fat body. *Nature*, **429**, 562–566.

Hye, A., Lynham, S., Thambisetty, M., Causevic, M., Campbell, J., Byers, H.L., Hooper, C., Rijsdijk, F., Tabrizi, S.J., Banner, S., et al. (2006) Proteome-based plasma biomarkers for Alzheimer's disease. *Brain*, **129**, 3042–3050.

Jackson, G.R., Salecker, I., Dong, X., Yao, X., Arnheim, N., Faber, P.W., MacDonald, M.E. and Zipursky, S.L. (1998) Polyglutamine-expanded human huntingtin transgenes induce degeneration of Drosophila photoreceptor neurons. *Neuron*, **21**, 633–642.

Jia, K., Hart, A.C. and Levine, B. (2007) Autophagy genes protect against disease caused by polyglutamine expansion proteins in Caenorhabditis elegans. *Autophagy*, **3**, 21–25.

Kaletta, T. and Hengartner, M.O. (2006) Finding function in novel targets: C. elegans as a model organism. *Nat Rev Drug Discov*, **5**, 387–398.

Kaltenbach, L.S., Romero, E., Becklin, R.R., Chettier, R., Bell, R., Phansalkar, A., Strand, A., Torcassi, C., Savage, J., Hurlburt, A., et al. (2007) Huntingtin interacting proteins are genetic modifiers of neurodegeneration. *PLoS Genet*, **3**, e82.

Kazantsev, A., Walker, H.A., Slepko, N., Bear, J.E., Preisinger, E., Steffan, J.S., Zhu, Y.Z., Gertler, F.B., Housman, D.E., Marsh, J.L., et al. (2002) A bivalent Huntingtin binding peptide suppresses polyglutamine aggregation and pathogenesis in Drosophila. *Nat Genet*, **25**, 25.

Khan, L.A., Bauer, P.O., Miyazaki, H., Lindenberg, K.S., Landwehrmeyer, B.G. and Nukina, N. (2006) Expanded polyglutamines impair synaptic transmission and ubiquitin-proteasome system in Caenorhabditis elegans. *J Neurochem*, **98**, 576–587.

Klamt, S., Saez-Rodriguez, J., Lindquist, J.A., Simeoni, L. and Gilles, E.D. (2006) A methodology for the structural and functional analysis of signaling and regulatory networks. *BMC Bioinformatics*, **7**, 56.

Kraemer, B.C., Zhang, B., Leverenz, J.B., Thomas, J.H., Trojanowski, J.Q. and Schellenberg, G.D. (2003) Neurodegeneration and defective neurotransmission in a Caenorhabditis elegans model of tauopathy. *Proc Natl Acad Sci U S A*, **100**, 9980–9985.

Kwok, T.C., Ricker, N., Fraser, R., Chan, A.W., Burns, A., Stanley, E.F., McCourt, P., Cutler, S.R. and Roy, P.J. (2006) A small-molecule screen in C. elegans yields a new calcium channel antagonist. *Nature*, **441**, 91–95.

Landles, C. and Bates, G.P. (2004) Huntingtin and the molecular pathogenesis of Huntington's disease. Fourth in molecular medicine review series. *EMBO Rep*, **5**, 958–963.

Lee, I., Lehner, B., Crombie, C., Wong, W., Fraser, A.G. and Marcotte, E.M. (2008) A single gene network accurately predicts phenotypic effects of gene perturbation in Caenorhabditis elegans. *Nat Genet*, **40**, 181–188.

Lee, S.S., Lee, R.Y., Fraser, A.G., Kamath, R.S., Ahringer, J. and Ruvkun, G. (2003) A systematic RNAi screen identifies a critical role for mitochondria in C. elegans longevity. *Nat Genet*, **33**, 40–48.

Levine, M.S., Cepeda, C., Hickey, M.A., Fleming, S.M. and Chesselet, M.F. (2004) Genetic mouse models of Huntington's and Parkinson's diseases: illuminating but imperfect. *Trends Neurosci*, **27**, 691–697.

Li, J.L., Hayden, M.R., Almqvist, E.W., Brinkman, R.R., Durr, A., Dode, C., Morrison, P.J., Suchowersky, O., Ross, C.A., Margolis, R.L., et al. (2003) A genome scan for modifiers of age at onset in Huntington disease: the HD MAPS Study. *Am J Hum Genet*, **73**, 682–687.

Lim, H.Y., Bodmer, R. and Perrin, L. (2006a) Drosophila aging 2005/06. *Exp Gerontol*, **41**, 1213–1216.

Lim, J., Hao, T., Shaw, C., Patel, A.J., Szabo, G., Rual, J.F., Fisk, C.J., Li, N., Smolyar, A., Hill, D.E., et al. (2006b) A protein-protein interaction network for human inherited ataxias and disorders of Purkinje cell degeneration. *Cell*, **125**, 801–814.

Lithgow, G.J. (2006) Why aging isn't regulated: a lamentation on the use of language in aging literature. *Exp Gerontol*, **41**, 890–893.

Liu, W., Li, D., Zhang, J., Zhu, Y. and He, F. (2006) SigFlux: a novel network feature to evaluate the importance of proteins in signal transduction networks. *BMC Bioinformatics*, **7**, 515.

Margolis, R.L., McInnis, M.G., Rosenblatt, A. and Ross, C.A. (1999) Trinucleotide repeat expansion and neuropsychiatric disease. *Arch Gen Psychiatry*, **56**, 1019–1031.

Margolis, R.L. and Ross, C.A. (2003) Diagnosis of Huntington disease. *Clin Chem*, **49**, 1726–1732.

Marsh, J.L., Walker, H., Theisen, H., Zhu, Y.Z., Fielder, T., Purcell, J. and Thompson, L.M. (2000) Expanded polyglutamine peptides alone are intrinsically cytotoxic and cause neurodegeneration in Drosophila. *Hum Mol Genet*, **9**, 13–25.

Morley, J.F., Brignull, H.R., Weyers, J.J. and Morimoto, R.I. (2002) The threshold for polyglutamine-expansion protein aggregation and cellular toxicity is dynamic and influenced by aging in Caenorhabditis elegans. *Proc Natl Acad Sci U S A*, **16**, 16.

Murphy, C.T., McCarroll, S.A., Bargmann, C.I., Fraser, A., Kamath, R.S., Ahringer, J., Li, H. and Kenyon, C. (2003) Genes that act downstream of DAF-16 to influence the lifespan of Caenorhabditis elegans. *Nature*, **424**, 277–283.

Natesan, S., Svensson, K.A., Reckless, G.E., Nobrega, J.N., Barlow, K.B., Johansson, A.M. and Kapur, S. (2006) The dopamine stabilizers (S)-(-)-(3-methanesulfonyl-phenyl)-1-propyl-piperidine [(-)-OSU6162] and 4-(3-methanesulfonylphenyl)-1-propyl-piperidine (ACR16) show high in vivo D2 receptor occupancy, antipsychotic-like efficacy, and low potential for motor side effects in the rat. *J Pharmacol Exp Ther*, **318**, 810–818.

Nollen, E.A., Garcia, S.M., van Haaften, G., Kim, S., Chavez, A., Morimoto, R.I. and Plasterk, R.H. (2004) Genome-wide RNA interference screen identifies previously undescribed regulators of polyglutamine aggregation. *Proc Natl Acad Sci U S A*, **101**, 6403–6408.

Olsen, A., Vantipalli, M.C. and Lithgow, G.J. (2006) Using Caenorhabditis elegans as a model for aging and age-related diseases. *Ann N Y Acad Sci*, **1067**, 120–128.

Outeiro, T.F., Kontopoulos, E., Altmann, S.M., Kufareva, I., Strathearn, K.E., Amore, A.M., Volk, C.B., Maxwell, M.M., Rochet, J.C., McLean, P.J., et al. (2007) Sirtuin 2 inhibitors rescue alpha-synuclein-mediated toxicity in models of Parkinson's disease. *Science*, **317**, 516–519.

Parker, J.A., Arango, M., Abderrahmane, S., Lambert, E., Tourette, C., Catoire, H. and Neri, C. (2005) Resveratrol rescues mutant polyglutamine cytotoxicity in nematode and mammalian neurons. *Nat Genet*, **37**, 349–350.

Parker, J.A., Connolly, J.B., Wellington, C., Hayden, M., Dausset, J. and Neri, C. (2001) Expanded polyglutamines in Caenorhabditis elegans cause axonal abnormalities and severe dysfunction of PLM mechanosensory neurons without cell death. *Proc Natl Acad Sci U S A*, **98**, 13318–13323.

Parker, J.A., Holbert, S., Lambert, E., Abderrahmane, S. and Neri, C. (2004) Genetic and pharmacological suppression of polyglutamine-dependent neuronal dysfunction in Caenorhabditis elegans. *J Mol Neurosci*, **23**, 61–68.

Parker, J.A., Metzler, M., Georgiou, J., Mage, M., Roder, J.C., Rose, A.M., Hayden, M.R. and Neri, C. (2007) Huntingtin-interacting protein 1 influences worm and mouse presynaptic function and protects Caenorhabditis elegans neurons against mutant polyglutamine toxicity. *J Neurosci*, **27**, 11056–11064.

Parry, D.H., Xu, J. and Ruvkun, G. (2007) A whole-genome RNAi Screen for C. elegans miRNA pathway genes. *Curr Biol*, **17**, 2013–2022.

Perrimon, N., Friedman, A., Mathey-Prevot, B. and Eggert, U.S. (2007) Drug-target identification in Drosophila cells: combining high-throughput RNAi and small-molecule screens. *Drug Discov Today*, **12**, 28–33.

Rast, J.P. (2003) Development gene networks and evolution. *J Struct Funct Genomics*, **3**, 225–234.

Ravikumar, B., Vacher, C., Berger, Z., Davies, J.E., Luo, S., Oroz, L.G., Scaravilli, F., Easton, D.F., Duden, R., O'Kane, C.J., et al. (2004) Inhibition of mTOR induces autophagy and reduces toxicity of polyglutamine expansions in fly and mouse models of Huntington disease. *Nat Genet*, **36**, 585–595.

Rubinsztein, D.C. (2002) Lessons from animal models of Huntington's disease. *Trends Genet*, **18**, 202–209.

Satyal, S.H., Schmidt, E., Kitagawa, K., Sondheimer, N., Lindquist, S., Kramer, J.M. and Morimoto, R.I. (2000) Polyglutamine aggregates alter protein folding homeostasis in Caenorhabditis elegans. *Proc Natl Acad Sci U S A*, **97**, 5750–5755.

Saudou, F., Finkbeiner, S., Devys, D. and Greenberg, M.E. (1998) Huntingtin acts in the nucleus to induce apoptosis but death does not correlate with the formation of intranuclear inclusions. *Cell*, **95**, 55–66.

Segalat, L. (2007) Invertebrate animal models of diseases as screening tools in drug discovery. *ACS Chem Biol*, **2**, 231–236.

Shelbourne, P.F., Keller-McGandy, C., Bi, W.L., Yoon, S.R., Dubeau, L., Veitch, N.J., Vonsattel, J.P., Wexler, N.S., Arnheim, N. and Augood, S.J. (2007) Triplet repeat mutation length gains correlate with cell-type specific vulnerability in Huntington disease brain. *Hum Mol Genet*, **16**, 1133–1142.

Sieburth, D., Ch'ng, Q., Dybbs, M., Tavazoie, M., Kennedy, S., Wang, D., Dupuy, D., Rual, J.F., Hill, D.E., Vidal, M., et al. (2005) Systematic analysis of genes required for synapse structure and function. *Nature*, **436**, 510–517.

Simmer, F., Moorman, C., Van Der Linden, A.M., Kuijk, E., Van Den Berghe, P.V., Kamath, R., Fraser, A.G., Ahringer, J. and Plasterk, R.H. (2003) Genome-wide RNAi of C. elegans using the hypersensitive rrf-3 strain reveals novel gene functions. *PLoS Biol*, **1**, E12.

Sipione, S. and Cattaneo, E. (2001) Modeling Huntington's disease in cells, flies, and mice. *Mol Neurobiol*, **23**, 21–51.

Stavrovskaya, I.G., Narayanan, M.V., Zhang, W., Krasnikov, B.F., Heemskerk, J., Young, S.S., Blass, J.P., Brown, A.M., Beal, M.F., Friedlander, R.M., et al. (2004) Clinically approved heterocyclics act on a mitochondrial target and reduce stroke-induced pathology. *J Exp Med*, **200**, 211–222.

Steffan, J.S., Agrawal, N., Pallos, J., Rockabrand, E., Trotman, L.C., Slepko, N., Illes, K., Lukacsovich, T., Zhu, Y.Z., Cattaneo, E., et al. (2004) SUMO modification of Huntingtin and Huntington's disease pathology. *Science*, **304**, 100–104.

Stilwell, G.E., Saraswati, S., Littleton, J.T. and Chouinard, S.W. (2006) Development of a Drosophila seizure model for in vivo high-throughput drug screening. *Eur J Neurosci*, **24**, 2211–2222.

Sulston, J.E. (1983) Neuronal cell lineages in the nematode Caenorhabditis elegans. *Cold Spring Harb Symp Quant Biol*, **48 Pt 2**, 443–452.

Tang, B.L. and Chua, C.E. (2008) SIRT1 and neuronal diseases. *Mol Aspects Med* **29**, 187–200.

Tickoo, S. and Russell, S. (2002) Drosophila melanogaster as a model system for drug discovery and pathway screening. *Curr Opin Pharmacol*, **2**, 555–560.

Toulmond, S., Tang, K., Bureau, Y., Ashdown, H., Degen, S., O'Donnell, R., Tam, J., Han, Y., Colucci, J., Giroux, A., et al. (2004) Neuroprotective effects of M826, a reversible caspase-3 inhibitor, in the rat malonate model of Huntington's disease. *Br J Pharmacol*, **141**, 689–697.

Tse, W. (2006) Optimizing pharmacotherapy: strategies to manage the wearing-off phenomenon. *J Am Med Dir Assoc*, **7**, 12–17.

Varma, H., Voisine, C., DeMarco, C.T., Cattaneo, E., Lo, D.C., Hart, A.C. and Stockwell, B.R. (2007) Selective inhibitors of death in mutant huntingtin cells. *Nat Chem Biol*, **3**, 99–100.

Vastenhouw, N.L., Fischer, S.E., Robert, V.J., Thijssen, K.L., Fraser, A.G., Kamath, R.S., Ahringer, J. and Plasterk, R.H. (2003) A genome-wide screen identifies 27 genes involved in transposon silencing in C. elegans. *Curr Biol*, **13**, 1311–1316.

Voisine, C., Varma, H., Walker, N., Bates, E.A., Stockwell, B.R. and Hart, A.C. (2007) Identification of potential therapeutic drugs for Huntington's disease using Caenorhabditis elegans. *PLoS ONE*, **2**, e504.

von Bohlen Und Halbach, O. (2005) Modeling neurodegenerative diseases in vivo review. *Neurodegener Dis*, **2**, 313–320.

Walker, F.O. (2007) Huntington's disease. *Lancet*, **369**, 218–228.

Ward, M. (2007) Biomarkers for Alzheimer's disease. *Expert Rev Mol Diagn*, **7**, 635–646.

Warrick, J.M., Chan, H.Y., Gray-Board, G.L., Chai, Y., Paulson, H.L. and Bonini, N.M. (1999) Suppression of polyglutamine-mediated neurodegeneration in Drosophila by the molecular chaperone HSP70. *Nat Genet*, **23**, 425–428.

Warrick, J.M., Paulson, H.L., Gray-Board, G.L., Bui, Q.T., Fischbeck, K.H., Pittman, R.N. and Bonini, N.M. (1998) Expanded polyglutamine protein forms nuclear inclusions and causes neural degeneration in Drosophila. *Cell*, **93**, 939–949.

Westphal, C.H., Dipp, M.A. and Guarente, L. (2007) A therapeutic role for sirtuins in diseases of aging? *Trends Biochem Sci* **32**, 555–560.

Wexler, N.S., Lorimer, J., Porter, J., Gomez, F., Moskowitz, C., Shackell, E., Marder, K., Penchaszadeh, G., Roberts, S.A., Gayan, J., et al. (2004) Venezuelan kindreds reveal that genetic and environmental factors modulate Huntington's disease age of onset. *Proc Natl Acad Sci U S A*, **101**, 3498–3503.

Zoghbi, H.Y. and Botas, J. (2002) Mouse and fly models of neurodegeneration. *Trends Genet*, **18**, 463–471.

7 Mouse Models for Validating Preclinical Candidates for Huntington's Disease

X. William Yang and Michelle Gray

CONTENTS

INTRODUCTION

CLINICAL FEATURES OF HUNTINGTON'S DISEASE

Ever since its original description by George Huntington in 1872, Huntington's disease (HD) has been known as one of the most devastating inherited neurodegenerative disorders afflicting the human brain. Currently in the United States, there are about 30,000 patients with HD and another 150,000 people who are at a genetic risk of developing the disease. HD is characterized by the clinical triad of late-onset motor disturbances (i.e., chorea and dystonia), psychiatric deficits (i.e., depression, irritability, and psychosis), and cognitive decline (Bates et al., 2002). The majority of HD patients experience onset of symptoms around the age of 40 (adult-onset HD), and the disease relentlessly progresses until the patient's death, which usually occurs within 10–20 years after disease onset. A small subset of HD patients experience the onset of symptoms before age 20 (juvenile HD), and these patients exhibit slightly different clinical features in that they tend to have more dystonia than chorea, as well as a higher incidence of epilepsy. Although the onset of HD is currently defined by the onset of motor deficits, recent studies using more sensitive motor studies, as well as cognitive studies, indicate that clinical manifestations of HD may occur years to decades before the onset of motor symptoms, and such deficits may correspond to the early and progressive cortical and striatal atrophy seen in presymptomatic HD patients (Aylward et al., 2004; Rosas et al., 2005, 2006).

HD NEUROPATHOLOGY

HD neuropathology is characterized by selective atrophy and neuronal loss, primarily targeting the striatum and cortex (Vonsattel and DiFiglia, 1998). The neostriatum (i.e., caudate and putamen) bears the brunt of the disease. In the striatum, the atrophy is accompanied by robust and highly selective loss of the striatal medium spiny neurons (MSNs), as well as reactive gliosis and microgliosis. Other striatal neuronal types, such as a variety of interneurons, may exhibit evidence of dysfunction but do not appear to degenerate in HD. Current neuropathological studies have provided very good clinical correlations between striatal atrophy and MSN degeneration with motor and certain psychiatric manifestations of HD.

Besides striatum, the HD cortex also suffers from atrophy and the loss of pyramidal projection neurons in the deep cortical layers (Hedreen et al., 1991; Heinsen et al., 1994). Similar to the striatum, the projection neurons in the cortex are more vulnerable to degeneration than the cortical interneurons. Until recently, the pathogenic significance of the cortex in HD remained poorly defined. Imaging studies in patients who demonstrate regional-selective cortical thinning, semi-independent of striatal atrophy, are significantly correlated with cognitive deficits in HD (Rosas et al., 2005). Moreover, the white matter abnormalities, particularly those connecting the cortical and subcortical structures such as the striatum, are affected early in

HD (Rosas et al., 2006). These emerging clinical studies highlight HD as a disease selectively targeting multiple components in the corticostriatal-thalamocortical circuit, a circuit mediating many of the basic functions affected in HD, including motor control, motor and reward learning, and cognition.

HD MOLECULAR GENETICS

A triumph in HD research is the identification of the HD mutation in 1993 (The Huntington's Disease Collaborative Research Group, 1993). HD is caused by a CAG repeat expansion mutation (>36 repeats) in the *Huntingtin (HTT)* gene, which translates into an expanded polyglutamine (polyQ) repeat. In HD and in eight other neurodegenerative disorders caused by the polyQ repeat expansion, the onset of clinical symptoms is inversely correlated with repeat length (Zoghbi and Orr, 2000). Although the mutant huntingtin (mhtt) protein is widely expressed in the brain and in other tissues outside the nervous system (Schilling et al., 1995; Sharp et al., 1995), the MSNs in the striatum and pyramidal neurons in the cortex (particularly in the deep cortical layers) are most vulnerable to degeneration (Vonsattel and DiFiglia, 1998). The cellular and molecular mechanisms underlying this selective neuropathology in HD remain unknown. Resolving this and other key HD pathogenic mechanisms may provide critical novel insights toward therapeutic development in HD.

ADVANTAGES OF USING MOUSE MODELS TO STUDY HD

A primary challenge in studying the pathogenesis and treatment of HD is to develop disease models that recapitulate both the genetic and phenotypic aspects of the disease, hence allowing one to go from genetic mutations to understanding disease mechanisms and ultimately to the development of effective therapeutics. Genetic models of HD have been generated using different model organisms to elucidate pathogenic mechanisms (Marsh and Thompson, 2006; Sipione and Cattaneo, 2001). Some of these genetic models (i.e., yeast, *Caenorhabditis elegans*, and *Drosophila*) are suitable for high-throughput target identification and/or validation and for drug screening (Hughes and Olson, 2001). It is generally believed, however, that an essential step in HD therapeutic discovery, before the pursuit of a full-scale human clinical trial, is the use of mammalian genetic models of HD for preclinical validation of the therapeutic targets and/or compounds.

The mouse is an ideal mammalian genetic model organism for modeling human neurodegenerative disorders. First, compared with the nonmammalian model organisms, mice have a genetic background more closely related to humans. Mice and humans diverged only about 75 million years ago compared with more than 600 million years of evolutionary divergence between *Drosophila* and humans. Because of such close evolutionary distance, mouse and human genomes are highly similar. About 90% of human genes have direct murine counterparts, and the overall genomic organization and gene expression are also similar between humans and mice (Mouse Genome Sequencing Consortium, 2002). Such genetic similarity increases the likelihood that the genomic response to the HD disease gene and/or to a therapeutic intervention may be comparable between HD mice and patients. Second, mice possess the basic neural circuits and neuronal cell types that are selectively targeted in HD, as well as a rich repertoire of behavior assays that may pinpoint the dysfunction and degeneration of neurons in the HD-relevant neural circuit (Watase and Zoghbi, 2003). This cellular/circuitry context permits the use of mouse models to study the cellular and molecular mechanisms underlying selective neuronal toxicity in HD, which could not be accomplished using other common nonmammalian models. Third, rich genetic resources and genetic manipulation tools are available in mice for sophisticated analyses of HD pathogenesis and treatment (Capecchi, 2005; Heintz, 2001). Mice are well known for their small size and short generation time (3 months). There are also many well-characterized inbred mouse strains, each with identical genetic background and well-defined characteristics. Furthermore, the mouse is readily amenable to genetic manipulations, including stable introduction of an exogenous gene into the mouse genome (transgenic mice), deletion of an endogenous gene in the mouse genome (knockout mice), and targeted replacement of an endogenous gene with an exogenous gene/sequence (knockin mice). Finally, more advanced mouse genetic strategies also permit conditional expression of a gene of interest (on or off) in specific cell types and/or at defined time points (Yamamoto et al., 2000). Thus, mice provide a powerful mammalian genetic model for the dissection of neural circuitry and molecular mechanisms underlying HD pathogenesis, as well as for validating and testing therapeutics for HD.

MOUSE GENETIC MODELS OF HD FOR PRECLINICAL STUDIES

Since the publication of the first HD mouse model (R6 mice) in 1996 (Mangiarini et al., 1996), the HD research field has produced more than a dozen different genetic mouse models of the disease (Levine et al., 2004; Menalled and Chesselet, 2002). Together, this rich repertoire of HD mouse models has been essential to unraveling the pathogenic mechanisms elicited by mhtt in mammalian organisms (Gusella and Macdonald, 2006; Landles and Bates, 2004; Orr and Zoghbi, 2007). Because recent excellent reviews on the subject can be found elsewhere, we will not attempt to comprehensively survey all the existing HD mouse models generated (Levine et al., 2004; Li and Li, 2006; Menalled, 2005; Menalled and Chesselet, 2002; Van Raamsdonk et al., 2007). Instead, we will focus on the few preclinical HD mouse models that are already being used or are intended for use in preclinical studies, that is, target validation and therapeutic compound testing (Table 7.1). We will address the genetic and phenotypic characteristics of each model, as well as important issues related to the use of these models for validating therapeutic targets in HD. The general principles discussed in this chapter will apply not only to the few HD mouse models discussed here but also to other future mouse models that may be used for HD preclinical studies.

VALIDATING PRECLINICAL MOUSE MODELS OF HD

The rationale for generating mouse genetic models of HD is based on the belief that these models can recapitulate, at least in part, some of the key disease phenotypes and/or pathogenic mechanisms that occur in patients; therefore, positive preclinical studies in these models may predict positive outcomes in clinical trials. In

TABLE 7.1

Summary of Preclinical Mouse Models of Huntington's Disease

Genetic Construct	R6/2	N171-82Q	Hdh[111]	CAG140	Hdh[CAG150]	YAC128	BACHD
Types of model	Transgenic	Transgenic	Knockin	Knockin	Knockin	Transgenic	Transgenic
Promoter	Human *HTT*	Murine prion	Murine Hdh	Murine Hdh	Murine Hdh	Human *HTT* locus	Human *HTT* locus
PolyQ repeat	150 CAG	82 CAG	111 CAG	140 CAG	150 CAG	128 CAG	97 CAA/CAG
PolyP region	Human	Human	Human	Human	Mouse	Human	Human
Protein context	Amino acids 1–82	Amino acids 1–171	Murine fl-Hdh	Murine fl-Hdh	Murine fl-Hdh	Human fl-mhtt	Human fl-mhtt
Protein expression level (Relative to endogenous Hdh)	75%	20%	50% or 100%	50% or 100%	100%	75%	150%
Repeat stability	Unstable	Unstable	Unstable	N.D.	N.D.	N.D.	Stable
Motor phenotypes							
Open field	+++ (8 wk.)			++ (4 wk.)	++ (8 wk.)	++ (8 wk.)	++ (8 wk.)
Rotarod	+++ (10–12 wk.)				100 wk.	++ (24 wk.)	+++ (24 wk.)
Grip strength	+++ (10–12 wk.)				100 wk.		
Gait			104 wk.	52 wk.			
Wheel running	+++ (4.5–5.5 wk.)						
Climbing	+++ (4.5–5.5 wk.)						
Cognitive phenotypes							
Reversal learning	++					++ (8 wk.)	
Morris water maze	++						
Instrumental learning							++ (24 wk.)
PPI	++					++ (52 wk.)	
Anxiety	++						++

continued

TABLE 7.1 (continued)
Summary of Preclinical Mouse Models of Huntington's Disease

Genetic Construct	R6/2	N171-82Q	Hdh^{111}	CAG140	Hdh^{CAG150}	YAC128	BACHD
Neuropathology							
Selective neuropathology	Nonselective	Nonselective	Selective	Selective	Selective	Selective	Selective
Brain weight	20% (12 wk.)	N.D.	N.D.	N.D.	N.D.	10% (52 wk.)	14% (52 wk.)
Striatal volume					40% (100 wk.)	15% (52 wk.)	28% (52 wk.)
Cortical volume						7% (52 wk.)	32% (52 wk.)
Striatal cell loss (stereology)					40% (100 wk.)	18% (52 wk.)	
Striatal dark neurons	+	+	3.5% (104 wk.)	N.D.	N.D.	N.D.	10–15% (52 wk.)
Mutant htt aggregates	+++	+++	++	++	++	++	+
Gliosis	+++	++					
Other phenotypes/ biomarkers							
Early lethality	12–13 wk.	—	—	—	—	—	
Body weight	Loss (7 wk.)	Loss (12 wk.)			Loss (70 wk.)	Gain (8 wk.)	Gain (8 wk.)
8-OHDG increase	+++						
NMR Spec./decrease NAA	+++						
Study duration	12–14 wk.	24 wk.			100 wk.	52 wk.	52 wk.
References	Hickey et al. (2005); Mangiarini et al. (1996)	Andreassen et al. (2001); Schilling et al. (1999)	Wheeler et al. (2000, 2002)	Menalled et al. (2003)	Heng et al. (2007); Lin et al. (2001); Woodman et al. (2007)	Slow et al. (2003); Van Raamsdonk et al. (2007)	Gray et al. (2008)

generating and analyzing these models, a key question that needs to be addressed is how valid these models are in recapitulating the salient genetic and clinical features of the human disease (Watase and Zoghbi, 2003). In this section, we will discuss some of the factors one should consider to determine the validity of a mouse genetic model of HD.

Construct Validity

Construct validity refers to the similarity of the *genetic context* used to develop the mouse model to that found in HD patients (Fleming et al., 2005). The genetic context should be viewed at multiple levels. At the DNA level, one should consider whether the types of mutations introduced (i.e., CAG repeat length) resemble those in the patients. In this regard, the targeted insertion of expanded CAG repeat and/or mhtt exon 1 into the endogenous murine huntingtin disease homolog (*Hdh*) locus in the mouse genome (i.e., knockin models) is the most genetically precise model of HD (Gusella and Macdonald, 2006). In the transgenic models of HD (with random insertion of transgenes into the mouse genome), one also needs to consider whether the genomic DNA context of the mutant transgene is similar to that in the patients. At the RNA level, one should consider whether the regulation of mhtt transcription is similar to the endogenous human htt, whether it is driven by an exogenous promoter, and whether endogenous mhtt mRNA processing such as splicing is preserved (i.e., models with an intact htt genomic locus). Finally, at the protein level, one needs to consider how similar the expressed mhtt protein in a given mouse model is to that in patients. Any divergence from the full-length human mhtt protein sequence may also reduce the construct validity of a model. For example, mouse models only expressing mhtt N-terminal fragments may have reduced construct validity at the protein level. Moreover, because the murine Hdh differs from its human counterpart in the polyproline region (polyP, a known protein–protein interaction domain of htt) and in another 273 amino acids outside the polyQ and polyP regions, these amino acid differences may subtly modify the disease processes mediated by the mutant Hdh compared with those by human mhtt.

Face Validity

Face validity refers to whether the overall phenotypes of a disease mouse model, both behavioral deficits and neuropathology, faithfully recapitulate those in HD patients. As described in more detail below, the phenotypes of the current preclinical HD mouse model can be studied at multiple levels to fully uncover its disease-relevant phenotypes. Commonly used behavioral phenotyping tools in HD mice include longitudinal behavioral observations (i.e., SHIRPA) (Rogers et al., 1997); motor behaviors such as rotarod performance (the ability to run a rotating wheel) and open field activity (automated measurement of spontaneous activity in an open field); cognitive tests that include a variety of learning tasks that may depend on the hippocampus or the corticostriatal circuit, such as swimming T mazes (Holmes et al., 2002) and instrumental conditioning (Balleine, 2005); and psychiatrically related behaviors such as anxiety and depressive behaviors and prepulse inhibition (PPI) (Arguello and Gogos, 2006; Holmes et al., 2002). Commonly used neuropathological studies in HD mice include stereological measurement of striatal and cortical volume and stereological counting of the striatal neurons (Menalled et al., 2002), counting striatal degenerating dark neurons as measured by toluidine blue staining of semithin brain sections (Gray et al., 2008; Hodgson et al., 1999; Turmaine et al., 2000), gliosis (Gu et al., 2005), and microgliosis (Simmons et al., 2007). In addition, weight loss and premature death are also used as secondary readouts in several preclinical mouse models. Each mouse model of HD, before entering the preclinical studies, has been extensively characterized to determine the extent of its phenotypic similarities to HD. Those models with relatively early-onset, progressive, and robust disease-like phenotypes (Table 7.1) are now being pursued for preclinical studies in HD.

Predictive Validity

An ultimate test for the validity of an HD mouse model is its *predictive validity*, which refers to whether key mechanistic findings and therapeutic efficacy in a given HD mouse model can predict positive clinical outcomes in human patients and *vice versa*. Currently, there is no effective disease-modifying therapy available for HD that can be used to validate the existing HD mouse models. Conversely, because the mouse genetic models of HD are widely used to identify pathogenic mechanisms and to test preclinical candidates, future clinical studies in HD patients using therapeutic leads originating from the mouse studies may be crucial to reveal the predictive validity of these models.

The concepts of construct validity, face validity, and predictive validity will be crucial to guide us in appreciating the strengths but also recognizing the weaknesses of the existing preclinical mouse models of HD. They may also be helpful in choosing the appropriate mouse models for preclinical studies. In the following section, we will describe the three major types of HD mouse models that are currently used or being prepared for use in HD preclinical studies. They include those expressing a small toxic N-terminal fragment of mhtt, those with precise genetic modification of the endogenous murine Hdh, and those human genomic locus transgenic mice expressing full-length human mhtt under its endogenous regulatory elements (Table 7.1). We will describe for each model its genetic construct and key phenotypic outcomes that may be used in preclinical studies.

N-TERMINAL MUTANT HUNTINGTIN FRAGMENT MODELS

R6/2 Mice

Genetic Construct and Expression

The R6/2 model is the first and one of the most influential HD mouse models for preclinical studies to date (Mangiarini et al., 1996) (Table 7.1). It was created on the (C57BL/6J × CBA) F1 mixed genetic background. R6/2 mice carry about 1 kilobase (kb) of the human huntingtin promoter, which drives the expression of mhtt exon 1 with ~150 CAG repeats. The transgene also contains ~262 base pair (bp) of human htt intron 1 sequence after the exon 1. R6/2 mice express the *mhtt-exon 1* transgene at about 75% of the endogenous Hdh level (Mangiarini et al., 1996). One issue

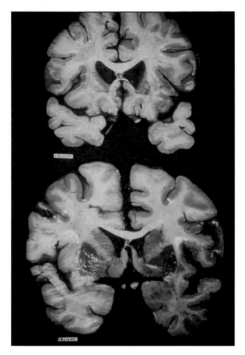

FIGURE 1.3 Gross neuropathology of HD. The top section is from an HD subject, whereas the lower section is from a control without neurologic disease. There is marked striatal atrophy with corresponding ventriculomegaly. There is also diffuse thinning of the cortical mantle. Although sections are not at precisely the same level, diffuse loss of cerebral volume, including loss of white matter, is evident in the HD specimen. (Courtesy of Dr. Andrew Lieberman, Department of Pathology, University of Michigan, Ann Arbor, MI.)

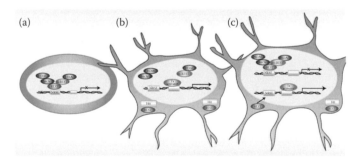

FIGURE 2.2 Model for polyglutamine-expanded huntingtin transcription interference of *BDNF* gene expression. Promoters containing a neuron-restrictive silencing element (a) are bound by transcription repressors that prevent expression of downstream genes, such as *BDNF*, in non-neural cell types. In normal neurons (b), normal huntingtin protein is localized to the cytosol and binds the transcription repressor REST there, allowing *BDNF* gene expression to occur. However, in HD neurons (c), polyglutamine-expanded huntingtin protein accumulates in the nucleus and does not sequester REST in the cytosol. REST thus inappropriately enters neuronal nuclei and represses neuronal expression of *BDNF* in HD. (From Thompson, L. M., *Nat Genet* 35, 13, 2003. With permission.)

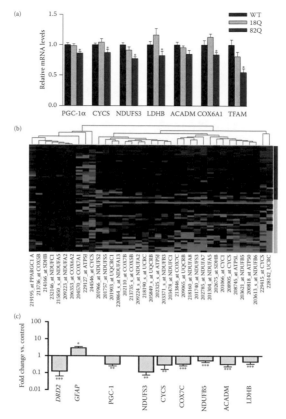

FIGURE 2.3 PGC-1α transcription interference in HD mice and HD patients. (a) Real-time RT-PCR analysis of striatal RNAs from HD 82Q mice (red), 18Q mice (gray), and wild-type mice (black) reveals decreased mitochondrial gene expression in the HD mouse model. (b) Microarray expression analysis of PGC-1α-regulated genes in human caudate. Here we see a heat map comparing the caudate nucleus expression of 26 PGC-1α target genes for 32 Grade 0–2 HD patients (adjacent to gold bar) and 32 matched controls (adjacent to blue bar). Most PGC-1α target genes are down-regulated. (c) Confirmation of expression reduction of PGC-1α-regulated genes in human caudate. We measured RNA expression levels for six PGC-1α targets (NDUFS3, CYCS, COX7C, NDUFB5, ACADM, and LDHB), PGC-1α, and two control genes (*GFAP* and *DRD2*). In this way, we confirmed significant expression reductions in PGC-1α targets and detected reduced PGC-1α in human HD striatum from early-grade patients. Statistical comparisons were performed with the *t*-test (*, $p < 0.05$; **, $p < 0.005$; ***, $p < 0.0005$). (From Weydt, P., *Cell Metab* 4, 349, 2006. With permission.)

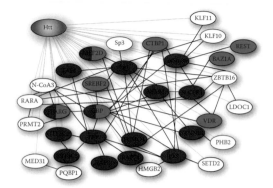

FIGURE 3.1 The interactions among transcription-related proteins that interact with Htt. As discussed in the text, evidence suggests the interaction of Htt with some partners is likely to be beneficial to the cell (green), whereas other interactions are most likely to be detrimental (red). Conflicting data exist for some of the proteins (half red, half green). (From the "Interactions" section of the online Entrez Gene database [http://www.ncbi.nlm.nih.gov/entrez/query.fcgi?db=gene; Maglott et al., 2005] of the National Center for Biotechnology Information website. The original publications describing each interaction can be viewed using the "Interactions" section "PubMed" links next to the appropriate interaction partner. Other sources for interactions not listed in Entrez Gene are referenced in the text.)

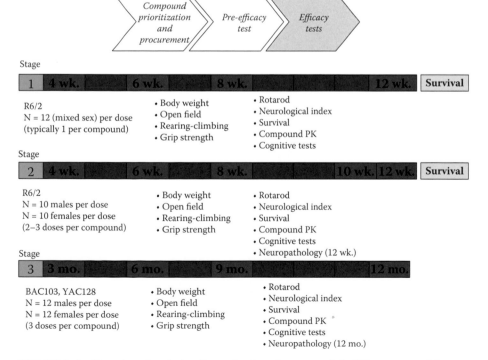

FIGURE 4.4 Staged testing design for evaluation of compounds in HD mouse models.

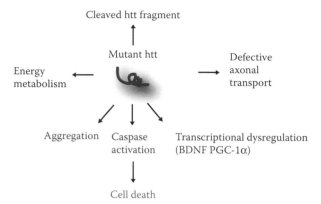

FIGURE 5.3 Multiple mechanisms implicated in HD pathology. The ones highlighted in red text are mechanisms that have been targeted in HTS assays, whereas the others are potential mechanisms for HTS assays.

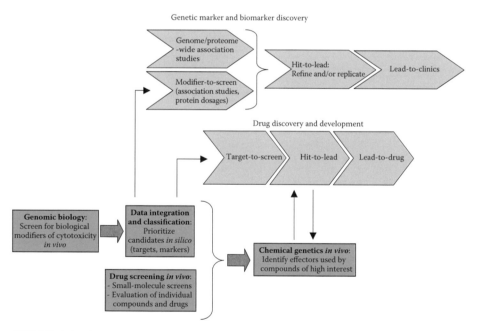

FIGURE 6.3 Implementation of *in vivo* biology and systems biology approaches into the drug and marker discovery process. Ways to combine invertebrate genetics and systems biology with the traditional components of drug and marker discovery are shown.

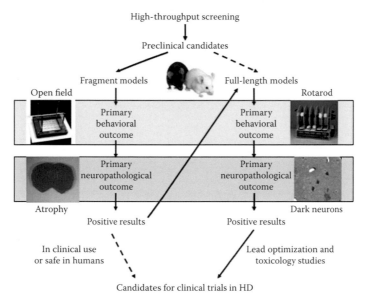

FIGURE 7.1 A dual-model scheme for preclinical studies using both the fragment models and full-length models of HD.

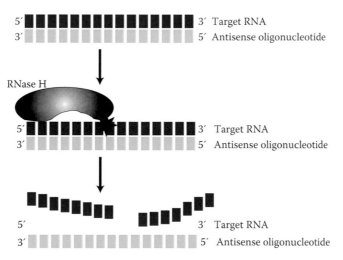

FIGURE 9.1 The most prominent mechanism of antisense oligonucleotide gene silencing—induction of RNase H endonuclease activity. The antisense oligonucleotide (typically a deoxyribonucleotide) binds the target RNA to form the heteroduplex substrate. RNase H binds via its binding domain at the 3′ antisense oligonucleotide/5′ RNA pole and cleaves the target RNA approximately 7 base pairs from its binding site. The target mRNA is degraded, whereas the antisense oligonucleotide remains intact, allowing it to form another heteroduplex substrate for induction of RNase H cleavage.

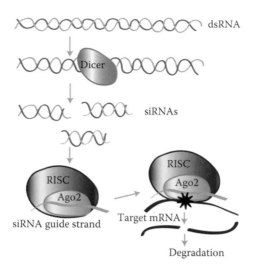

FIGURE 9.2 siRNA-mediated RNAi. dsRNAs are bound by the enzyme Dicer and cleaved by Dicer's catalytic component (RNase III) into siRNAs—21-nucleotide dsRNA with 2-nucleotide 3′ overhangs. With assistance of R2D2, the guide strand is loaded into the RISC. Within RISC, the guide strand binds to Ago2. After complementary base pairing between siRNA guide strand and target mRNA, the endonuclease Ago2 cleaves the target mRNA at the scissile phosphate. The 5′ end of the target is degraded in the exosome, and the 3′ end is degraded in the cytoplasm. Ago2 is a multiple turnover enzyme, and one Ago-bound guide strand directs cleavage of hundreds of target mRNAs.

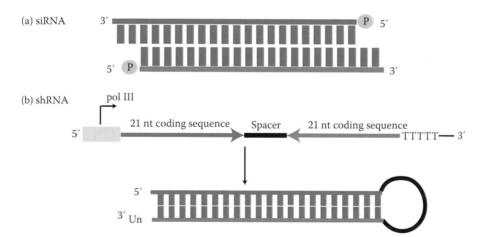

FIGURE 9.3 (a) siRNAs: siRNAs can be synthesized to mimic endogenous RNase III Dicer cleavage products—21-nucleotide dsRNA with 5′ phosphates and 2-nucleotide 3′ unphosphorylated overhangs. (b) shRNAs: vectors introduced into the cell contain the DNA template with the pol III promoter, which drives expression of the ssRNA transcript through the polyT termination sequence. The sense and antisense strand on the ssRNA associate in cis to form the shRNA—a dsRNA separated by a loop (Paul et al., 2002; Sui et al., 2002; Yu et al., 2002). The shRNA is processed by Dicer into an siRNA (Dykxhoorn et al., 2003).

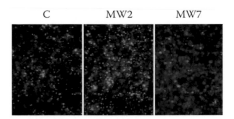

FIGURE 10.3 MW7 prevents whereas MW2 promotes aggregation of mutant HDx1-enhanced green fluorescent protein (EGFP) in PC12 cells. MW7 and MW2 cDNAs were cloned into ecdysone-inducible vectors and transfected into PC12 cells that were engineered to express HDx1 in response to ecdysone [26]. Selected PC12 cell clones were then treated with ecdysone to induce simultaneous expression of HDx1 and the scFv. A luciferase construct was used as control (far left panel).

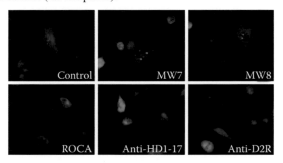

FIGURE 10.5 The anti-huntingtin antibodies/intrabodies, anti-N1–17, MW7, and MW8, stain living striatal cells with a punctate pattern (red) similar to an anti-dopamine D2 receptor (D2R) antibody. The striatal ST-14 cell line was transduced with HDx1-enhanced green fluorescent protein (EGFP) (PQ103) lentivirus, and live cells were incubated with either control antibodies (mouse Ig2b) and a non-neuronal anti-CD9 (ROCA) or anti-Htt antibodies/intrabodies as indicated. A polyclonal antibody against D2R was used as positive control for cell surface staining. Alexa 568-conjugated secondary antibody was used to visualize staining (red); the green fluorescence is native HDx1-EGFP.

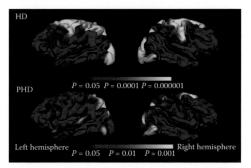

FIGURE 11.1 Cortical changes occur early and are extensive in HD. Top, maps of statistically significant thinning in 33 patients with early HD compared with age- and sex-matched subjects. Bottom, maps of statistically significant thinning during the premanifest period, a total of 35 subjects. The most yellow areas correspond to loss of the cortical gray matter more than 20% thinner than controls. The distribution of cortical change and their correspondence with clinical symptoms suggest that cortical atrophy may play an important role in the clinical manifestations of HD.

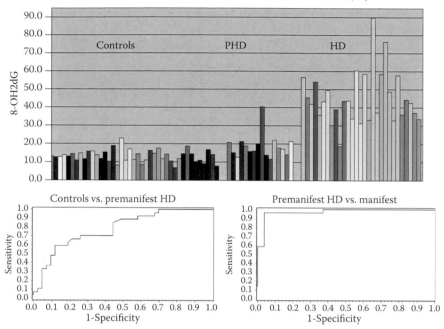

FIGURE 11.2 Individual 8-hydroxy-2-deoxyguanosine (8-OHDG) levels in plasma from age- and gender-matched controls (mean = 13.5), premanifest HD (PHD) subjects (mean = 18.1), and early manifest HD subjects (mean = 45.3). Levels are higher in PHD subjects than in controls ($P = .003$) and almost fourfold higher in manifest HD subjects ($P = 5.013E-13$). Partial discrimination of controls from premanifest HD subjects (area under the receiver operating characteristic curve = 0.784) suggests that 8-OHDG elevates during the HD prodrome. Complete discrimination of premanifest HD from manifest HD subjects (area under the receiver operating characteristic curve = 0.973) suggests that 8-OHDG elevations accelerate as individuals become symptomatic. 8-OHDG has potential as a biomarker of a disease activity (DNA damage), as a diagnostic biomarker of phenoconversion, and as a pharmacodynamic biomarker because it can be suppressed by antioxidant and energy-buffering treatments.

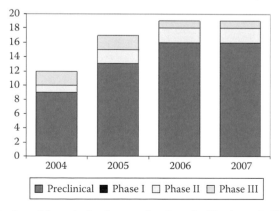

FIGURE 12.1 Number of drugs in development by phase for Huntington's disease, 2004–2007.

with R6/2 is its genetic heterogeneity as a result of the CAG repeat instability in the germ line, which can range from low 100s to >300 repeats. Such dramatic repeat alterations may significantly alter R6/2 mouse phenotypes. Therefore, in a preclinical study, the repeat length for the R6/2 mice should be closely monitored to avoid genetic drift of the repeat lengths during the study.

Behavioral Phenotypes

The earliest *motor deficits* in R6/2 mice were detected at 4.5 weeks and include decreased running wheel activities in mice individually housed, climbing a wired cylinder, and hypoactivity in the open field test (Hickey et al., 2005). At 7–8 weeks, grip strength and rotarod deficits can be readily detected, and these deficits increase until the animal's premature death around 16 weeks of age (Carter et al., 1999; Lione et al., 1999; Murphy et al., 2000). In addition, R6/2 mice also exhibit a variety of other motor phenotypes, including changes in circadian rhythms (Morton et al., 2005), clasping (a dystonic posture when the mice were suspended by the tail), tremor, involuntary jerky movements, and seizures (Li et al., 2005; Mangiarini et al., 1996).

Before the onset of overt motor deficits (at 8 weeks of age), R6/2 mice already exhibit a variety of *cognitive deficits* reminiscent of HD patients. For example, HD patients exhibit perseverance of learned tasks indicating frontal-striatal inhibition deficits (Aron et al, 2003). R6/2 mice also demonstrate difficulty in reversing pre-learned tasks in the two-choice swim tank, T maze (Lione et al., 1999), and an automated video-based reversal task (Morton et al., 2006). R6/2 mice also exhibit deficits in spatial learning in the Morris water-maze task (Murphy et al., 2000) and in contextual fear conditioning (Bolivar et al., 2003). In the psychiatrically related deficits, R6/2 mice exhibit impairment in PPI, a sensorimotor gating abnormality also seen in the HD patients (Swerdlow et al., 1995) and in patients with schizophrenia (Swerdlow et al., 1994).

Neuropathology

R6/2 mice demonstrate robust brain atrophy (about 20% brain weight loss at 12 weeks). Striatal atrophy in these mice can be readily measured using unbiased stereology (Li et al., 2005). R6/2 mice exhibited only moderate dark neuron degeneration in the late stage (Mangiarini et al., 1996; Turmaine et al., 2000), but the striatal neuronal atrophy can be readily quantified as a preclinical outcome. Another frequently used pathological endpoint in R6/2 mice is the presence of mhtt-containing nuclear inclusions (NIs) and neuropil aggregates (NAs), some of which can be detected with antibodies against ubiquitin or against the N-terminal region of huntingtin, for example, EM48 antibody (Davies et al., 1997; Gutekunst et al., 1999; Li et al., 1999). The aggregates in this model are much more widespread than those observed in patients using the same EM48 antibody (DiFiglia et al., 1997; Gutekunst et al., 1999). Moreover, because the significance of mhtt aggregates to the pathogenesis of HD neuronal dysfunction and degeneration remains unclear (Bates, 2003), and there is accumulating evidence suggesting that the large nuclear inclusions may be neuroprotective (Arrasate et al., 2004), mhtt aggregates should not be used as the sole pathological outcome in a preclinical study.

Survival and Weight Loss

Two commonly used preclinical outcomes in R6/2 mice are weight loss and premature death. Because these phenotypes can be readily modified by environmental enrichment (Hockly et al., 2002), which may confound any therapeutic study, it is now recommended that standardized living conditions with a moderate level of enriched environment should be implemented for this model (Hockly et al., 2003). Because the underlying etiology for weight loss and death in R6/2 mice remains unclear, these phenotypes are often only used as surrogate markers to support findings from the primary behavioral and neuropathological outcomes (Bates and Hockly, 2003; Hockly et al., 2003).

Other Phenotypes

R6/2 mice also have a set of molecular readouts, such as gene expression changes, that could be readily used in preclinical studies. A recent microarray analysis comparing gene expression changes in HD mice and HD patients revealed that many of the gene expression changes seen in the R6/2 striatum significantly overlap with those occurring in patients (Kuhn et al., 2007). This result supports the idea of using gene expression signatures as a biomarker for preclinical efficacy in R6/2 mice. Other promising biomarkers in R6/2 mice include using NMR spectroscopy to detect a robust (53%) nonlinear decrease in *in vivo N*-acetyl aspartate (NAA) levels beginning at 6 weeks of age, a brain energetic deficit also seen in patients (Jenkins et al., 2000). Finally, 8-hydroxy-2-deoxyguanosine (8-OHDG), a biomarker for oxidized DNA damage, which is elevated in HD patient serum (Hersch et al., 2006), is also elevated in R6/2 and R6/1 mice (a model with similar construct to R6/2) (Bogdanov et al., 2001; Kovtun et al., 2007). These latter biomarkers (NAA and 8-OHDG), once validated in HD patients, may be used as relatively noninvasive and unbiased endpoints for preclinical and clinical studies in HD (Hersch et al., 2006).

N171-82Q Mice

Genetic Construct

Another commonly used N-terminal fragment mouse model of HD in preclinical studies is the N171-82Q mouse model. N171-82Q mice express an N-terminal fragment of human htt with 82 polyglutamine repeats and the first 171 amino acids of htt in all the neurons of the brain driven by a mouse prion protein promoter (Schilling et al., 1999). The N171-82Q line was created on the C3H/HeJ × C57BL/6J background. This model expresses mhtt fragment at about 20% of levels of the endogenous murine Hdh protein (Schilling et al., 1999).

Behavioral Phenotypes

N171-82Q mice exhibit progressive motor deficits starting at 12 weeks of age, including tremors, uncoordination, hypoactivity, abnormal gait, and clasping (Schilling et al., 1999). These symptoms are progressive until premature death at approximately 5–6 months of age. The most used quantifiable motor phenotype in this model is rotarod impairment, beginning at approximately 11 weeks of age (Andreassen et al., 2001; Schilling et al., 2004b). The behavioral deficits in N171-82Q mice appear to be more variable compared with R6/2 mice; thus, to attain adequate statistical power, a typical preclinical study with N171-82Q mice would require a large number of mice (i.e., minimum of 20) than a study with R6/2 (Hersch and Ferrante, 2004).

Neuropathology

N171-82Q mice have several pathological features mimicking HD. First, they exhibit brain weight loss at 120 days of age (Andreassen et al., 2001; Hersch and Ferrante, 2004), striatal atrophy (as measured by stereological measurement of the ventricular volume), and striatal neuronal atrophy (Gardian et al., 2005). Moreover, these mice exhibit other neuropathological features more similar to HD patients, including more mhtt aggregates in the cortex than in the striatum, robust reactive gliosis at 4–5 months of age, and apoptotic neuronal degeneration can be detected in the cortex and striatum at 4.5 months of age (Yu et al., 2003).

Survival and Weight Loss

Similar to R6/2 mice, N171-82Q mice also exhibit progressive weight loss (beginning at approximately 90 days of age) and shortened lifespan (average survival of about 130 days). These phenotypes are frequently used as surrogate markers of therapeutic efficacy in the preclinical studies. Because environmental enrichment can also significantly modify these phenotypes (Schilling et al., 2004a), standardized housing conditions are critical in preclinical trials using this model.

Other Phenotypes

N171-82Q mice exhibit hyperglycemia (i.e., increased baseline fasting glucose level and impaired glucose tolerance test) at about 80 days of age, which could be used as an outcome measure (Andreassen et al., 2001). The utility of the test is relatively limited because of its unclear etiology and high variability.

Full-Length Murine Huntington's Disease Homolog KI Mouse Models

Knockin (KI) models of HD were generated by targeting an expanded polyglutamine repeat and/or adjacent human mhtt exon 1 sequences (including the polyP region) to replace the corresponding sequences in the endogenous murine Hdh; hence the mutant Hdh is expressed from the endogenous *Hdh* locus in a manner similar to the expression of mhtt in patients. Thus, Hdh-KI mice are generally considered the most precise genetic mouse model of HD (Gusella and Macdonald, 2006). Existing Hdh-KI mice carry expanded mutant CAG repeats up to 150 (Menalled, 2005). Compared with the mhtt N-terminal fragment models, Hdh-KI mice exhibit slow progression and relatively mild phenotypes, and their lifespans are usually normal. Three Hdh-KI models—Hdh^{Q111}, CAG140, and Hdh$^{(CAG)150}$—are currently being developed for preclinical studies and will be discussed in more detail here.

Hdh^{Q111} Mice

Genetic Construct

The Hdh^{Q111} model is a KI mouse model of HD generated on a CD1 × 129Sv background. This model targeted a chimeric murine Hdh/human mhtt exon 1 into the endogenous *Hdh* locus, and the human mhtt portion includes 111 CAG repeat and human polyP region (Wheeler et al., 1999).

Behavioral Phenotypes

The behavioral phenotypes of Hdh^{Q111} are very mild. Gait abnormalities are detected at 24 months of age, and no rotarod, clasping, or open field abnormalities were detected in heterozygous or homozygous Hdh^{Q111} mice up to 17 months of age (Wheeler et al., 2002).

Neuropathology

Hdh^{Q111} homozygous mice display selective accumulation of nuclear mhtt at 2.5 months of age, as stained by EM48 antibody (Wheeler et al., 2000). This phenotype progresses as EM48-positive puncta are detected at 5 months of age. Nuclear inclusion formation is first detected at 10 months of age (Wheeler et al., 2000, 2002). At 24 months, these mice also demonstrate reactive gliosis in the striatum and a small number (about 3.5%) of striatal dark neurons as stained by toluidine blue (Wheeler et al., 2002). Brain atrophy has not been reported in this model.

Other Phenotypes

Hdh^{Q111} mice exhibit a variety of molecular phenotypes, including but not limited to increased expression of a ribosomal signaling protein Rrs1, somatic Hdh CAG-repeat instability in the striatum, and low mitochondrial ATP levels (Fossale et al., 2002; Lloret et al., 2006; Seong et al., 2005; Wheeler et al., 2003).

CAG140 Mice

Genetic Construct

A second preclinical Hdh-KI mouse model has targeted replacement of the endogenous murine Hdh exon 1 with a chimeric mouse and human exon 1 with 140 CAG repeats (Menalled et al., 2003). They were generated on a 129Sv × C57BL/6J background.

Behavioral Phenotypes

CAG140 mice exhibit early hyperactivity as measured by an increase in rearing at 1 month of age, followed by hypoactivity at 4 months of age (Menalled et al., 2003). This pattern of locomotor activity changes is similar but with an earlier onset compared with another Hdh-KI model, CAG94 (Menalled et al., 2002). CAG140 mice exhibit gait abnormality at 12 months of age, with a decrease in stride length (Menalled et al., 2003).

Neuropathology

CAG140 mice exhibit selective striatal and cortical nuclear staining of mhtt microaggregates by EM48-positive antibody (Menalled et al., 2003). Nuclear inclusions are present in the striatum at 4 months of age and do not appear in the cortex until 6 months of age. At this age, nuclear staining and nuclear microaggregates are also seen in the cerebellum, an area relatively spared in HD. Neuropil microaggregates are also present in the striatum, globus pallidus, and piriform cortex at 2 months of age and increase with age. To date, no overt cell loss or brain atrophy has been reported for this model. However, CAG94, which is a model very similar to CAG140, exhibits significant striatal atrophy (15%) but no neuronal loss at 18–26 months of age as measured by unbiased stereology (Menalled et al., 2002). Therefore, it is very likely that CAG140 may exhibit significant striatal atrophy phenotype that could facilitate its utility in a preclinical study.

Hdh$^{(CAG)150}$ Mice

Genetic Construct

The third preclinical Hdh-KI model is Hdh$^{(CAG)150}$, which replaced the short CAG repeat in the murine Hdh exon 1 with a repeat of 150 CAG (Lin et al., 2001). These mice, created in a 129/Ola × C57BL/6J background, were analyzed either in this mixed genetic background (Heng et al., 2007) or in a (CBA × C57BL/6) F1 background generated from C57BL/6 and CBA congenic lines (i.e., backcrossed for >12 generations) (Woodman et al., 2007). The latter breeding strategy ensures the homogeneity of the genetic background in the mutant and control mice (Silva, 1997).

Behavioral Phenotypes

Homozygous Hdh$^{(CAG)150}$ mice were extensively studied using a battery of behavioral paradigms that revealed several slowly progressive motor abnormalities (Heng et al., 2007; Woodman et al., 2007). In the mixed genetic background, a mild exploratory deficit is present in Hdh$^{CAG(150)}$ mice at 70 weeks of age, with more robust deficits seen at 100 weeks (Heng et al., 2007). In the rotarod test, homozygous mice display a significant deficit on the rotarod at 40 weeks (Lin et al., 2001) and 100 weeks of age (Heng et al., 2007). At the latter time point, these mice also exhibit gait abnormalities, difficulty in traversing balance beams, and clasping.

In the (CBA × C57BL/6) F1 background (Woodman et al., 2007), homozygous Hdh$^{CAG(150)}$ mice also have onset of significant and progressive rotarod deficits at 18 months of age. A grip-strength deficit was observed at 6 months of age, but it is nonprogressive until 10 months of age. No locomotor deficits were detected in the mutant mice up to 18 months of age, suggesting this phenotype may depend heavily on the genetic background of the mutant mice.

Neuropathology

Mutant huntingtin aggregates can be detected in the striatum and hippocampus in the homozygous Hdh$^{CAG(150)}$ mice at 6 months of age and become widespread in the brain at 10 months of age (Tallaksen-Greene et al., 2005). Similar to what was observed in the patients, the recent study reveal that at 100 weeks, Hdh$^{CAG(150)}$ homozygous mice exhibit reduced striatal D1 and D2 dopamine receptor ligand binding and reduced dopamine transporter (Dat) ligand binding (Heng et al., 2007). Using stereological counting of NeuN(+) striatal neurons, the same group found that the Hdh$^{CAG(150)}$ mice exhibit 40% reduction in striatal volume and 43% reduction in striatal NeuN(+) neurons. This latter stereological result is very promising to use striatal atrophy and/or neuronal loss as preclinical outcomes in this model. However, because these results were obtained from a relatively small sample size, it would be important for the study to be replicated using a much larger sample size, and preferably also in recombinant inbred F1 background.

Survival and Body Weight

Hdh$^{CAG(150)}$ mice do not have a shortened lifespan. However, in both mixed and F1 genetic backgrounds, the mutant mice exhibit significant and progressive weight loss at 70 weeks of age (Heng et al., 2007) and at 52 weeks in the pure F1 background (Woodman et al., 2007).

Molecular Phenotypes

Brain gene expression profiling in the Hdh$^{CAG(150)}$ F1 mice revealed parallel changes in 22-month-old Hdh$^{CAG(150)}$ mice and 12-week-old R6/2 mice (Woodman et al., 2007). Decreases in the expression of chaperones (i.e., heat shock protein 70, heat shock cognate 70) can be confirmed by Western blots in both models. Such parallels in the molecular phenotypes between the two different types of HD mouse models support the further exploration of gene expression changes as biomarkers in preclinical studies.

FULL-LENGTH HUMAN HUNTINGTIN TRANSGENIC MOUSE MODELS

Multiple full-length human huntingtin transgenic mouse models (fl-mhtt Tg) have been generated to study the dominant toxicity elicited by expanded polyglutamine repeat in the context of full-length mhtt (fl-mhtt). This group of fl-mhtt transgenic models include one cDNA transgenic mouse driven by the CMV promoter (Reddy et al., 1998) and two genomic transgenic models expressing fl-mhtt from the human genomic locus on a yeast artificial chromosome (YAC) (Hodgson et al., 1999; Slow et al., 2003; Van Raamsdonk et al., 2007) or on a bacterial artificial chromosome (BAC) (Gray et al., 2008). As a group, the fl-mhtt Tg mouse models demonstrate earlier and relatively robust motor deficits (i.e., rotarod deficits), as well as selective atrophy and/or neurodegeneration in the striatum and cortex. Therefore, fl-mhtt Tg mouse models, particularly YAC and BAC models, are currently being used or developed as preclinical mouse models in HD. In this review, we will focus our discussion on the two latter models, YAC128 and BACHD.

YAC128 Mice

Genetic Construct

A series of YAC transgenic models of HD expressing full-length human mhtt with 18, 46, 72, and 128 CAG repeats (i.e., YAC18, YAC46, YAC72, and YAC128) were generated in the inbred FvB background (Hodgson et al., 1999; Slow et al., 2003; Van Raamsdonk et al., 2007). In these models, the YAC transgenes carry the entire 170-kb genomic locus of human htt gene plus 25 kb of the 5′ flanking sequence and about 120 kb of 3′ flanking sequence. YAC128 mice are the latest of the series and exhibit by far the most robust phenotypes among all the YAC models; therefore, YAC128 is used as a preclinical model in HD (Slow et al., 2003). At the protein level, the YAC128 line expresses human mhtt at about 75% of the level of the endogenous murine Hdh (Slow et al., 2003).

Behavioral Deficits

YAC128 mice exhibit hyperactivity at 2 months of age and hypoactivity at 8–12 months of age. They also exhibit rotarod deficit initially at 4 months of age but become more prominent at 6 months of age (Graham et al., 2006; Van Raamsdonk et al., 2005c). YAC128 mice demonstrate a variety of cognitive deficits, including motor learning deficits on rotarod beginning at 2 months, a reversal learning deficit in a swimming T maze beginning at 2 months, PPI deficits at 12 months, open-field habituation test at 9 months, and a linear swimming test at 8 months (Van Raamsdonk et al., 2005e).

Neuropathology

The earliest pathological marker in YAC128 mice is the selective nuclear localization of EM48-positive mhtt in the striatum at 1–2 months of age, which intensifies at 3 months of age (Van Raamsdonk et al., 2005a). EM48-positive mhtt nuclear accumulation is present in the cortex and hippocampus at 3 months of age, and EM48-positive nuclear inclusions appear in the striatum by 18 months of age (Van Raamsdonk et al., 2005a). The YAC model is the first HD model to demonstrate selective atrophy in the striatum and cortex but not in the cerebellum, a pattern reminiscent of that in HD (Van Raamsdonk et al., 2005a). Stereological studies reveal striatal volume loss first detected at 9 months of age, as well as a significant decrease in striatal volume (10%), cortical volume (8.6%), and globus pallidus (10%) at 12 months of age but no change in the hippocampus or cerebellar volumes (Van Raamsdonk et al., 2005a). The striatal volume loss is associated with a loss of NeuN-positive neurons (18%) (Van Raamsdonk et al., 2005a).

Survival and Body Weight

The lifespan of YAC128 mice is slightly decreased. YAC128 mice gain weight between 2 and 6 months of age (about 27%), and the weight gain occurs in all organs except the brain and testis, where mhtt is particularly toxic and causes weight loss (Van Raamsdonk et al., 2006). The weight gain phenotype appears to be caused by the overexpression of human htt because YAC18 mice overexpressing wild-type htt also exhibit similar weight gain phenotype. Because HD patients exhibit weight loss rather than weight gain, the weight gain phenotype in the YAC128 model is not a feature of the disease and should not be used as a preclinical endpoint in this model.

BACHD Mice

Genetic Construct

BACHD is a novel transgenic mouse model of HD generated and maintained in the FvB inbred background. BACHD mice express full-length human mhtt from its own regulatory elements on a 240-kb BAC, which contains the intact 170-kb human htt locus plus about 20 kb of 5′ flanking genomic sequence and 50 kb of 3′ sequence (Gray et al., 2008). The BAC was engineered to include an mhtt exon 1 containing a mixed CAA/CAG repeat encoding an intact polyglutamine stretch (Kazantsev et al., 1999; Yang et al., 1997). A recent more precise sizing of the BACHD CAA/CAG repeat by Laragen (Los Angeles, CA), using both direct sequencing and GeneMapper methods, shows that the BACHD CAA/CAG repeat length is 97. Unlike Hdh-KI and R6/2 mice, the CAA/CAG repeat length in BACHD mice appears stable in the germline over many generations. Another feature of the BACHD transgene design is the inclusion of two LoxP sites flanking mhtt-exon 1, which permits the selective removal of mhtt expression in any cell types expressing the Cre recombinase (Branda and Dymecki, 2004; M. Gray and X. W. Yang, unpublished data) Therefore, this model is particularly useful to study the cell autonomous toxicity and pathological cell–cell interactions elicited by mhtt (Gu et al., 2005). BACHD mice have five copies of the transgene integrated and express fl-mhtt protein at about 1.5- to 2-fold of the endogenous Hdh level. The expression of mhtt is in an endogenous pattern and functional because the BACHD transgene can rescue the embryonic lethality of the murine Hdh knockout mice (Gray et al., 2008; Zeitlin et al., 1995).

Behavioral Phenotypes

BACHD mice exhibit mild but significant rotarod deficits at 2 months, but repeated testing revealed that the rotarod deficits in these mice are progressive and become very pronounced at 6 and 12 months of age. At 6 months, BACHD mice also exhibit hypoactivity in the open field test. In the cognitive domain, a preliminary study reveals BACHD mice exhibit deficits in an instrumental reward learning paradigm, in which these mice are unable to efficiently learn the relationships between stimulus, action, and the rewarding outcomes (B. Balleine, unpublished data). The BACHD mice also exhibit significant impairment in acquiring the swimming T-maze task at 4–6 months of age (L. Menalled and D. Howland, personal communication). Finally, BACHD mice also exhibit several phenotypes in the psychiatric-like behavioral domain (Holmes et al., 2002) starting at about 6 months of age, including enhanced anxiety in the light-dark box and elevated plus maze, enhanced depressive-like behavior in forced swim test, and altered PPI (Menalled et al., 2009; M. Gray and X. W. Yang, unpublished data).

Neuropathology

At 6 months of age, BACHD mouse brains are indistinguishable from the wild-type controls and have comparable cortical and striatal volumes. At 12 months of age, BACHD brains are visibly atrophic with 20% reduction in the forebrain weight but normal cerebellar weight compared with the wild-type mice (Gray et al., 2008). Stereological measurement reveals a 28% reduction in the striatal volume and 32% reduction in the cortical volume compared with the wild types. Furthermore, using toluidine blue staining of striatal semithin sections, we found the 12-month-old BACHD mice contain about 15% dark degenerating neurons in the lateral striatum compared with only 0.3% in the control littermates. One distinguishing feature of BACHD mice from the other full-length huntingtin mouse models is the lack of early nuclear localization of EM48-positive mhtt in the striatum and cortex. Instead, at 12 months of age, EM48 staining reveals large and predominantly neuropil mhtt aggregates in the deep cortical layers and very few small aggregates in the striatum. This distribution pattern of mhtt aggregates is reminiscent of that in adult-onset HD (DiFiglia et al., 1997; Gutekunst et al., 1999), albeit the abundance of the aggregates in BACHD mice appears to be less than that in the patients.

Survival and Body Weight

BACHD mice have a normal lifespan compared with their wild-type littermates. Similar to YAC128 mice, BACHD mice also exhibit significant weight gain between 2 and 6 months and maintain the weight differential without further weight gain until 12 months. As discussed above, this phenotype may be related to the overexpression of human huntingtin. Importantly, rotarod performance and body weight at 6 months demonstrate that body weight in BACHD mice is not correlated with their poor rotarod performance; furthermore, a subset of BACHD mice within the normal weight range still exhibit robust rotarod deficits (Gray et al., 2008; Menalled et al., 2009). These results suggest that the neuronal dysfunction and degeneration in this model are related to the toxicity of mhtt in the brain and are unrelated to the body weight changes.

SUMMARY OF PRECLINICAL HD MOUSE MODELS

In this section, we have described the three major types of preclinical HD mouse models. These models are used in various preclinical studies because of their particular strengths in genetic and/or phenotypic characteristics. The mhtt fragment transgenic models exhibit early-onset and rapidly progressing behavioral and neuropathological phenotypes associated with significant weight loss and premature death. Furthermore, these models also exhibit distinct molecular changes (i.e., gene expression changes and 8-OHDG level) that are also replicated in the patients. The advantage of the full-length Hdh-KI and mhtt transgenic models is that they may better resemble the pathogenic mechanisms that occur in the patients (Gusella and MacDonald, 2006; Van Raamsdonk et al., 2007). The shared disadvantage of these models, compared with the fragment models, is that they are slowly progressive models, and quantifiable motor and pathological deficits usually do not occur until 6–12 months of age. Hdk-KI mice are the most precise genetic model of HD and hence are valuable to study molecular mechanisms and therapeutic interventions requiring such precise level of mutant Hdh expression in the endogenous genomic context. The motor and pathological phenotypes of the Hdh-KI mice are slowly progressing and mild, but recent studies of CAG140 (Menalled et al., 2003) and Hdh$^{(CAG)150}$ (Heng et al., 2007; Woodman et al., 2007) mice demonstrate that these models may exhibit motor and striatal atrophy phenotypes that are relatively robust and could be used as preclinical outcomes. Finally, the human fl-mhtt Tg mice, BACHD and YAC128, both exhibit progressive rotarod and open field deficits, late onset, and selective neuropathology reminiscent of HD. In these latter two models, rotarod deficits and striatal or cortical atrophy may be used as outcome measures in a preclinical study. In the next section, we will discuss how these HD mouse models can be used for preclinical studies in HD.

PRECLINICAL TRIALS WITH MOUSE MODELS OF HD

From a genetic perspective, HD is a relatively simple single gene disorder with full penetrance; hence the study of HD from the very beginning has been heralded as a model genetic brain disorder to uncover a rational route from etiology to treatment (Wexler et al., 1991). Our current understanding of HD pathogenic mechanisms remains incomplete, and a large number of pathogenic mechanisms are implicated in HD pathogenesis. These include proteolysis of mhtt to generate toxic mhtt fragments, transcriptional dysregulation, impairment of the proteasome, mitochondrial dysfunction and energetic deficit, excitotoxicity, deficits in vesicular trafficking/transport, and finally, mhtt toxic conformation change (i.e., aggregation) (Gatchel and Zoghbi, 2005; Li and Li, 2006). In addition to targeting to these specific mechanisms, many HD neuroprotective therapeutics are designed to target broadly the processes of cell death (i.e., apoptosis) or increase cellular "healthiness" (Beal and Ferrante, 2004). Based on the broad set of potential HD pathogenic mechanisms and even broader set of therapeutic strategies, a large number of preclinical candidates are being developed in the HD therapeutic pipeline, particularly from the high-throughput screening using cellular and small organism models (i.e., *C. elegans* or *Drosophila* models) (Hughes and Olson, 2001). Because only a small subset of these candidates can possibly be moved into human clinical trials, preclinical studies in the HD mouse models will play a central role in prioritizing the lead candidates for the very expensive next stage of drug discovery (i.e., chemical lead optimization, toxicity studies, and clinical trials). In the remainder of this chapter, we will focus on addressing several important issues related to the use of preclinical HD mouse models in the current large-scale HD drug discovery process. Because of space limitations, we will not be able to discuss in detail the specific preclinical compounds that have already been tested in HD mice. Excellent recent reviews on the subject can be found elsewhere (Beal and Ferrante, 2004; Butler and Bates, 2006; Di Prospero and Fischbeck, 2005; Hersch and Ferrante, 2004).

A DUAL-MODEL APPROACH TO TEST HD PRECLINICAL CANDIDATES

Because of the rich repertoire of HD mouse models available, a critical first question is which mouse model or models should be used in a preclinical study of a therapeutic candidate. An ideal HD mouse model should have (1) a genetic construct that is similar if not identical to the patients; (2) robust behavioral deficits and selective neuropathology mimicking the patients; and (3) these phenotypes are early onset, rapidly progressing, easily quantifiable, and have limited variability between mice. From our description of the preclinical HD mouse models, it is apparent that each of the current models only partially satisfies such requirements. The fragment models (i.e., R6/2) have very rapid onset and progression of disease, as well as low phenotypic variability (10–20 per treatment group in preclinical trials); weight loss and early lethality in these models can be used as surrogate markers of disease. Thus, these fragment models are particularly suitable for screening a relatively large number of candidates and have already made significant contributions in identifying promising preclinical leads in HD (Beal and Ferrante, 2004). However, there are also some concerns with the use of mhtt fragment models to determine the preclinical efficacy of therapeutic candidates (Beal and Ferrante, 2004; Hersch and Ferrante, 2004). First, the neuropathology in the fragment models is more widespread and relatively nonselective. Second, because the disease progression in patients is usually very slow, another concern is that a potential efficacious therapeutic mechanism or candidate in human HD (i.e., with a more slowly progressing disease process) may not be effective in the fragment model. Third, the fragment models only express a small portion of mhtt regulatory elements and proteins; thus, potential pathogenic interventions requiring the intact mhtt genomic context (i.e., RNA interference based on the full-length human htt mRNA) or the protein context (i.e., proteolysis inhibition) could not be tested in such a model.

The availability of several full-length mhtt mouse models (i.e., full-length models; Table 7.1) provides new opportunities to test HD preclinical candidates in model systems that (1) have genetic, genomic, and protein context more similar to human HD; (2) have a slowly progressing disease process that may be more closely related to the process in the human disease; and (3) have neuropathology that is selective to the striatum and cortex. However, the main concern in the use of the full-length models in preclinical trials is cost, which is relatively high because of the length of the trial (i.e., up to 12 months in YAC128 and BACHD mice and up to 1–2 years in Hdh-KI mice), as well as because of the potential variability of their outcomes, which may require a larger number of mice for each study (see below). However, recent studies using YAC128 mice in preclinical studies provide some examples of full-length models being effectively used to identify promising preclinical candidates in HD (Van Raamsdonk et al., 2005b, 2005d).

It is clear that both the mhtt fragment models and the full-length models have their potential strengths and weaknesses in the preclinical studies. Without unequivocal proof that any of these HD models can predict the clinical outcome in patients (i.e., predictive validity), and faced with the daunting task of prioritizing a potentially large number of preclinical candidates for very costly and lengthy clinical studies, we and many others in the HD research community favor the *dual-model* approach of using both a fragment model and a full-length huntingtin mouse model for the preclinical study (Bates and Hockly, 2003; Beal and Ferrante, 2004). In this scheme (see Figure 7.1), a large number of preclinical candidates coming out of the screening assays will be first tested, if the therapeutic mechanism permits, in a fragment mouse model of HD. R6/2 mice may be a preferred model for this purpose because of the model's rapid disease course and superior statistical power to detect efficacy (Hersch and Ferrante, 2004; Hockly et al., 2003). Such a therapeutic screen, with primary outcomes focusing on behavioral improvements (i.e., rotarod and grip strength) and neuroprotection (i.e., striatal atrophy), can provide relatively rapid information on the potential therapeutic efficacy of a large number of preclinical candidates at a reasonable cost. After demonstrating therapeutic efficacy in the fragment model (defined as both behavioral and pathological improvements), the next decision point is whether the compound should go directly to the clinical study or whether a second mouse trial in a full-length model is warranted. At this point, if the compound has proven safety in humans and/or is already in clinical use, direct clinical study in HD patients may be warranted. If a compound would require more extensive medicinal chemistry studies and toxicity and tolerability

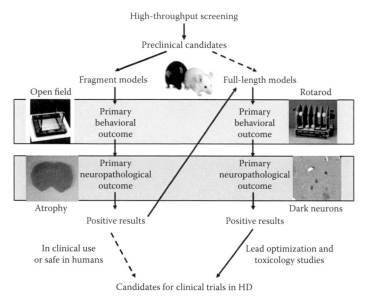

High-throughput screening

Preclinical candidates

Fragment models   Full-length models
Open field   Rotarod

Primary   Primary
behavioral   behavioral
outcome   outcome

Primary   Primary
neuropathological   neuropathological
outcome   outcome

Atrophy   Dark neurons
Positive results   Positive results

In clinical use   Lead optimization and
or safe in humans   toxicology studies

Candidates for clinical trials in HD

FIGURE 7.1   (See color insert following page 172.) A dual-model scheme for preclinical studies using both the fragment models and full-length models of HD.

studies in animals and humans, which are time consuming and costly, we prefer the strategy of performing a second preclinical study in a full-length mouse model of HD before these compounds enter the clinical phase (Figure 7.1). Any compounds demonstrating therapeutic efficacy in such dual-model trials will be the best clinical candidates for further therapeutic development toward clinical trials. Of course, preclinical studies in a full-length mhtt model could also be initiated without a fragment model trial or despite a negative result from such trial if (1) the mechanism of therapy requires the genomic or protein context of the full-length mhtt (i.e., small interfering RNA against human full-length mRNA, or proteolysis inhibitor targeting mhtt regions beyond those in the fragment models); or (2) the therapeutic modality favors a more slowly progressive and/or selective disease process (i.e., aging-related therapeutic targets). In summary, in the bottleneck process of selecting preclinical candidates for further clinical studies, we believe the dual-model approach in which first screening with a fragment model of HD and then following up with a second study using a full-length model is the most rational and prudent approach to identify promising clinical candidates in HD.

DESIGNING AND CONDUCTING MOUSE PRECLINICAL TRIALS

Preclinical trials in HD mice are an expensive and, in the case of a full-length model, relatively lengthy process. Therefore, before the initiation of such a study, careful design of the testing paradigms and rigorous standardization of the testing conditions and data analyses within a laboratory and across different laboratories are critical to produce accurate and reproducible results. The importance of mouse trial design and standardization in HD is highlighted in a Hereditary Disease Foundation-sponsored workshop at Cardiff in 2002 (see the workshop report at http://www.hdfoundation.org/workshops/200207Report.php) and reviewed by others (Bates and Hockly, 2003; Hersch and Ferrante, 2004; Hockly et al., 2003). In this section, we will discuss some of the salient points related to the design and execution of a mouse trial for HD.

Power Analyses to Determine the Sample Size for the Study

A very important consideration in the design of preclinical trials is to perform *power analyses* for a given preclinical outcome to determine the number of HD mice needed for a given set of outcomes in a preclinical trial (Bates and Hockly, 2003). Similar to the human clinical trials, insufficiently powered studies can be costly and may not provide the necessary information to assess the efficacy of a compound. Performing power analyses before the mouse trial will help one to determine how many animals will be needed in a study to have a predetermined reasonable chance (i.e., 80% or 90%) of detecting a significant improvement at a predetermined magnitude of improvement (i.e., 30% improvement). The basic formula for sample size estimation is the following (adapted from Hockly et al., 2003):

$$n = [(SD1^2 + SD2^2) (y + z)^2] / (\mu1 - \mu2)^2$$

In this formula, n represents the number of mice in each treatment group; $\mu1$ and $\mu2$ are the means of the two groups; and SD1 and SD2 are the standard deviations in the two groups. Furthermore, y is 1.96 for $P = 0.05$ (two-sided normal distribution); and z is 1.28 for power of 90% and 0.84 for power of 80%.

Because of cost concerns, power analyses are crucial for minimizing cost by selecting outcomes and/or HD mouse models that may require a relatively smaller number of mice per treatment group to detect a significant therapeutic benefit. For example, the use of R6/2 mice normally requires only about 10 mice per group for most of the phenotypic outcomes, whereas the use of N171-82Q mice may require 20 mice per group (Hersch and Ferrante, 2004). For the full-length models, most of the potential preclinical outcomes in these models should be analyzed by power analyses to determine whether they are suitable for preclinical studies. In the case of the YAC128 model, striatal atrophy is a particularly good readout because only eight animals per group are needed to provide 80% power to detect a significant ($P < 0.05$) treatment benefit that is 33% or greater (Slow et al., 2003). In BACHD mice, both our analyses (Gray et al., 2008) and an independent analysis at PsychoGenics (Tarrytown, NY) (L. Menalled and D. Howland, personal communications) reveal that the rotarod deficit at 6 months is a particularly robust phenotype outcome: only 22 BACHD mice are needed per group to provide 80% power to detect significant ($P < 0.05$) improvement of 33% or better. Thus, in both fragment and full-length models of HD, power analyses should be done for all the major outcomes so that one can begin the study with sufficient power to detect a therapeutic benefit.

Rigorous Standardization of Preclinical Mouse Trials

A key to reducing the variability in a mouse trial is the rigorous standardization of the testing condition (see HDF Cardiff Workshop Report; also Bates and Hockly, 2003; Hockly et al., 2003). There are several areas to which the investigator should pay particular attention. First, the genetics of a given HD mouse model has to be carefully controlled. If possible, an inbred mouse background should be used for a study. We do not recommend that all the mouse trials for all the different types of HD models should be done in the same genetic background, although it is well known that certain behavioral readouts may be more robust in one inbred background than another (Crawley et al., 1997). We do recommend, however, within the same type of model, a consistent and relatively pure genetic background (i.e., inbred or recombinant F1) should be used, and preferably a genetic background in which a given model exhibits the most robust phenotypes that will be used as preclinical outcomes (i.e., YAC128 model in FvB background) (Van Raamsdonk et al., 2007). For the fragment models such as R6/2, which can only be maintained in a mixed F1 background, one should consistently split transgenic littermates between the treatment group and control group, and use litters from different breeding pairs to avoid potential breeder effects (Hockly et al., 2003). Because CAG repeat size can vary in certain mouse models, such as R6/2 and Hdh-KI mice, the mice used in the study should be routinely genotyped to rule out any mice with dramatic expansion or contraction of the repeats, which may influence the phenotypes. Second, mouse housing and diet conditions should also be standardized. There are several reports that demonstrate environmental enrichment and dietary influences, which may significantly improve outcomes in HD mice (Hockly et al., 2002; Schilling et al., 2004a; Spires et al., 2004). For R6/2 mice, a moderately enriched housing condition is recommended (Hockly et al., 2003). For other HD mouse models, such as the full-length models, the housing conditions should be maintained constant in different trials and comparable across different laboratories. Finally, the outcome testing conditions and statistical analyses should also be the same for a given model in different trials and different laboratories. For example, the equipment and methods for behavioral testing and neuropathological studies (i.e., stereological counting and dark neuron quantization) may be different from laboratory to laboratory. Therefore, it is critical that those laboratories using the same preclinical mouse model reach a consensus on a set of standardized protocols for various outcome studies and make such standardized protocols accessible to others in the community. For the phenotypic studies, the investigators conducting the tests should be blind to the genotypes. Finally, statistical analyses should be performed in the same way for a given outcome in a given model, so one can readily compare the preclinical results of different compounds and their ability to be replicated by different laboratories.

Improving the Outcome Measures in Preclinical Mouse Studies

In the majority of preclinical trials with HD mice, the primary phenotypic outcomes are rotarod for motor behaviors and stereological measurement of the striatal atrophy. Secondary outcomes, such as weight loss and lethality, are only possible for the fragment models, and their relevance to HD remains unclear. However, in HD patients, the clinical features are far more complex and consist of motor, psychiatric, and cognitive deficits. Because we are currently unclear whether this clinical triad is caused by neurodegeneration or neuronal dysfunction in HD, and whether different brain regions may mediate these distinct deficits, the exclusive use of motor and striatal pathology outcomes may be insufficient, particularly in evaluating candidates that may selectively improve cognitive- and/or psychiatric-related phenotypes (which may be highly relevant to HD). For example, a recent preclinical study with R6/2 mice showed that the imposition of sleep with alprazolam selectively improves the cognitive deficits in this model but not motor deficits (Pallier et al., 2007). Currently, there is a concerted effort to develop more sensitive potentially automatable cognitive behavioral outcomes for the preclinical models of HD (Lione et al., 1999; Morton et al., 2006; Van Raamsdonk, et al., 2005e). In our view, the use of robust cognitive measures, especially those affecting the same corticostriatal neural circuit as in patients, should be evaluated carefully and should be included when possible as primary outcome measures in HD preclinical studies.

Finally, another concerted effort in the HD therapeutic field is the attempt to discover biomarkers that may reflect the disease process in patients and may be used in future clinical studies. Some of these potential biomarkers, such as gene expression alterations and/or specific molecular changes in the serum such as increase in 8-OHDG (a marker for oxidized damage to DNA), have already been elucidated in patients (Dalrymple et al., 2007; Hersch et al., 2006; Runne et al., 2007). Importantly, in a recent clinical study with creatine, 8-OHDG was shown to dramatically normalize with the treatment, thus indicating it may be a useful peripheral biomarker that can predict therapeutic benefit (Hersch et al., 2006). If similar biomarkers also are altered in preclinical mouse models, they could be incorporated in the mouse trial to provide the opportunity to noninvasively monitor preclinical outcomes in mice and clinical outcomes in patients.

TRANSLATING PRECLINICAL STUDIES TO THE CLINIC

The ultimate goal for the preclinical studies with HD mice is to develop clinical candidates that can eventually succeed in human clinical trials. This goal will not only validate a specific mouse model and/or other nonmammalian models of HD but also validate the entire rational approach we are undertaking to develop an effective treatment or a cure for HD. Currently, with the predominant use of fragment models of HD, a large set of compounds have already been demonstrated to have preclinical efficacy (Beal and Ferrante, 2004). Several of these compounds, including creatine, coenzyme Q, α-lipoic acid, cystamine, and histone deacetylase inhibitors, have been consistently beneficial in multiple HD mouse models and in trials done by different laboratories (Beal and Ferrante, 2004). Furthermore, cystamine remains the only compound that satisfies the dual-model testing paradigm, demonstrating some efficacy in both fragment models (Beal and Ferrante, 2004) and in the full-length YAC128 model (Van Raamsdonk et al., 2005b).

If the role of a genetic mouse model in a rational development of therapy for HD or other brain disorders remains unproven, we may learn some lessons from another field in which 30 years of intensive research has brought some visible successes in the clinic. In the cancer field, mouse genetic models have provided critical mechanistic insights into the disease pathogenic process (Frese and Tuveson, 2007; Van Dyke and Jacks, 2002); some lessons about the difference in the biology between the mouse and human, which allowed the building of better mouse models (Rangarajan and Weinberg, 2003); and eventually a few success stories—in the case for acute promyelocytic leukemia (APL), a mouse model of APL provided promising preclinical results (Lallemand-Breitenbach et al., 1999), which subsequently translated into effective new treatment for patients (Soignet and Maslak, 2004). In the HD field, a relatively encouraging Phase III clinical trial demonstrated coenzyme Q10, which has reproducible efficacy in the preclinical models of HD (i.e., R6/2 and N171-82Q mice), also reveals a nonsignificant trend (14%) of improvement in HD patients. With the cancer field in mind, we are optimistic that the preclinical HD mouse models will not only teach us about the intricate pathogenic mechanisms in HD but will also play a critical role in identifying promising therapeutic candidates that can be translated into the clinic.

ACKNOWLEDGMENTS

We apologize for our omission, in part because of space limitations, of citations or topics. We thank the National Institute of Neurodegenerative Disorders and Stroke/National Institutes of Health (Grants 5R01NS049501 and 1R21NS047391), Hereditary Disease Foundation, High Q Foundation, and CHDI Foundation, Inc. (to X. W. Yang) for their support. We also thank Nancy Wexler, Carl Johnson, Allan Tobin, and Ethan Signer for their support. We thank Bernard Balleine at UCLA, Liliana Menalled and Daniela Brunner at PsychoGenics, and David Howland at High Q Foundation for sharing unpublished data. Finally, we thank Xiaohong Lu, Xiaofeng Gu, Binh Vuong, and other members of the Yang laboratory for discussions and for help in preparing the chapter manuscript.

REFERENCES

Andreassen, O. A., A. Dedeoglu, R. J. Ferrante, B. G. Jenkins, K. L. Ferrante, M. Thomas, A. Friedlich, S. E. Browne, G. Schilling, D. R. Borchelt, et al. 2001. Creatine increase survival and delays motor symptoms in a transgenic animal model of Huntington's disease. *Neurobiol Dis* 8 (3):479–91.
Arguello, P. A., and J. A. Gogos. 2006. Modeling madness in mice: one piece at a time. *Neuron* 52 (1):179–96.
Aron, A. R., B. J. Sahakian, and T. W. Robbins. 2003. Distractibility during selection-for-action: differential deficits in Huntington's disease and following frontal lobe damage. *Neuropsychologia* 41 (9):1137–47.
Arrasate, M., S. Mitra, E. S. Schweitzer, M. R. Segal, and S. Finkbeiner. 2004. Inclusion body formation reduces levels of mutant huntingtin and the risk of neuronal death. *Nature* 431 (7010):805–10.

Aylward, E. H., B. F. Sparks, K. M. Field, V. Yallapragada, B. D. Shpritz, A. Rosenblatt, J. Brandt, L. M. Gourley, K. Liang, H. Zhou, et al. 2004. Onset and rate of striatal atrophy in preclinical Huntington disease. *Neurology* 63 (1):66–72.

Balleine, B. W. 2005. Neural bases of food-seeking: affect, arousal and reward in corticostriatolimbic circuits. *Physiol Behav* 86 (5):717–30.

Bates, G. 2003. Huntingtin aggregation and toxicity in Huntington's disease. *Lancet* 361 (9369):1642–44.

Bates, G. P., and E. Hockly. 2003. Experimental therapeutics in Huntington's disease: are models useful for therapeutic trials? *Curr Opin Neurol* 16 (4):465–70.

Bates, G. P., P. S. Harper, and A. L. Jones, eds. 2002. *Huntington's Disease*. Oxford: Oxford University Press.

Beal, M. F., and R. J. Ferrante. 2004. Experimental therapeutics in transgenic mouse models of Huntington's disease. *Nat Rev Neurosci* 5 (5):373–84.

Bogdanov, M. B., O. A. Andreassen, A. Dedeoglu, R. J. Ferrante, and M. F. Beal. 2001. Increased oxidative damage to DNA in a transgenic mouse model of Huntington's disease. *J Neurochem* 79 (6):1246–49.

Bolivar, V. J., K. Manley, and A. Messer. 2003. Exploratory activity and fear conditioning abnormalities develop early in R6/2 Huntington's disease transgenic mice. *Behav Neurosci* 117 (6):1233–42.

Branda, C. S., and S. M. Dymecki. 2004. Talking about a revolution: the impact of site-specific recombinases on genetic analyses in mice. *Dev Cell* 6 (1):7–28.

Butler, R., and G. P. Bates. 2006. Histone deacetylase inhibitors as therapeutics for polyglutamine disorders. *Nat Rev Neurosci* 7 (10):784–96.

Capecchi, M. R. 2005. Gene targeting in mice: functional analysis of the mammalian genome for the twenty-first century. *Nat Rev Genet* 6 (6):507–12.

Carter, R. J., L. A. Lione, T. Humby, L. Mangiarini, A. Mahal, G. P. Bates, S. B. Dunnett, and A. J. Morton. 1999. Characterization of progressive motor deficits in mice transgenic for the human Huntington's disease mutation. *J Neurosci* 19 (8):3248–57.

Crawley, J. N., J. K. Belknap, A. Collins, J. C. Crabbe, W. Frankel, N. Henderson, R. J. Hitzemann, S. C. Maxson, L. L. Miner, A. J. Silva, et al. 1997. Behavioral phenotypes of inbred mouse strains: implications and recommendations for molecular studies. *Psychopharmacology (Berl)* 132 (2):107–24.

Dalrymple, A., E. J. Wild, R. Joubert, K. Sathasivam, M. Bjorkqvist, A. Petersen, G. S. Jackson, J. D. Isaacs, M. Kristiansen, G. P. Bates, et al. 2007. Proteomic profiling of plasma in Huntington's disease reveals neuroinflammatory activation and biomarker candidates. *J Proteome Res* 6 (7):2833–40.

Davies, S. W., M. Turmaine, B. A. Cozens, M. DiFiglia, A. H. Sharp, C. A. Ross, E. Scherzinger, E. E. Wanker, L. Mangiarini, and G. P. Bates. 1997. Formation of neuronal intranuclear inclusions underlies the neurological dysfunction in mice transgenic for the HD mutation. *Cell* 90 (3):537–48.

Di Prospero, N. A., and K. H. Fischbeck. 2005. Therapeutics development for triplet repeat expansion diseases. *Nat Rev Genet* 6 (10):756–65.

DiFiglia, M., E. Sapp, K. O. Chase, S. W. Davies, G. P. Bates, J. P. Vonsattel, and N. Aronin. 1997. Aggregation of huntingtin in neuronal intranuclear inclusions and dystrophic neurites in brain. *Science* 277 (5334):1990–93.

Fleming, S. M., P. O. Fernagut, and M. F. Chesselet. 2005. Genetic mouse models of parkinsonism: strengths and limitations. *NeuroRx* 2 (3):495–503.

Fossale, E., V. C. Wheeler, V. Vrbanac, L. A. Lebel, A. Teed, J. S. Mysore, J. F. Gusella, M. E. MacDonald, and F. Persichetti. 2002. Identification of a presymptomatic molecular phenotype in Hdh CAG knock-in mice. *Hum Mol Genet* 11 (19):2233–41.

Frese, K. K., and D. A. Tuveson. 2007. Maximizing mouse cancer models. *Nat Rev Cancer* 7 (9):654–58.

Gardian, G., S. E. Browne, D. K. Choi, P. Klivenyi, J. Gregorio, J. K. Kubilus, H. Ryu, B. Langley, R. R. Ratan, R. J. Ferrante, et al. 2005. Neuroprotective effects of phenylbutyrate in the N171-82Q transgenic mouse model of Huntington's disease. *J Biol Chem* 280 (1):556–63.

Gatchel, J. R., and H. Y. Zoghbi. 2005. Diseases of unstable repeat expansion: mechanisms and common principles. *Nat Rev Genet* 6 (10):743–55.

Graham, R. K., E. J. Slow, Y. Deng, N. Bissada, G. Lu, J. Pearson, J. Shehadeh, B. R. Leavitt, L. A. Raymond, and M. R. Hayden. 2006. Levels of mutant huntingtin influence the phenotypic severity of Huntington disease in YAC128 mouse models. *Neurobiol Dis* 21 (2):444–55.

Gray, M., D. I. Shirasaki, C. Cepeda, V. M. Andre, B. Wilburn, X.-H. Lu, J. Tao, I. Yamazaki, S.-H. Li, Y. E. Sun, et al. 2008. Full-length human mutant huntingtin with a stable polyglutamine repeat can elicit progressive and selective neuropathogenesis in BACHD mice. *J Neurosci* 28 (24):6182–95.

Gu, X., C. Li, W. Wei, V. Lo, S. Gong, S. H. Li, T. Iwasato, S. Itohara, X. J. Li, I. Mody, et al. 2005. Pathological cell-cell interactions elicited by a neuropathogenic form of mutant Huntingtin contribute to cortical pathogenesis in HD mice. *Neuron* 46 (3):433–44.

Gusella, J. F., and M. E. Macdonald. 2006. Huntington's disease: seeing the pathogenic process through a genetic lens. *Trends Biochem Sci* 31 (9):533–40.

Gutekunst, C. A., S. H. Li, H. Yi, J. S. Mulroy, S. Kuemmerle, R. Jones, D. Rye, R. J. Ferrante, S. M. Hersch, and X. J. Li. 1999. Nuclear and neuropil aggregates in Huntington's disease: relationship to neuropathology. *J Neurosci* 19 (7):2522–34.

Hedreen, J. C., C. E. Peyser, S. E. Folstein, and C. A. Ross. 1991. Neuronal loss in layers V and VI of cerebral cortex in Huntington's disease. *Neurosci Lett* 133 (2):257–61.

Heinsen, H., M. Strik, M. Bauer, K. Luther, G. Ulmar, D. Gangnus, G. Jungkunz, W. Eisenmenger, and M. Gotz. 1994. Cortical and striatal neurone number in Huntington's disease. *Acta Neuropathol (Berl)* 88 (4):320–33.

Heintz, N. 2001. BAC to the future: the use of bac transgenic mice for neuroscience research. *Nat Rev Neurosci* 2 (12):861–70.

Heng, M. Y., S. J. Tallaksen-Greene, P. J. Detloff, and R. L. Albin. 2007. Longitudinal evaluation of the Hdh(CAG)150 knock-in murine model of Huntington's disease. *J Neurosci* 27 (34):8989–98.

Hersch, S. M., and R. J. Ferrante. 2004. Translating therapies for Huntington's disease from genetic animal models to clinical trials. *NeuroRx* 1 (3):298–306.

Hersch, S. M., S. Gevorkian, K. Marder, C. Moskowitz, A. Feigin, M. Cox, P. Como, C. Zimmerman, M. Lin, L. Zhang, et al. 2006. Creatine in Huntington disease is safe, tolerable, bioavailable in brain and reduces serum 8OH2′dG. *Neurology* 66 (2):250–52.

Hickey, M. A., K. Gallant, G. G. Gross, M. S. Levine, and M. F. Chesselet. 2005. Early behavioral deficits in R6/2 mice suitable for use in preclinical drug testing. *Neurobiol Dis* 20 (1):1–11.

Hockly, E., P. M. Cordery, B. Woodman, A. Mahal, A. van Dellen, C. Blakemore, C. M. Lewis, A. J. Hannan, and G. P. Bates. 2002. Environmental enrichment slows disease progression in R6/2 Huntington's disease mice. *Ann Neurol* 51 (2): 235–42.

Hockly, E., B. Woodman, A. Mahal, C. M. Lewis, and G. Bates. 2003. Standardization and statistical approaches to therapeutic trials in the R6/2 mouse. *Brain Res Bull* 61 (5):469–79.

Hodgson, J. G., N. Agopyan, C. A. Gutekunst, B. R. Leavitt, F. LePiane, R. Singaraja, D. J. Smith, N. Bissada, K. McCutcheon, J. Nasir, et al. 1999. A YAC mouse model for Huntington's disease with full-length mutant huntingtin, cytoplasmic toxicity, and selective striatal neurodegeneration. *Neuron* 23 (1):181–92.

Holmes, A., C. C. Wrenn, A. P. Harris, K. E. Thayer, and J. N. Crawley. 2002. Behavioral profiles of inbred strains on novel olfactory, spatial and emotional tests for reference memory in mice. *Genes Brain Behav* 1 (1):55–69.

Hughes, R. E., and J. M. Olson. 2001. Therapeutic opportunities in polyglutamine disease. *Nat Med* 7 (4):419–23.

The Huntington's Disease Collaborative Research Group. 1993. A novel gene containing a tri-nucleotide repeat that is expanded and unstable on Huntington's disease chromosomes. *Cell* 72 (6):971–83.

Jenkins, B. G., P. Klivenyi, E. Kustermann, O. A. Andreassen, R. J. Ferrante, B. R. Rosen, and M. F. Beal. 2000. Nonlinear decrease over time in N-acetyl aspartate levels in the absence of neuronal loss and increases in glutamine and glucose in transgenic Huntington's disease mice. *J Neurochem* 74 (5):2108–19.

Kazantsev, A., E. Preisinger, A. Dranovsky, D. Goldgaber, and D. Housman. 1999. Insoluble detergent-resistant aggregates form between pathological and nonpathological lengths of polyglutamine in mammalian cells. *Proc Natl Acad Sci U S A* 96 (20):11404–09.

Kovtun, I. V., Y. Liu, M. Bjoras, A. Klungland, S. H. Wilson, and C. T. McMurray. 2007. OGG1 initiates age-dependent CAG trinucleotide expansion in somatic cells. *Nature* 447 (7143):447–52.

Kuhn, A., D. R. Goldstein, A. Hodges, A. D. Strand, T. Sengstag, C. Kooperberg, K. Becanovic, M. A. Pouladi, K. Sathasivam, J. H. Cha, et al. 2007. Mutant huntingtin's effects on striatal gene expression in mice recapitulate changes observed in human Huntington's disease brain and do not differ with mutant huntingtin length or wild-type huntingtin dosage. *Hum Mol Genet* 16 (15):1845–61.

Lallemand-Breitenbach, V., M. C. Guillemin, A. Janin, M. T. Daniel, L. Degos, S. C. Kogan, J. M. Bishop, and H. de The. 1999. Retinoic acid and arsenic synergize to eradicate leukemic cells in a mouse model of acute promyelocytic leukemia. *J Exp Med* 189 (7):1043–52.

Landles, C., and G. P. Bates. 2004. Huntingtin and the molecular pathogenesis of Huntington's disease. Fourth in molecular medicine review series. *EMBO Rep* 5 (10):958–63.

Levine, M. S., C. Cepeda, M. A. Hickey, S. M. Fleming, and M. F. Chesselet. 2004. Genetic mouse models of Huntington's and Parkinson's diseases: illuminating but imperfect. *Trends Neurosci* 27 (11):691–97.

Li, H., S. H. Li, A. L. Cheng, L. Mangiarini, G. P. Bates, and X. J. Li. 1999. Ultrastructural localization and progressive formation of neuropil aggregates in Huntington's disease transgenic mice. *Hum Mol Genet* 8 (7):1227–36.

Li, J. Y., N. Popovic, and P. Brundin. 2005. The use of the R6 transgenic mouse models of Huntington's disease in attempts to develop novel therapeutic strategies. *NeuroRx* 2 (3):447–64.

Li, S., and X. J. Li. 2006. Multiple pathways contribute to the pathogenesis of Huntington disease. *Mol Neurodegener* 1:19.

Lin, C. H., S. Tallaksen-Greene, W. M. Chien, J. A. Cearley, W. S. Jackson, A. B. Crouse, S. Ren, X. J. Li, R. L. Albin, and P. J. Detloff. 2001. Neurological abnormalities in a knock-in mouse model of Huntington's disease. *Hum Mol Genet* 10 (2):137–44.

Lione, L. A., R. J. Carter, M. J. Hunt, G. P. Bates, A. J. Morton, and S. B. Dunnett. 1999. Selective discrimination learning impairments in mice expressing the human Huntington's disease mutation. *J Neurosci* 19 (23):10428–37.

Lloret, A., E. Dragileva, A. Teed, J. Espinola, E. Fossale, T. Gillis, E. Lopez, R. H. Myers, M. E. MacDonald, and V. C. Wheeler. 2006. Genetic background modifies nuclear mutant huntingtin accumulation and HD CAG repeat instability in Huntington's disease knock-in mice. *Hum Mol Genet* 15 (12):2015–24.

Mangiarini, L., K. Sathasivam, M. Seller, B. Cozens, A. Harper, C. Hetherington, M. Lawton, Y. Trottier, H. Lehrach, S. W. Davies, et al. 1996. Exon 1 of the HD gene with an expanded CAG repeat is sufficient to cause a progressive neurological phenotype in transgenic mice. *Cell* 87 (3):493–506.

Marsh, J. L., and L. M. Thompson. 2006. Drosophila in the study of neurodegenerative disease. *Neuron* 52 (1):169–78.

Menalled, L. B. 2005. Knock-in mouse models of Huntington's disease. *NeuroRx* 2 (3):465–70.

Menalled, L. B., and M. F. Chesselet. 2002. Mouse models of Huntington's disease. *Trends Pharmacol Sci* 23 (1):32–39.

Menalled, L. B., F. El-Khodor, M. Patry, M. Suarez-Farinas, S. J. Orenstein, B. Zahasky, C. Leahy, V. Wheeler, X. W. Yang, M. MacDonald, et al. 2009. Systematic behavioral evaluation of Huntington's disease transgenic and knock-in mouse models. *Neurobiol Dis* 35 (3):319–36.

Menalled, L. B., J. D. Sison, I. Dragatsis, S. Zeitlin, and M. F. Chesselet. 2003. Time course of early motor and neuropathological anomalies in a knock-in mouse model of Huntington's disease with 140 CAG repeats. *J Comp Neurol* 465 (1):11–26.

Menalled, L. B., J. D. Sison, Y. Wu, M. Olivieri, X. J. Li, H. Li, S. Zeitlin, and M. F. Chesselet. 2002. Early motor dysfunction and striosomal distribution of huntingtin microaggregates in Huntington's disease knock-in mice. *J Neurosci* 22 (18):8266–76.

Morton, A. J., E. Skillings, T. J. Bussey, and L. M. Saksida. 2006. Measuring cognitive deficits in disabled mice using an automated interactive touchscreen system. *Nat Methods* 3 (10):767.

Morton, A. J., N. I. Wood, M. H. Hastings, C. Hurelbrink, R. A. Barker, and E. S. Maywood. 2005. Disintegration of the sleep-wake cycle and circadian timing in Huntington's disease. *J Neurosci* 25 (1):157–63.

Mouse Genome Sequencing Consortium. 2002. Initial sequencing and comparative analysis of the mouse genome. *Nature* 420 (6915):520–62.

Murphy, K. P., R. J. Carter, L. A. Lione, L. Mangiarini, A. Mahal, G. P. Bates, S. B. Dunnett, and A. J. Morton. 2000. Abnormal synaptic plasticity and impaired spatial cognition in mice transgenic for exon 1 of the human Huntington's disease mutation. *J Neurosci* 20 (13):5115–23.

Orr, H. T., and H. Y. Zoghbi. 2007. Trinucleotide repeat disorders. *Annu Rev Neurosci* 30:575–621.

Pallier, P. N., E. S. Maywood, Z. Zheng, J. E. Chesham, A. N. Inyushkin, R. Dyball, M. H. Hastings, and A. J. Morton. 2007. Pharmacological imposition of sleep slows cognitive decline and reverses dysregulation of circadian gene expression in a transgenic mouse model of Huntington's disease. *J Neurosci* 27 (29):7869–78.

Rangarajan, A., and R. A. Weinberg. 2003. Opinion: Comparative biology of mouse versus human cells: modelling human cancer in mice. *Nat Rev Cancer* 3 (12):952–59.

Reddy, P. H., M. Williams, V. Charles, L. Garrett, L. Pike-Buchanan, W. O. Whetsell, Jr., G. Miller, and D. A. Tagle. 1998. Behavioural abnormalities and selective neuronal loss in HD transgenic mice expressing mutated full-length HD cDNA. *Nat Genet* 20 (2):198–202.

Rogers, D. C., E. M. Fisher, S. D. Brown, J. Peters, A. J. Hunter, and J. E. Martin. 1997. Behavioral and functional analysis of mouse phenotype: SHIRPA, a proposed protocol for comprehensive phenotype assessment. *Mamm Genome* 8 (10):711–13.

Rosas, H. D., N. D. Hevelone, A. K. Zaleta, D. N. Greve, D. H. Salat, and B. Fischl. 2005. Regional cortical thinning in preclinical Huntington disease and its relationship to cognition. *Neurology* 65 (5):745–47.

Rosas, H. D., D. S. Tuch, N. D. Hevelone, A. K. Zaleta, M. Vangel, S. M. Hersch, and D. H. Salat. 2006. Diffusion tensor imaging in presymptomatic and early Huntington's disease: selective white matter pathology and its relationship to clinical measures. *Mov Disord* 21 (9):1317–25.

Runne, H., A. Kuhn, E. J. Wild, W. Pratyaksha, M. Kristiansen, J. D. Isaacs, E. Regulier, M. Delorenzi, S. J. Tabrizi, and R. Luthi-Carter. 2007. Analysis of potential transcriptomic biomarkers for Huntington's disease in peripheral blood. *Proc Natl Acad Sci U S A* 104 (36):14424–29.

Schilling, G., M. W. Becher, A. H. Sharp, H. A. Jinnah, K. Duan, J. A. Kotzuk, H. H. Slunt, T. Ratovitski, J. K. Cooper, N. A. Jenkins, et al. 1999. Intranuclear inclusions and neuritic aggregates in transgenic mice expressing a mutant N-terminal fragment of huntingtin. *Hum Mol Genet* 8 (3):397–407.

Schilling, G., A. V. Savonenko, M. L. Coonfield, J. L. Morton, E. Vorovich, A. Gale, C. Neslon, N. Chan, M. Eaton, D. Fromholt, C. A. Ross, et al. 2004a. Environmental, pharmacological, and genetic modulation of the HD phenotype in transgenic mice. *Exp Neurol* 187 (1):137–49.

Schilling, G., A. V. Savonenko, A. Klevytska, J. L. Morton, S. M. Tucker, M. Poirier, A. Gale, N. Chan, V. Gonzales, H. H. Slunt, et al. 2004b. Nuclear-targeting of mutant huntingtin fragments produces Huntington's disease-like phenotypes in transgenic mice. *Hum Mol Genet* 13 (15):1599–610.

Schilling, G., A. H. Sharp, S. J. Loev, M. V. Wagster, S. H. Li, O. C. Stine, and C. A. Ross. 1995. Expression of the Huntington's disease (IT15) protein product in HD patients. *Hum Mol Genet* 4 (8):1365–71.

Seong, I. S., E. Ivanova, J. M. Lee, Y. S. Choo, E. Fossale, M. Anderson, J. F. Gusella, J. M. Laramie, R. H. Myers, M. Lesort, et al. 2005. HD CAG repeat implicates a dominant property of huntingtin in mitochondrial energy metabolism. *Hum Mol Genet* 14 (19):2871–80.

Sharp, A. H., S. J. Loev, G. Schilling, S. H. Li, X. J. Li, J. Bao, M. V. Wagster, J. A. Kotzuk, J. P. Steiner, A. Lo, et al. 1995. Widespread expression of Huntington's disease gene (IT15) protein product. *Neuron* 14 (5):1065–74.

Silva, E., Simpson, J. Takahashi, H. Lipp, S. Nakanishi, J. Wehner, K. Giese, T. Tully, T. Abel, and P. Chapman. 1997. Mutant mice and neuroscience: recommendations concerning genetic background. *Neuron* 19 (4):755–59.

Simmons, D. A., M. Casale, B. Alcon, N. Pham, N. Narayan, and G. Lynch. 2007. Ferritin accumulation in dystrophic microglia is an early event in the development of Huntington's disease. *Glia* 55 (10):1074–84.

Sipione, S., and E. Cattaneo. 2001. Modeling Huntington's disease in cells, flies, and mice. *Mol Neurobiol* 23 (1):21–51.

Slow, E. J., J. van Raamsdonk, D. Rogers, S. H. Coleman, R. K. Graham, Y. Deng, R. Oh, N. Bissada, S. M. Hossain, Y. Z. Yang, et al. 2003. Selective striatal neuronal loss in a YAC128 mouse model of Huntington disease. *Hum Mol Genet* 12 (13): 1555–67.

Soignet, S., and P. Maslak. 2004. Therapy of acute promyelocytic leukemia. *Adv Pharmacol* 51:35–58.

Spires, T. L., H. E. Grote, S. Garry, P. M. Cordery, A. Van Dellen, C. Blakemore, and A. J. Hannan. 2004. Dendritic spine pathology and deficits in experience-dependent dendritic plasticity in R6/1 Huntington's disease transgenic mice. *Eur J Neurosci* 19 (10):2799–807.

Swerdlow, N. R., D. L. Braff, N. Taaid, and M. A. Geyer. 1994. Assessing the validity of an animal model of deficient sensorimotor gating in schizophrenic patients. *Arch Gen Psychiatry* 51 (2):139–54.

Swerdlow, N. R., J. Paulsen, D. L. Braff, N. Butters, M. A. Geyer, and M. R. Swenson. 1995. Impaired prepulse inhibition of acoustic and tactile startle response in patients with Huntington's disease. *J Neurol Neurosurg Psychiatry* 58 (2):192–200.

Tallaksen-Greene, S. J., A. B. Crouse, J. M. Hunter, P. J. Detloff, and R. L. Albin. 2005. Neuronal intranuclear inclusions and neuropil aggregates in HdhCAG(150) knockin mice. *Neuroscience* 131 (4):843–52.

Turmaine, M., A. Raza, A. Mahal, L. Mangiarini, G. P. Bates, and S. W. Davies. 2000. Nonapoptotic neurodegeneration in a transgenic mouse model of Huntington's disease. *Proc Natl Acad Sci U S A* 97 (14):8093–97.

Van Dyke, T., and T. Jacks. 2002. Cancer modeling in the modern era: progress and challenges. *Cell* 108 (2):135–44.

Van Raamsdonk, J. M., W. T. Gibson, J. Pearson, Z. Murphy, G. Lu, B. R. Leavitt, and M. R. Hayden. 2006. Body weight is modulated by levels of full-length huntingtin. *Hum Mol Genet* 15 (9):1513–23.

Van Raamsdonk, J. M., Z. Murphy, E. J. Slow, B. R. Leavitt, and M. R. Hayden. 2005a. Selective degeneration and nuclear localization of mutant huntingtin in the YAC128 mouse model of Huntington disease. *Hum Mol Genet* 14 (24):3823–35.

Van Raamsdonk, J. M., J. Pearson, C. D. Bailey, D. A. Rogers, G. V. Johnson, M. R. Hayden, and B. R. Leavitt. 2005b. Cystamine treatment is neuroprotective in the YAC128 mouse model of Huntington disease. *J Neurochem* 95 (1):210–20.

Van Raamsdonk, J. M., J. Pearson, D. A. Rogers, N. Bissada, A. W. Vogl, M. R. Hayden, and B. R. Leavitt. 2005c. Loss of wild-type huntingtin influences motor dysfunction and survival in the YAC128 mouse model of Huntington disease. *Hum Mol Genet* 14 (10):1379–92.

Van Raamsdonk, J. M., J. Pearson, D. A. Rogers, G. Lu, V. E. Barakauskas, A. M. Barr, W. G. Honer, M. R. Hayden, and B. R. Leavitt. 2005d. Ethyl-EPA treatment improves motor dysfunction, but not neurodegeneration in the YAC128 mouse model of Huntington disease. *Exp Neurol* 196 (2):266–72.

Van Raamsdonk, J. M., J. Pearson, E. J. Slow, S. M. Hossain, B. R. Leavitt, and M. R. Hayden. 2005e. Cognitive dysfunction precedes neuropathology and motor abnormalities in the YAC128 mouse model of Huntington's disease. *J Neurosci* 25 (16):4169–80.

Van Raamsdonk, J. M., S. C. Warby, and M. R. Hayden. 2007. Selective degeneration in YAC mouse models of Huntington disease. *Brain Res Bull* 72 (2–3):124–31.

Vonsattel, J. P., and M. DiFiglia. 1998. Huntington disease. *J Neuropathol Exp Neurol* 57 (5):369–84.

Watase, K., and H. Y. Zoghbi. 2003. Modelling brain diseases in mice: the challenges of design and analysis. *Nat Rev Genet* 4 (4):296–307.

Wexler, N. S., E. A. Rose, and D. E. Housman. 1991. Molecular approaches to hereditary diseases of the nervous system: Huntington's disease as a paradigm. *Annu Rev Neurosci* 14:503–29.

Wheeler, V. C., W. Auerbach, J. K. White, J. Srinidhi, A. Auerbach, A. Ryan, M. P. Duyao, V. Vrbanac, M. Weaver, J. F. Gusella, et al. 1999. Length-dependent gametic CAG repeat instability in the Huntington's disease knock-in mouse. *Hum Mol Genet* 8 (1):115–22.

Wheeler, V. C., C. A. Gutekunst, V. Vrbanac, L. A. Lebel, G. Schilling, S. Hersch, R. M. Friedlander, J. F. Gusella, J. P. Vonsattel, D. R. Borchelt, et al. 2002. Early phenotypes that presage late-onset neurodegenerative disease allow testing of modifiers in Hdh CAG knock-in mice. *Hum Mol Genet* 11 (6):633–40.

Wheeler, V. C., L. A. Lebel, V. Vrbanac, A. Teed, H. te Riele, and M. E. MacDonald. 2003. Mismatch repair gene Msh2 modifies the timing of early disease in Hdh(Q111) striatum. *Hum Mol Genet* 12 (3):273–81.

Wheeler, V. C., J. K. White, C. A. Gutekunst, V. Vrbanac, M. Weaver, X. J. Li, S. H. Li, H. Yi, J. P. Vonsattel, J. F. Gusella, et al. 2000. Long glutamine tracts cause nuclear localization of a novel form of huntingtin in medium spiny striatal neurons in HdhQ92 and HdhQ111 knock-in mice. *Hum Mol Genet* 9 (4):503–13.

Woodman, B., R. Butler, C. Landles, M. K. Lupton, J. Tse, E. Hockly, H. Moffitt, K. Sathasivam, and G. P. Bates. 2007. The Hdh(Q150/Q150) knock-in mouse model of HD and the R6/2 exon 1 model develop comparable and widespread molecular phenotypes. *Brain Res Bull* 72 (2–3):83–97.

Yamamoto, A., J. J. Lucas, and R. Hen. 2000. Reversal of neuropathology and motor dysfunction in a conditional model of Huntington's disease. *Cell* 101 (1):57–66.

Yang, X. W., P. Model, and N. Heintz. 1997. Homologous recombination based modification in Escherichia coli and germline transmission in transgenic mice of a bacterial artificial chromosome. *Nat Biotechnol* 15 (9):859–65.

Yu, Z. X., S. H. Li, J. Evans, A. Pillarisetti, H. Li, and X. J. Li. 2003. Mutant huntingtin causes context-dependent neurodegeneration in mice with Huntington's disease. *J Neurosci* 23 (6):2193–202.

Zeitlin, S., J. P. Liu, D. L. Chapman, V. E. Papaioannou, and A. Efstratiadis. 1995. Increased apoptosis and early embryonic lethality in mice nullizygous for the Huntington's disease gene homologue. *Nat Genet* 11 (2):155–63.

Zoghbi, H. Y., and H. T. Orr. 2000. Glutamine repeats and neurodegeneration. *Annu Rev Neurosci* 23:217–47.

8 Pharmaceutical Development for Huntington's Disease

Richard J. Morse, Janet M. Leeds,
Douglas Macdonald, Larry Park,
Leticia Toledo-Sherman, and Robert Pacifici

CONTENTS

INTRODUCTION

Huntington's disease (HD) is a devastating neurological disorder caused by mutations in the human gene encoding the huntingtin protein, htt. Since the identification of the HD gene nearly 15 years ago, there has been enormous progress in understanding the molecular features of HD pathology at the cellular level, as well the development of a plethora of HD models in the mouse and other experimental organisms. These studies have led to the beginnings of a systematic process of target selection and validation in the service of formulating rational strategies for pharmaceutical development of HD therapeutics (see Chapter 4, this volume). Some of these targets have matured to the point of being the subjects of directed translational research programs in academia, biotechnology, and pharma organizations. This chapter will describe some of the features that have made the identification of treatments for HD a challenge and how the field is moving forward in the face of such challenges. In particular, we will focus on the activities of the CHDI Foundation, Inc. to bring disease-modifying treatments to the clinic.

PEOPLE WITH HD HAVE FEW THERAPEUTIC OPTIONS

HD's devastating and relentless progression motivates patients, families, and their physicians to try almost any potential disease-modifying treatment. A few treatments alleviate symptoms, including antidepressants, which improve mood and function; dopamine antagonists, which suppress involuntary movements; and sleeping medication. Only one compound, coenzyme Q (CoQ), has shown even a modest trend toward slowing disease progression [1]. Several clinical trials have attempted to evaluate the effectiveness of putative disease-modifying treatments, but thus far, no successful drug is available. The Huntington Study Group (HSG)—a nonprofit group of clinical investigators from the United States, Canada, Europe, and Australia—has conducted many of the HD clinical trials, including CoQ, ethyl-eicosapentaenoate (ethyl-EPA), tetrabenazine, memantine, phenylbutyrate, dimebolin hydrochloride, and minocycline, but none of these compounds was successful. Members of the European Huntington's Disease Network (EHDN)—a similar group based in Europe—have evaluated riluzole, adding it to the list of unsuccessful drugs. (See Chapter 12, this volume, for more discussion of the clinical trial pipeline in HD.)

    To date, the rationale for clinical trials has been the perceived ability to address some aspect of HD pathogenesis. For example, CoQ, available over the counter as a nutraceutical, is a naturally occurring and endogenous enzymatic cofactor that also functions as an antioxidant. CoQ participates in the mitochondrial respiratory chain and thereby contributes to ATP generation. In a small trial, CoQ demonstrated benefit in patients in the early phase of Parkinson's disease, apparently slowing disease progression [2]. Because energy deficits and oxidative stress may also be important in HD pathogenesis and because CoQ is safe and readily available, HSG investigators set out to evaluate it first for safety, tolerability, and availability and then for possible benefit. However, in an open-label dose-escalating trial, HSG found only a trend toward benefit after 20 weeks of relatively high doses of CoQ [3]. A subsequent 30-month study reported a positive trend that did not reach statistical significance [3]. The CoQ study was the only trial to use disease progression, rather than chorea, as an endpoint. Similarly, a double-blinded placebo-controlled trial investigated the potential benefit of ethyl-EPA, a compound with multiple relevant putative mechanisms of action [4]. (See Chapter 2, this volume, for additional discussion of pathogenic mechanisms in HD.) Research subjects received 1 g/day of ethyl-EPA. A positive trend was noted in the total motor score among patients who followed the protocol, but the study did not meet the statistical significance endpoint [5].

Bonelli and Hofmann [1] have systematically reviewed all the published HD clinical evaluations and classified them according to study design. In their nomenclature, a "Level 1" study is a randomized controlled study with at least 2-week duration and more than 10 HD subjects enrolled for the entire study. Of the published Level 1 studies, only two—tetrabenazine and amantadine—showed clinical benefit. In both cases, the benefit was symptomatic. Tetrabenazine—an inhibitor of vesicular dopamine transport—is approved in Europe for the treatment of chorea, but the side effects of this compound are so severe that its use remains controversial. Amantadine—an inhibitor of N-methyl-D-aspartic acid (NMDA)-type glutamate receptors—showed a modest positive effect on chorea at 400 mg/day [6], but a subsequent study at a slightly lower dose (300 mg/day) found no evidence of decreased chorea [7].

All the compounds used so far in HD clinical trials fall into two classes, representing two strategies for drug discovery: nutraceuticals and pharmaceuticals developed for other indications. Each of these approaches has serious limitations but nonetheless merit exploration absent more rational alternatives. Nutraceuticals and other natural products often have no clear mechanism of action or molecular target, making it difficult to study structure–activity relationships (SARs) and to optimize their structures for HD therapy. Further, because nutraceutical manufacture is not regulated as is that of pharmaceuticals, the product quality is not consistent and the concentration of active ingredient may vary greatly. Still, because there is no available treatment for HD, a systematic empirical evaluation of nutraceuticals could benefit HD patients and families. If a nutraceutical were safe and well-tolerated, even with only a modest benefit, it would certainly be recommended by clinicians. From the perspective of drug development, pharmaceuticals developed for other indications are a more attractive route. If such a compound showed efficacy in HD, we could build on previous knowledge about its specific molecular target and its mode of action to improve its utility for HD. Further, such a compound, known to modulate a specific molecular target, can serve as a "validating ligand," thereby providing a sharp tool with which to look for alternate points of intervention. Ultimately, none of the compounds used in clinical trials addressed a specific molecular target known to participate in HD pathogenesis. Even if one of these compounds had shown benefit for HD patients, no clear path for optimization would be obvious without knowing its mode of action.

THE LANDSCAPE IS FAVORABLE FOR HD DRUG DISCOVERY

As the population ages, neurodegenerative diseases have become particularly important to society. Both government and industry have increased their support for relevant drug development. HD is a "paradigmatic" neurodegenerative disease, offering many advantages to researchers and clinicians seeking new therapeutic routes. HD results from an autosomal dominant allele that can be identified long before disease onset, and its pathogenesis is likely to involve the same pathogenic processes implicated in other neurodegenerative disorders—excitotoxicity, mitochondrial dysfunction, apoptosis, and transcriptional dysregulation [8]. In addition, basic research into HD pathophysiology has already led to the development of powerful tools, including antibodies to htt, transgenic and knockin mice and rats, cell-based models of mutant htt-induced pathogenesis, invertebrate animal models, and new compounds that address HD-relevant pathogenic mechanisms. Cell-based HD models have been the focus of much effort to identify novel compounds and targets that can modify HD toxicity.

CELL-BASED MODELS SUGGEST NEW THERAPEUTIC TARGETS

Heterologous expression of mutant *htt* in yeast [9] has led to the identification of genes that modify pathogenesis and, using two-hybrid technology, to the elucidation of the network of proteins that interact directly and indirectly with Htt [10,11]. This elaborated Htt "interactome" has already suggested new potential targets for drug discovery. Other cellular models include transiently transfected non-neuronal and neuronal cell lines, primary neuronal cultures, and inducible, stably transfected, neuron-like cell lines [12], as well as cell lines derived from murine HD models. Of the stable neuron-like cell lines, the pheochromocytoma 12 (PC12) [13] and the striatally derived ST14A [14] are the most frequently used. Another cell line of particular interest is STHdhQ111 [15], generated by the conditional immortalization of striatal neuron precursors from embryos of the Hdh^{Q111} knockin mouse. These tools are not only contributing to basic research on HD pathogenesis but they are also providing platforms for screening chemical libraries.

IN VIVO MODELS OF MUTANT HTT ACTION ARE AVAILABLE IN RATS, MICE, WORMS, AND FLIES

In vivo studies of HD pathogenesis have been problematic. Several invertebrate HD models have come from the roundworm *Caenorhabditis elegans* [16] and the fly *Drosophila melanogaster* [17]. Although these models are particularly convenient for genetic studies and perhaps for compound evaluation, their direct clinical relevance to human pathogenesis remains uncertain. HD researchers have also used at least 10 transgenic and knockin mouse models and one transgenic rat model [18,19].

The most frequently used rodent models are the R6/2 and R6/1 transgenic mouse models [20], which each produce an exon 1-derived fragment of mutant Htt, with approximately 150 glutamine residues. Until now, most *in vivo* drug evaluation has used R6/2 because it develops neurological signs most rapidly. Although its rapid pathogenesis has led to its extensive characterization, it is less than an ideal model: (1) it contains only a small fragment of the entire *htt* gene; (2) it is genetically unstable, with the number of glutamine repeats varying from about 100 to more than 300; (3) it has a heterogeneous genetic background that segregates modifying genes; (4) unlike human HD, it shows no substantial neuronal loss in the striatum or cortex; and (5) its altered behavior and cognition do not fully mimic the human condition—not surprising for a mouse [21]. Full-length transgenic mouse models are also available. One model, now no longer available, derived its full-length *htt* from a full-length complementary DNA (cDNA) [22]. Others derived their mutant *htt* from a yeast artificial chromosome (YAC) and some from a bacterial artificial chromosome (BAC) [23–25]. None of these models, however, develop abnormal behavior as rapidly as the R6/2 fragment model. (See Chapter 7, this volume, for extended discussion of transgenic mouse HD models.)

HTT-EXPRESSING CELLS ALLOW HIGH-THROUGHPUT "PHENOTYPIC SCREENING"

Traditional drug discovery depends on target identification and validation, followed by the development of target-directed cell-free assays. Cell-based models can also provide "phenotypic assays" that allow screening of compound libraries. In this complementary "biology-driven HTS" (high-throughput screening) [26] approach, the drug developer seeks compounds that modulate phenotype in disease-relevant biological assays, for example, mutant *htt*-dependent cell death, that incorporate biological context and complexity upfront.

There are reports of several drug-like small molecules that rescue a mutant *htt*-dependent phenotype. One of these, called C2-8, emerged from a screen for small molecules that inhibited the aggregation of mutant exon 1-derived Htt fragments in PC12 cells [27]. Another compound, B2, emerged from a screen of compounds that induced aggregation but mitigated proteasome dysfunction [28]. Another screen found 29 molecules that prevented mutant *htt* exon 1 cytotoxicity in ST14A cells, a neuron-like cell line derived from E14 striatal primordia [29].

Since the discovery of the *huntingtin* gene, basic research has suggested more than 300 molecular targets where intervention could slow or stop pathogenesis [30]. During this time, the biotechnology and pharmaceutical industries, as well as government and university laboratories, have developed powerful new pharmacological tools under the rubric of "chemical genomics"—the application of the tools of small molecule drug discovery to the development of research tools that can improve our understanding of protein and cell function. Particularly useful products of this activity are "validating ligands," which specifically interact with individual receptors, enzymes, and other proteins [31]. These new pharmacological and biological tools will be increasingly useful for the identification and validation of molecular targets that we can then address by traditional target-based discovery (see Chapter 4, this volume).

WHAT IS THE FUNCTION OF WILD-TYPE HTT?

Despite the publication of more than 1,000 articles dealing with some aspect of htt, we still have little idea of its normal function(s). Htt is required for embryonic development, and it plays a role in regulating gene expression, especially in the transcription of the gene encoding brain-derived neurotrophic factor (BDNF) [32,33]. Htt also participates in intracellular trafficking [34–37]. HD has long served as a textbook example of a disease that results from a dominant allele, and research has focused on understanding a toxic "gain of function." Some investigators, however, have argued that the disease also results from a loss of normal function. Evidence for loss of function comes not only from studies in animal models but also from comparisons with other polyglutamine expansion diseases, such as the spinocerebellar ataxias.

In the most frequently used cell and animal models of HD, mutant exon 1-derived Htt fragments form nuclear and perinuclear aggregates. These aggregates that contain ubiquitinated Htt appear in many cases to increase sensitivity to toxic insults [38]. Similar aggregates—containing the mutant androgen receptor—accumulate in cell and animal models of spinal and bulbar muscular atrophy (SBMA) [39].

MUTANT HTT AFFECTS SOME CELLS MORE THAN OTHERS, BUT WE DO NOT KNOW WHICH AFFECTED CELLS CAUSE HD PATHOGENESIS

Htt itself, with 3,144 amino acid residues, has limited homology to any family of proteins. Although the presence of a long polyglutamine tract near its amino terminus is certainly the cause of HD, Htt is not obviously amenable to targeting with traditional drugs. Nor does Htt appear to have a specific molecular function that could be modulated by a protein biopharmaceutical. Therefore, the drug developer's task is to look for intervention points both "upstream" (e.g., in transcription and translation) and "downstream" (e.g., in the signaling cascades affected by mutant Htt). Among the many "downstream" processes affected by mutant Htt are (1) mitochondrial dysfunction and impaired energy metabolism (decreased ATP levels), (2) inflammatory processes [40] and excitotoxicity [41], (3) caspase-mediated apoptosis [42,43], and (4) oxidative stress and damage [44,45].

The effects of mutant Htt vary greatly with cell type. In postmortem HD brains, the medium-sized spiny neurons (MSNs) of the striatum are the most affected [46,47]. Although many researchers have adopted MSN vulnerability as the defining feature of HD [42], several recent studies suggest that MSN loss is not cell autonomous but rather results from the deprivation of trophic support from the cortex, specifically from the loss of BDNF [48,49].

The question of cell autonomy is central to the development of appropriate cell models for drug development. Every HD drug developer seeks a cell system in which the cell autonomous action of mutant Htt recapitulates the disease process, but a single cell may never provide an accurate model. The most useful cell-based assay may instead require coculture of two or more cell types.

Essentially all cells make Htt, and despite substantial striatal cell loss, HD pathology may occur in the periphery and elsewhere in the brain. In particular, inflammation may contribute to early pathogenesis both in HD and in mouse models, with activation of microglia in the brain and movement of microglial precursors from the periphery. Microglia produce a number of enzymes in the kynurenine pathway, which is responsible for the degradation of tryptophan. Among the enzymes up-regulated in early HD is kynurenine monooxygenase (KMO), which converts kynurenine into 3-hydroxykynurenine [50], a precursor of quinolinic acid, a naturally occurring neurotoxin. Increased levels of 3-hydroxykynurenine and quinolinate are present in several transgenic mouse models of HD [51].

In fact, in a yeast model of mutant Htt toxicity, KMO deletion protects yeast from the cytotoxicity of mutant exon 1-derived Htt fragment [52]. Furthermore, treating R6/2 mice with Ro-61-8049, a potent inhibitor of KMO, leads to improved motor behavior in both the rotarod and open field tests. Although the results with the Ro-61-8049 inhibitor of KMO are interesting and suggest the involvement of inflammatory cells in HD pathogenesis, distribution studies indicate that the compound does not cross the blood–brain barrier (BBB). Thus, the observed effect has three possible explanations: (1) an active metabolite of Ro-61-8049 passes through the BBB and acts either on the KMO pathway or on an entirely different pathway; (2) the compound inhibits KMO in the periphery, altering metabolites in the body; or (3) the Ro-61-8049 exerts its effect systemically—either via the KMO pathway or an alternative pathway.

"Phenotypic" Cell-Based Screens Have Identified Molecules That Modulate Htt Cytotoxicity

Because of the lack of tractable validated molecular targets, cell-based drug discovery has focused on "phenotypic assays," in which mutant *htt* expression induces a cytotoxic cascade. One advantage of such a phenotypic assay is that the researcher can simultaneously address multiple targets in a single assay. For such a phenotypic assay to be meaningful, however, the choice of the cell line used is crucial. Usually the cells used for such assays are easily manipulated cell lines, but few of these lines are of neuronal origin, possibly limiting their relevance to neuropathogenesis. Because HD's most prominent pathology is the loss of MSNs in the striatum and of pyramidal cells in the cortex, a better strategy would use cells that derive from those cells or at least share as many characteristics as possible with them.

A cell-based phenotypic assay also requires a choice of pathogenic *htt*. Among the constructs used so far are pure polyglutamine, *htt* exon 1, otherwise truncated *htt*, and full-length *htt* [16,53–57]. Cells that express mutant exon 1 show a more rapidly developing phenotype than those expressing longer fragments, but they do not allow the assessment of effects that depend on more carboxyl-terminal portions of Htt. Because mutant Htt is cytotoxic, the cell lines that constitutively express mutant *htt* die, unless they are resistant to such cytotoxicity. Such resistance can result from cellular adaptations, possibly from increased production of cellular machinery that sequesters toxic proteins, or from selection of mutant *htt*-resistant mutants. To minimize these problems, investigators have used inducible expression systems, enabling them to initiate the pathogenic cascade at will. Examples of cell lines used in HD drug discovery include PC12 cells with ecdysone-dependent expression of *htt* exon 1, primary embryonic striatal neurons transfected with *htt* exon 1, and a striatal cell line derived from a knockin mouse model [53,58,59]. Selecting a measurable readout poses yet another problem for a phenotypic assay. Choosing a single readout, such as cell death, may miss earlier and more subtle events of the cytotoxic cascade, but these early events may or may not be relevant to the cell loss that characterizes HD. Therefore, CHDI and our collaborators are developing and using high-content technologies, which can monitor multiple readouts simultaneously with the hope of a fuller understanding of cellular pathology.

Drug developers in companies and in universities have used several cell-based assay systems for HD phenotypic screening. For example, Trophos SA (Marseille, France) has driven cytotoxicity in rat embryo primary striatal neurons with a trans-fected 480-amino acid fragment of mutant *htt*. Their readout was fluorescence from an Htt carboxyl-terminal green fluorescent protein tag used as a marker of neuronal viability [60]. Cellumen's (Pittsburgh, PA) high-content assay measures aggregation of exon 1-derived Htt fragments after ecdysone [53] induction in PC12 cells. CombinatoRx's (Cambridge, MA) assay follows the subcellular distribution of mutant Htt in Stl11 cells, a line derived from a homozygous knockin model with 111 glutamine residues. Finally, Varma et al. [61] have used the ST14A cell line in cytotoxicity screens.

TRANSGENIC MICE ALLOW *IN VIVO* EVALUATION OF COMPOUNDS

The R6/2 mouse carries exon 1 of the human *htt* with between 105 and 225 CAG repeats [20]. Because the R6/2 develops abnormal behavior by 4–6 weeks, CHDI and others have used it most frequently to evaluate potential drugs. However, even when investigators have studied the same compounds in the R6/2 mouse model, they have found notable differences in outcome. These differences appear to arise from the different protocols for behavioral testing and from different husbandry [62]. High Q Foundation and CHDI, working with PsychoGenics (Tarrytown, NY), have developed standardized protocols for husbandry and testing to reduce variability and to increase the robustness of behavioral testing.

Furthermore, because R6/2 contains only *htt* exon 1, it is not useful for testing all candidate therapeutics, for example, compounds expected to inhibit the cleavage of full-length Htt or to alter the phosphorylation of residues not encoded by exon 1. Still, the R6/2's rapid development recommends it, and its pathology resembles that of the Q150 knockin mouse, suggesting that full-length Htt is not necessary for the initiation of disease and that testing in R6/2 is not irrelevant [63]. Still lacking, however, are compounds that can serve as positive controls in cells and in several animal models. Validation of a compound—showing that it slows pathology in HD and in animal and cell models—would also validate those models, but this goal still appears distant. At least seven full-length mouse models are also available. These differ in behavior, pathology, and genetic background and thus far have had limited use for drug evaluation [25,64]. (See Chapter 7, this volume, for extended discussion of transgenic mouse HD models.)

POTENTIAL HD THERAPIES HAVE PARTICULARLY SEVERE SAFETY REQUIREMENTS

Because genetic testing can reveal the disease-causing allele long before clinical signs, we must anticipate long-term dosing of healthy young adults. Any HD therapeutic agent must be extremely safe and well tolerated, with "tolerable side effects" determined by the profile of the people who receive the therapy. Physicians and drug developers faced similar issues in treating chronic psychiatric illnesses and HIV-positive individuals. People with HIV, for example, often lead long and productive lives using the current triple therapy, but many of these drugs can cause serious side effects that reduce the quality of life. Physicians treating HIV-positive subjects assess the tolerability of medications by evaluating the impact of the side effects on daily living and on perceived quality of life [65]. Compounds used for HD treatment will require a better safety profile than compounds used for cancer because untreated illness so often leads to rapid death and patients receive medication for shorter times. The challenge for HD will be to identify disease-modifying compounds with a good therapeutic window.

THE DRUG DISCOVERY PROCESS

CHDI's MISSION IS TO "RAPIDLY DISCOVER AND DEVELOP DRUGS THAT PREVENT OR SLOW HUNTINGTON'S DISEASE"

CHDI is unusual among drug developers: we are a not-for-profit foundation, and our motivation is time. Our strategy embraces both biology-based screening and traditional target-based drug discovery, and we are agnostic about therapeutic modality. We are pursuing—in parallel—options that range from small molecules, to proteins, to gene- and cell-based therapies. Notwithstanding our unusual strategy, we use industry standards to pursue our goals, to track our successes, and to determine the appropriate level of resource commitment. Our metrics for pharmaceutical development, from target to investigational new drug (IND) (illustrated in Figure 8.1), are most applicable to small molecules but are easily adapted to our modality-agnostic approach. We adapt our definitions of common drug discovery terms (such as "hit," "lead," or "preclinical candidate") from the *ISOA/ARF Drug Development Tutorial* [66] and the *Assay Guidance Manual Version 4.1* (commonly called the *NIH Quantitative Biology Manual*) [67].

There are two kinds of screening strategies: "target based" and "phenotypic." To help us prioritize potential targets for HD therapeutics, CHDI has developed a target validation (TV) scale, described in Chapter 4, this volume. Once we have decided

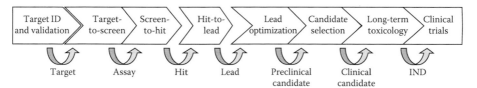

FIGURE 8.1 Representation of the pharmaceutical development process from target to IND.

that a target has a sufficiently high TV score, our next step is to develop an assay in which we can identify molecules that modulate that target. We then adapt the assay to a high-throughput format, usually a 386-well plate.

We have several metrics for a successful assay: (1) it should be inexpensive, costing less than \$0.50 per well for a biochemical assay and less than \$2.00 per well for a cell-based assay; (2) it must be robust, with an assay quality factor (Z') greater than 0.5 (Z' is 1.0, less three times the ratio of the sum of the standard deviations of signals from positive and negative controls to the difference of their means); (3) it must be reproducible, with a coefficient of variation of less than 20%; and (4) it must be quick, allowing a maximum turnaround for a set of tested compounds of 1 week and a total time for lead generation of 3 to 6 months. Once we have an assay that meets these criteria, we can initiate the production of reagents and the launching of the screen. At the same time, we start a program to study the selectivity and the mechanism of action of potential hits.

Figure 8.1 illustrates the stages of compound progression from "active" to "hit" to "lead." A compound is called "active" when it meets a threshold level of activity in the primary screen. A compound becomes a "hit" only when it satisfies the following criteria: (1) confirmed identity and purity; (2) a reproducible dose–response curve, with a biochemical IC_{50} less than 5 μM or a cellular EC_{50} less than 30 μM; (3) confirmed mechanism and specificity; (4) a relationship between structure and activity across chemically related analogues; and (5) chemical tractability—the molecule must have no obvious intrinsic chemical liabilities. Declaration of a compound as a hit triggers the next sequence of activities: (1) selection of appropriate secondary biochemical and cell-based assays; (2) mechanistic studies; (3) synthesis of more chemical analogues to explore SAR; and (4) studies of absorption, distribution, metabolism, and excretion (ADME), as well as solubility, metabolic stability, cell permeability, and cytotoxicity.

Advancing a hit to a "lead" requires the satisfaction of additional criteria: (1) a series of analogues of the lead compound must show selectivity and *in vitro* SAR consistent with the target; (2) the compound and its active analogues must be active *in vivo*; (3) at least one compound must have biochemical IC_{50} less than 100 nM or cellular EC_{50} less than 1 μM; (4) potential issues associated with physiochemical properties, *in vivo* ADME, and biopharmaceutical properties of the lead series must be identified; and (5) we must be able to plot a clear, data-supported path to lead optimization. Only when we declare that a compound is a lead do we commit resources for lead optimization. At that point, we again scale up the production of several related molecules for *in vivo* pharmacokinetic (PK) and (pseudo)efficacy studies. Lead optimization involves more ADME studies (tier 2 ADME) and toxicology, using appropriate surrogates, for broad-based pharmacological profiling, *in vitro* toxicological studies, protein binding, and the induction of cytochrome P450, which can selectively metabolize potential lead compounds.

The next step in our scheme is progression from lead to *preclinical candidate*, which requires that at least one molecule satisfy all the following criteria: (1) *in vitro* EC_{50} less than 100 nM, ideally in a cellular assay; (2) at least 100-fold selectivity compared with the most closely related target (including isoforms); (3) a useable formulation; (4) *in vivo* efficacy (or pseudoefficacy) in at least one (transgenic) animal model; (5) at least 25% bioavailability at 30 mg/kg; (6) a brain exposure greater than the EC_{50} with the selected route of administration and formulation, preferably in two animal species; (7) no genetic toxicity at 50 μM; and (8) acceptable *in vivo* safety in a 7-day rodent study.

Declaration of a compound as a preclinical candidate triggers an effort to determine whether it is an acceptable "clinical candidate," in particular its synthesis at a scale that can support both *in vivo* efficacy studies and 14-day rodent toxicology. A backup molecule from the same series must also be selected at this point. Before filing an application for approval as an IND, a compound must meet the additional criteria of a "clinical candidate": (1) reproducible and statistically significant positive results in at least one relevant outcome measure in at least one mammalian HD model; (2) demonstrated mechanism *in vivo*, as determined by direct or surrogate measurements; (3) acceptable subchronic toxicity in a 14-day non-Good Laboratory Practice (GLP) rodent and nonrodent toxicity study, with acceptable toxicokinetics; (4) acceptable ADME profile; and (5) low liability for cardiotoxicity as determined by QT interval prolongation in dog.

Declaration of clinical candidacy triggers scaled-up synthesis of the compound under conditions of GLP for the studies required for IND approval. Once the preferred formulation of the molecule is chosen, the chemical, manufacturing, and controls studies are initiated, including two safety studies of at least 28 days, with at least one study in a nonrodent species. Depending on the specific nature of the clinical candidate and its potential toxicities, the compound may also require additional *in vitro* evaluation and *in vivo* studies.

The Current Pipeline

Five distinct streams of candidate therapeutics, described below, currently feed CHDI's *in vivo* testing program, which is hosted at several contract research organizations: (1) we have assembled a list of "validating ligands" for molecular targets of high interest, starting with those identified through the community-based Systematic Evaluation of Therapeutics for Huntington's Disease (SET-HD) initiative [68]; we have added compounds used for other neurodegenerative disorders, including many that address specific molecular targets on our extended list; (2) we are working to improve a set of compounds identified in previous high-throughput screens, supported by the CHDI Foundation, that used several biology-driven phenotypic screens; (3) we are also attempting to improve compounds, such as CoQ, which are the subject of ongoing clinical trials, using medicinal chemistry to improve the therapeutic window; (4) we have assembled a portfolio of new target-based drug discovery programs, as well as programs that use new technologies; and (5) we have initiated several projects for nontraditional therapeutics, including antisense technology or gene therapy. We are presently using the R6/2 transgenic mouse as a frontline *in vivo* model to evaluate and prioritize candidate molecules.

Many of the validating ligands that CHDI has evaluated *in vivo* are molecules already used in clinical studies. The rationale for these choices was that, if a compound should exhibit efficacy, the regulatory hurdle to initiate a clinical study would be low, and clinical studies could be initiated rapidly. Among the 60 compounds already evaluated, about half come from the SET-HD process. Before choosing the compounds for *in vivo* testing, we reviewed each nominated compound for appropriate physical, chemical, PK, and safety properties. This process led us to reject SET-HD compounds such as geldanamycin, which is prohibitively expensive and had significant PK and toxicological liabilities [69]. In such cases, however, we sought alternative compounds that could address the same putative mechanism of action.

In addition to the SET-HD compounds, CHDI scientists nominated best-in-class ligands for specific targets and mechanisms. For example, to address the role of oxidative stress in HD pathogenesis, we chose to evaluate the free radical scavenger NXY-059 (Cerovive, AstraZeneca, Wilmington, DE), which had progressed to Phase III clinical trials for cerebrovascular ischemia [44]. Similarly, to examine the potential pathogenic role of adenosine 2A (A2A) receptors, we evaluated the A2A receptor antagonist KW-6002, which is in Phase III clinical trials for Parkinson's disease [70].

Phenotypic cell-based screens have provided another list of compounds for *in vivo* evaluation. In 1998 the Cure Huntington Disease Initiative (predecessor of the current CHDI), initiated an HTS campaign, mostly using mutant exon 1 cytotoxicity assays in PC12 cells, *Drosophila*, *C. elegans*, and yeast. This primary screening campaign found 181 active compounds with EC_{50} less than 10 μM from a total of 815,000 compounds. These initial hits, together with commercially available analogues, were the input to a secondary screening campaign whose purpose was to confirm the original activity and to begin the elaboration of SAR required for lead optimization.

The first step in winnowing the original 181 active compounds and analogues to 12 compound series was computational filtering. Filtering keeps identified non-productive types of compounds ("chemotypes") from advancing further, thereby saving valuable chemistry resources. The computational filters identified molecules known to aggregate; to bind other molecules promiscuously; to have inappropriate, non-drug-like characteristics or reactive groups; and to be metabolically unstable. To facilitate the SAR studies we sometimes allowed some non-drug-like compounds, such as flavonoids (known to be difficult to optimize for therapeutic use), to pass the filters. However, when we ranked the ensuing compounds, they had low priorities.

The secondary screening campaign used seven assays to find 165 compounds, from 14 chemical classes, that showed activity in at least one assay. Of these, 45 exhibited activity in two assays, and five compounds were active in three assays. Four of these five represented chemotypes that were active in one or more assays. One singleton compound, with a chemotype unique among the original 165, was active in three assays. The structure of this compound predicts that it can penetrate the BBB and that it may function as a prodrug, converted to an active compound by enzymes present in the assay. Another singleton was an adenosine receptor A2B antagonist.

Thirteen compound classes showed promising SARs and underwent additional evaluation with the following criteria: (1) amenability to chemical modification at multiple sites; (2) potential to improve PK and ADME; (3) ease of analogue generation; (4) parallel chemistry synthesis; (5) drug-like characteristics; and (6) predicted ability to penetrate the BBB [71].

We classified the remaining compounds into chemical series based on common chemical scaffolds and listed 16 series and 30 singletons. The computational filters discussed above allowed us to reject compounds with reactive functionality, promiscuous binding, tendency to aggregate, and non-drug-like structures. We then grouped the resulting compounds into classes, prioritized them according to the activity of each class in multiple HD-relevant assays, and synthesized libraries of appropriate analogues for testing in PC12 and ST14A cell-based assays. The results of these ongoing tests will determine which analogues will move forward to *in vivo* evaluation.

Medicinal Chemistry May Be Able to Improve Compounds Already Evaluated in Clinical Trials

Several compounds—notably creatine and CoQ—have shown positive trends toward efficacy in HD clinical trials but only at extremely high doses. CHDI has tried to improve the efficacy and bioavailability of these two compounds. The treatment of HD patients with creatine and CoQ means to repair defects in energy metabolism and mitochondrial dysfunction associated with HD. Creatine is the metabolic precursor of phosphocreatine, as a temporary energy for ATP-derived high-energy phosphate, whereas CoQ participates in the mitochondrial respiratory chain. Both molecules have substantial deficiencies as drugs, and neither readily crosses the BBB. CoQ, with cLogP greater than 20, has very poor tissue distribution characteristics [72,73]. Creatine, at the high concentrations used in clinical trials, may well saturate BBB's creatine transporter [74,75].

CHDI has initiated two medicinal chemistry collaborations to improve these molecules, with the goal of shifting their therapeutic windows and allowing more definitive studies of the value of enhancing mitochondrial function in HD. We are working with Edison Pharmaceuticals (Mountain View, CA) to generate CoQ analogs that both are more bioavailable and have higher redox potential, and with XenoPort, Inc. (Santa Clara, CA) to develop a BBB-permeable prodrug that will generate phosphocreatine.

Edison has adopted a dual approach to re-engineering CoQ: (1) modifying the *para*-benzoquinone headgroup to optimize redox potential, and (2) simultaneously modifying the lipophilic tail to improve ADME properties. Edison has assembled a collection of CoQ analogues based on their performance in separate assays of redox potential and their ability to rescue cells from acute oxidative stress. To address the specific insult of mutant htt, Edison measures the ability of their CoQ analogues to rescue human HD fibroblasts producing Htt with 69 glutamine residues from an acute oxidative challenge. Although this assay allows them to rank compounds, the ordering may not be relevant to HD itself: the mitochondrial activities of dividing fibroblasts almost certainly differ from those of postmitotic neurons. Despite this reservation, Edison has identified a CoQ analogue with a cellular $EC_{50} < 30$ nM and is now testing this and other compounds *in vivo*.

XenoPort's program is attempting to produce a prodrug of creatine phosphate that readily crosses the BBB. Their strategy is to attach nonpolar moieties around the highly charged phosphate group to produce a much less polar molecule whose high-energy phosphate is protected from plasma phosphatases. The attached moieties, which are also substrate BBB transporters, are designed to be cleaved by esterases within the brain. As a proof of principle, XenoPort has already shown that such prodrugs can rescue cells subjected to metabolic insult.

CHDI Is Developing Small Molecule Drugs
to Address High-Priority Targets

In Chapter 4, this volume, our CHDI colleagues discuss our metrics of TV. The resulting ranking has allowed us to: (1) quantify our confidence that pharmacological intervention at a given target will modify disease progression; (2) specify additional experiments that can increase or decrease a target's rank; and (3) prioritize targets for prosecution. Table 4.2 lists the highest ranking targets from those efforts (see Chapter 4, this volume). We expect that some of these targets are most amenable to small molecule therapeutics, and CHDI is trying to develop drugs that address these targets, which include (1) histone deacetylases (HDACs), especially class I and II; (2) caspase-6; (3) caspase-1; and (4) transglutaminase 2. Other targets on the list, however, including Htt and growth factor receptors (for BDNF, glial cell derived neurotrophic factor [GDNF], and fibroblast growth factor [FGF] 2) will probably require alternative modalities.

To choose among individual caspases and HDACs, CHDI has initiated a collaboration with Amphora Discovery Corporation (Research Triangle Park, NC). Amphora has profiled a collection of diverse compounds against individual caspases and HDACs, and has developed a database of the selectivity of each compound against all the enzymes tested [76]. By mining this database, Amphora has identified chemical scaffolds that can serve as seeds for subsequent library generation and additional screening, resulting in a dramatic shortening of hit-to-lead timelines for individual family members. Currently, CHDI's HDAC and caspase projects are exploiting several identified scaffolds as lead series in the search for isoform-selective ligands with the requisite central nervous system (CNS) exposure needed for *in vivo* studies.

CHDI and CombinatoRx are exploring the existing pharmacopoeia (and some new chemical entities) for combinations that inhibit cytotoxicity in cell-based phenotypic assays. These empirical studies, which do not rely on previous knowledge of target interactions, have discovered unexpected synergies—the pharmacological equivalent of synthetic lethal interactions found in yeast and other model organisms [77]. Fully understanding these synergies will require a marriage of chemogenomics (using small molecules as biological probes) and systems biology (the study of complex interactions).

CombinatoRx argues that its platform may also dramatically reduce the time to the clinic because its starting compounds are already Food and Drug Administration (FDA)-approved drugs. Even if this program does not result immediately in an HD therapeutic, data about pair-wise interactions can lead to new insights relevant to our efforts to identify and validate HD drug targets [78].

LARGE MOLECULE THERAPEUTICS REQUIRE UNCONVENTIONAL DELIVERY

Mutant Htt is the only fully validated target for HD, and reducing its concentration should suffice to slow or prevent HD. With this goal in mind, CHDI is exploring two routes to reduce Htt production—antisense oligodeoxynucleotides (ASO) and short-hairpin RNA interference [79,80] (see Chapter 9, this volume). Our ASO approach, in partnership with Isis Pharmaceuticals (Carlsbad, CA), uses technology developed by Isis—single-stranded modified oligonucleotides complementary to *htt* mRNA. Once an ASO binds to a portion of the target mRNA, the mRNA becomes a substrate for nuclease digestion, thereby preventing its translation. Isis has the capacity to rapidly identify and optimize ASO drug candidates, as well as the means to determine efficacious dosing in humans and experimental animals. The company has previously characterized the subchronic toxicities of ASOs, and they have studied tissue (and brain) distribution and PK and pharmacodynamics (PD) after intracerebroventricular (ICV) or intrathecal delivery.

Isis's work to date is promising. They have identified candidate *htt* ASOs from screens in human and murine cell lines, and they have confirmed the *in vitro* silencing of both wild-type and mutant *htt*. *In vivo* toxicology and pharmacology studies are complete in mice, and efficacy studies are underway in a transgenic HD model. Optimization of backbone chemistries and gap sizes for our CNS application of the ASOs are still pending, as are longer safety studies, but this program is on a fast track toward further IND-enabling studies.

Although the Isis ASO program now envisions direct injection into the brain or into cerebrospinal fluid, we are also working on methods that would allow oral or parenteral drugs to cross the BBB. The BBB, which protects the brain from blood-borne infectious agents, prevents circulating proteins and most small molecules from entering the brain, thereby posing an enormous challenge for neurotherapeutic agents [81]. Small molecules that passively cross the BBB—lipophilic compounds with molecular masses less than 500—represent but a tiny fraction of the chemical universe and do not include proteins and oligonucleotides. Some xenobiotics do cross the BBB, however, because they contain functional groups recognized by BBB transporters. The XenoPort phosphocreatine prodrug program, discussed previously, attempts to exploit such carrier-mediated transport.

We are seeking to exploit carrier-mediated transport to overcome the BBB's barrier to large molecules such as proteins. In the meantime, we are exploring more invasive modalities to seek a proof of concept for potential therapeutics. A number of such techniques are available: (1) ICV injection, with convection enhanced diffusion; (2) transient chemical disruption of the BBB; (3) focal BBB disruption with high intensity focused ultrasound; and (4) transnasal delivery (for a restricted class of molecules). Such invasive modalities, however, cannot be used to treat a condition that may require life-long dosing. The much preferred route exploits the BBB's own receptor-mediated transport (RMT), which is responsible for delivering selected protein or peptide cargoes to the brain. The best characterized of these transport systems involves the transferrin receptor [82]. This receptor can bind either transferrin or peptidomimetic monoclonal antibodies on the outside of the endothelial cells that comprise the BBB. These cells then move the receptor–ligand complex through the

BBB by endocytosis. The RMT system acts as "molecular Trojan horses" to ferry drugs, proteins, and nonviral gene medicines across the BBB [81].

Investigators have developed several molecular *targeting vectors*, often monoclonal antibodies, that can carry macromolecules and immunoliposomes—phospholipid vesicles, about 90 nm in diameter, coated with targeting vectors—across rodent and nonhuman primate BBBs [83–88]. The monoclonal antibodies, which bind to the transferrin (or other) receptor, are attached to polyethylene glycol (PEG) that decorates the liposome exterior and protects it from metabolization. Therefore, the liposomes are said to be "PEGylated." PEGylated immunoliposomes bind to the receptor and then move into the endothelial cells by endocytosis.

CHDI and Hermes Biosciences (South San Francisco, CA) have entered into a collaboration to explore the utility of immunoliposomes for HD drug delivery. Hermes has extensive experience with liposomal formulations for oncology, and it has demonstrated the ability to produce targeted immunoliposomal therapeutics with the standards required for human clinical trials [89,90]. This collaboration has three goals: (1) delivery of small molecule validating ligands that would not otherwise be available in the brain; (2) delivery of therapeutic proteins, via nonviral gene therapy [91]; and (3) delivery of ASO therapeutics. Success with this platform would allow CHDI to explore the possible therapeutic use of neurotrophic factors such as BDNF, FGF2, and GDNF (and the genes encoding them).

A Validating Ligand Must Bind to Its Target *In Vivo*

To use a validating ligand to test a molecular target, that compound must engage the target *in vivo*. Such a demonstration requires adequate exposure in the tissue of interest, as well as PD markers of target engagement. Such measurements may be direct, for example, positron emission tomography (PET) measurements of receptor occupancy by radioligand displacement, or indirect, for example, histone acetylation after HDAC inhibition. In the antisense program, the PD marker is the level of Htt or of *htt* mRNA, whereas in the KMO program, the markers are the downstream metabolites of the kynurenine pathway. In the caspase program, the PD readout is the amount of Htt fragments, determined immunologically.

Although measures of target engagement at the preclinical efficacy stage have been stressed, their absence once a compound is on a clinical trajectory can prove costly; the failure of the NMDA receptor glycine site antagonist gavestinel may have been one such example [92].

The first requirement for pharmacological TV is a molecule that is sufficiently potent and selective *in vitro*. *In vivo* validation then requires identification of a suitable vehicle and the determination of a route for administration that gives appropriate exposure at the desired site of action. The design of TV experiments must be sensitive to mode of action: for example, it is pointless to evaluate a caspase-6 inhibitor in the R6/2 model because the exon 1-encoded Htt fragment does not contain the putative caspase-6 cleavage site. Similarly, the exon 1-derived *htt* mRNA in R6/2 does not contain the sequence complementary to the antisense oligonucleotides that Isis discovered to be most effective in reducing Htt levels *in vitro*; therefore, such studies must use a full-length (or at least sufficiently long) mutant transgene.

IN VIVO EVALUATION REQUIRES ATTENTION TO SCALED-UP SYNTHESIS AND FORMULATION

Formulation, PK, and *in vivo* testing require much more compound than *in vitro* studies. Some compounds may be purchased, but others require contract synthesis. Once a sufficient quantity of a compound is acquired, it then enters CHDI's chemical repository, where it is registered and checked for identity and purity (>95%). Next, we dispatch the compound for pre-efficacy formulation, whose goal is to produce a formulation and dosing regimen with an acceptable safety profile to test in mice for several months.

Pre-efficacy formulation starts with the determination of the aqueous solubility of the compound at pH 4.5, 7, and 8, the tolerable pH range in mouse models. The next step is to identify suitable formulations for both parenteral and oral administration. Ideally the compound should dissolve at concentrations of 20 mg/ml, enough to allow relatively high doses. In practice, however, we may need to settle for concentrations of 10 mg/ml or even 1 mg/ml. Some compounds have proven insoluble even at 1 mg/ml and require oral dosing as a methylcellulose suspension.

After pre-efficacy formulation, the next steps are to determine the route of administration and the compound's ability to traverse the BBB. This evaluation, which uses a single-dose PK study in wild-type mice, measures the total plasma and brain exposure (the area under the curve, representing the integral of concentration over time) after intravenous (IV), subcutaneous, intraperitoneal, and oral administration. For a compound whose half-life is 12 hours or less, this study extends for 24 hours, during which time its concentration is measured eight times in the plasma and four times in the brain.

Single-dose PK studies have established dosing, route, and formulation for some 60 compounds of interest to CHDI, but 10 of these did not appear to penetrate the BBB. That is, these compounds reached a concentration in unperfused brain of less than 2%–4% that in the plasma, which is consistent with the contribution of blood in the unperfused brain [93]. Such a failure can prevent *in vivo* study, but not always, because some compounds act peripherally. For example, a pan-caspase inhibitor can modulate the inhibition of interleukin 1 induction in the periphery, thus reducing peripheral inflammatory response. This effect may also extend to immunomodulatory cells (macrophages and microglia) that may contribute to neuroinflammation within the CNS [94].

Compounds that are expected to act only within the CNS but that do not cross the BBB require re-evaluation. The single-dose PK protocol uses 5 mg/kg for IV administration and 10 mg/kg for other routes, and some compounds do not have significant CNS exposure at these doses. If the literature suggests that higher doses are tolerated and efficacious in disease models, then we move to a higher dose with the hope of finding sufficient compound exposure in the brain. In some cases, however, a compound does not penetrate the BBB even at higher concentrations, and our next step is to search for an active metabolite that does enter the brain. In the case of tetrabenazine, for example, a BBB-penetrant metabolite reduces chorea in HD, although tetrabenazine itself has little CNS exposure [95].

Even if a metabolite enters the brain, it may not act in the CNS. Peripheral exposure to the parent compound or its metabolite may be responsible for the pharmacological effect. For example, systemic administration of a compound may change the level of endogenous peripheral metabolites, and these compounds may cross the BBB and lead to effects within the CNS. To overcome such ambiguity requires both PK studies and PD readouts, as described previously. A compound may advance toward *in vivo* efficacy studies only if a target engagement correlates with plasma levels of the test compound, even in the absence of CNS exposure. A PD readout is required for such a conclusion, but such measurements are not available for every compound, and unfortunately we have not always been able to establish PK/PD correlations, especially because of the limited number of tools (such as PET radioligands) available for studying CNS effects.

After determining the optimal route and dosing regimen, the next goal is to establish tolerability, usually starting with a single-day toxicity study at three doses, typically at half-log steps. A second study follows, usually for 2 weeks with three doses. To prevent the selection of a dose that may confound the efficacy study, the tolerability study must encompass the range of doses contemplated for the long-term efficacy trial. The dose range follows from a review of the literature and an examination of the single-dose PK data. During the 2-week study, we monitor neurological signs with a modified Irwin test every 3 days [96], and we collect plasma and brain samples 24 hours after the last dose to determine the accumulation of compound.

INITIAL EFFICACY SCREENING *IN VIVO*

For each compound that jumps the hurdles discussed above, we now do efficacy testing in the R6/2 mouse using the determined formulation and dosing. The mice begin treatment at 4.5 weeks (about 10 days after weaning), and treatment continues until death. Our usual study design uses three cohorts, with two doses of each compound plus vehicle alone. Although three or more doses would give more information, such a design would significantly limit the number of tested compounds.

In collaboration with our colleagues at PsychoGenics, we have decided to stage our efficacy trials to increase the number of evaluated compounds. In this design, the first cohorts of mice ("stage 1") are sufficient to detect a trend ($0.05 < P < 0.3$) of a 20% improvement in motor function or survival. With the robust tests now used, 12 mice per cohort are sufficient. When a positive trend appears, testing proceeds to "stage 2," which is powered to achieve statistical significance ($P < 0.05$) for a 20% effect. Stage 2 includes a larger array of motor readouts than stage 1, as well as cognitive measurements and histological analysis.

Stage 1 evaluates multiple motor behaviors, including the rotarod, grip strength, and open field behavior (e.g., distance traveled, rearing, and speed) (Figure 8.2). Our current design uses separate cohorts to evaluate cognitive performance and histopathology, but—having discovered that the two types of tests do not interfere with one another—we are now in the process of combining motor and cognitive testing in the same stage 2 cohorts. If any of the stage 1 outcome measures exhibit a positive trend, the compound is considered for stage 2 testing. The decision to advance a compound to stage 2 depends partly on a comparison of stage 1 data with *in vivo* data from other compounds with similar molecular targets or mechanisms of action.

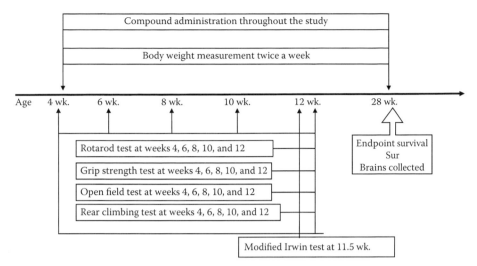

FIGURE 8.2 The R6/2 stage 1 testing paradigm that primarily measures motor and survival endpoints.

The evaluation of compounds in full-length models now takes close to a year, limiting the practical utility of these animals for initial compound screening. If stage 2 studies confirm a compound's efficacy in R6/2 mice, however, we then test it in one of the more slowly developing full-length transgenic models, either in YAC128 or BAC-HD mice. Efforts are underway to establish early and robust readouts from these full-length models.

Compounds Entering the Clinic Must Be Safe

Demonstration of efficacy in one or more animal models still does not necessarily mean that a compound is an appropriate candidate for clinical evaluation. Any compound that enters the clinic must be safe, meaning that it must satisfy multiple *in vitro* and *in vivo* safety pharmacology studies before nomination as a clinical candidate. Generally, compounds that have already been in clinical studies for non-HD indications will be the easiest to advance to clinical studies for HD. Even FDA-approved compounds, however, may be unsuitable for HD clinical studies. For example, one group has reported that minocycline, an antibiotic, increases the lifetime of R6/2 mice, although not all reports concur [97–99]. The long-term safety of minocycline in humans has not been established because the drug is usually used acutely. Rodent studies suggest that its long-term use may increase autoimmune disorders, provoke serious hypersensitivity, and induce hyperpigmentation [100].

The FDA requires safety studies both before the initiation of clinical studies (before approval of an IND application) and again before the approval of a therapeutic agent (before approval of a new drug application [NDA]). The required length of these studies depends on the length of time over which the compound will be administered. Because HD treatment will require long-term administration, compounds approved for short-term use will require longer safety studies before initiating HD clinical trials.

SAFETY AND EFFICACY STUDIES MUST ADDRESS ISSUES IDENTIFIED BY REGULATORY AGENCIES

Dialogue with the U.S. FDA and European Medicines Agency (EMEA) can identify issues likely to be important in clinical studies well in advance, such as specific adverse effects known to be associated with a particular class of compound. One FDA guidance document, *ICH Guidance for Industry S7A Safety Pharmacology Studies for Human Pharmaceuticals* [101], underscores the need to evaluate such off-target adverse events as cardiac effects (QT interval prolongation), which may be independent of the primary PD effect. A compound can also have on-target or off-target adverse effects that derive from its binding to receptors or enzymes, necessitating broad surveys of potential targets. The scope and range of new safety studies must respond to concerns of regulatory agencies, as well as to the results of previous safety studies, secondary PD studies, and previous clinical investigations. Safety studies must address the potential for specific organ toxicity, including cardiovascular, respiratory, CNS, renal, and gastrointestinal systems, focusing on irreversible effects. (Transient disruption of these systems may also result from specific drug action, with no irreversible effects.)

The choice of species for preclinical safety studies depends on the potential of each species to reflect the likely human response to a given compound. Species may differ, for example, in the level of cell- and organ-specific expression of individual targets and in the distribution and metabolism of individual compounds. Safety studies must define the dose–response relationship for any observed adverse effect, and they should also encompass a dose range that achieves efficacious pharmacological exposure.

Discussions with regulatory agencies must also address the proper outcome measures and endpoints for efficacy trials in humans. Because HD is a late-onset, slowly progressing disease, clinical trials could benefit from the use of a surrogate endpoint defined by an appropriate biomarker [102]. Longitudinal observational studies, such as PREDICT-HD [103,104], Pharos [105], TRACK-HD, and the HD Neuroimaging Initiative, are in the process of evaluating the biochemical, structural, clinical, cognitive, and behavior changes that occur before HD diagnosis, and these studies may also identify useful biomarkers and surrogate that will accelerate HD clinical trials.

CASTING A WIDE NET FOR POSSIBLE HD THERAPIES

HD undoubtedly results from the expression of mutant *htt*, but the supporting evidence for any of the competing pathogenic models is not compelling, and each model implicates many potential molecular targets. Therefore, CHDI's drug development program cannot be limited to a single molecular target or a single point of intervention. Rather, the rapid identification of a disease-modifying HD therapeutic agent will require a portfolio of parallel programs.

CHDI is presently working on 18 therapeutic programs (see Table 8.1). Some of these are defined by traditional molecular targets such as the "Adenosine A2a receptor program." However, others encompass a broader effort, such as the "*In vivo* program,"

TABLE 8.1

Current CHDI Therapeutic Programs

Project Number	Name and Description
CHDI 144001	HDAC platform
CHDI 144002	Htt-induced cytotoxicity
CHDI 144003	Screening in primary striatal cultures
CHDI 144004	Existing pharmacopoeia
CHDI 144005	Cellular aggregation
CHDI 144006	Toxic fragment production
CHDI 144007	Stress and folding
CHDI 144008	Transglutaminase
CHDI 144009	RNA-based therapies
CHDI 144010	Trophic factors
CHDI 144011	Excitotoxicity
CHDI 144013	*In vivo* portfolio
CHDI 144014	Adenosine A2a antagonism
CHDI 144015	KMO
CHDI 144016	mGluR platform
CHDI 144017	Antioxidants
CHDI 144018	Model organism screening

which encompasses our *in vivo* testing effort, including compound procurement or synthesis, formulation, PK, and dosing. Our objective for these parallel programs is accelerating the entrance of several molecules into clinical evaluation with the hope of identifying at least one disease-modifying treatment. We expect to evaluate more than one clinical candidate before we find a disease-modifying treatment, and we also expect that successful treatment of HD will require the combination of more than one therapy, as is the case for other serious, chronic illnesses, such as AIDS.

Clinical evaluation of potential therapies will require a synergistic collaboration of translational scientists and clinicians to ensure two-way traffic of observations and ideas. Such interaction will be particularly important in the development and evaluation of potential biomarkers and surrogate endpoints, whose incorporation into the drug discovery program will greatly improve its potential for success.

The goal of most of CHDI's current efforts is to bring candidate therapeutics to the clinic, but we have begun to devote more resources to preparing for clinical evaluation. Our preclinical work increasingly attends to the magnitudes and correlations of therapeutic effects. Information from such preclinical studies can inform estimates of the number of subjects needed to observe a parallel effect in a clinical trial. Slowing or stopping HD will almost certainly require the treatment of pre-manifest gene carriers, people who carry the mutant *htt* gene but who have not developed the diagnostic signs. For these subjects, the Unified Huntington Disease Rating Scale (UHDRS) is useless because they appear clinically normal by almost every UHDRS measure [106]. One possible readout for a trial in premanifest subjects is "phenoconversion," the development of HD-specific signs, particularly of chorea and other characteristic motor signs, but such a trial would require more than 3 years and more than 500 subjects. With these limitations in mind, CHDI is supporting the four studies mentioned previously—PREDICT, Pharos, TRACK-HD, and the HD Neuroimaging Initiative—to identify biomarkers and potential surrogate endpoints. Recent reports show increases in a number of inflammatory biomarkers, including microglial activation and inflammatory cytokine levels, in premanifest HD [40,106]. Several neuroimaging studies also suggest that the size of brain, caudate, and putamen and the thinning of the cortex might be useful biomarkers [107,108].

For HD, as for any chronic illness, the ideal medication would be a daily oral tablet with few or no side effects. Because no disease-modifying treatments are available, patients and their families are also likely to be receptive to almost any treatment modality, including implanted systems for direct CNS administration, such as those used in the treatment of chronic pain. Although such devices may overcome problems of brain exposure, potential side effects and adverse events will need further study. Maintaining compliance will be a challenge for any disease-modifying treatment that significantly reduces the quality of life for people with no overt symptoms or signs. Almost any long-term treatment regimen, however, raises significant issues of safety and compliance, and these issues will be magnified when there are significant side effects [109].

CHDI's mission is to find treatments that slow or stop HD. Achievement of this mission will require close collaboration not only among the translational scientists and clinicians within CHDI but also among the scientists, clinicians, patients, and families of the entire HD community.

REFERENCES

1. Bonelli RM and Hofmann P. A systematic review of the treatment studies in Huntington's disease since 1990. *Expert Opin Pharmacother* (2007) 8: pp. 141–153.
2. Shults CW, Oakes D, Kieburtz K, Beal MF, Haas R, Plumb S, Juncos JL, Nutt J, Shoulson I, Carter J, et al. Effects of coenzyme q10 in early Parkinson disease: evidence of slowing of the functional decline. *Arch Neurol* (2002) 59: pp. 1541–1550.
3. Huntington Study Group. A randomized, placebo-controlled trial of coenzyme q10 and remacemide in Huntington's disease. *Neurology* (2001) 57: pp. 397–404.
4. Murck H and Manku M. Ethyl-epa in Huntington disease: potentially relevant mechanism of action. *Brain Res Bull* (2007) 72: pp. 159–164.
5. Puri BK, Leavitt BR, Hayden MR, Ross CA, Rosenblatt A, Greenamyre JT, Hersch S, Vaddadi KS, Sword A, Horrobin DF, et al. Ethyl-epa in Huntington disease: a double-blind, randomized, placebo-controlled trial. *Neurology* (2005) 65: pp. 286–292.
6. Verhagen Metman L, Morris MJ, Farmer C, Gillespie M, Mosby K, Wuu J, and Chase TN. Huntington's disease: a randomized, controlled trial using the NMDA-antagonist amantadine. *Neurology* (2002) 59: pp. 694–699.
7. O'Suilleabhain P and Dewey RBJ. A randomized trial of amantadine in Huntington disease. *Arch Neurol* (2003) 60: pp. 996–998.
8. Leegwater-Kim J and Cha JJ. The paradigm of Huntington's disease: therapeutic opportunities in neurodegeneration. *NeuroRx* (2004) 1: pp. 128–138.
9. Meriin AB, Zhang X, He X, Newnam GP, Chernoff YO, and Sherman MY. Huntington toxicity in yeast model depends on polyglutamine aggregation mediated by a prion-like protein RNQ1. *J Cell Biol* (2002) 157: pp. 997–1004.

10. Kaltenbach LS, Romero E, Becklin RR, Chettier R, Bell R, Phansalkar A, Strand A, Torcassi C, Savage J, Hurlburt A, et al. Huntingtin interacting proteins are genetic modifiers of neurodegeneration. *PLoS Genet* (2007) 3: p. e82.
11. Goehler H, Lalowski M, Stelzl U, Waelter S, Stroedicke M, Worm U, Droege A, Lindenberg KS, Knoblich M, Haenig C, et al. A protein interaction network links git1, an enhancer of huntingtin aggregation, to Huntington's disease. *Mol Cell* (2004) 15: pp. 853–865.
12. Sipione S and Cattaneo E. Modeling Huntington's disease in cells, flies, and mice. *Mol Neurobiol* (2001) 23: pp. 21–51.
13. Li SH, Cheng AL, Li H, and Li XJ. Cellular defects and altered gene expression in pc12 cells stably expressing mutant huntingtin. *J Neurosci* (1999) 19: pp. 5159–5172.
14. Cattaneo E and Conti L. Generation and characterization of embryonic striatal conditionally immortalized st14a cells. *J Neurosci Res* (1998) 53: pp. 223–234.
15. Trettel F, Rigamonti D, Hilditch-Maguire P, Wheeler VC, Sharp AH, Persichetti F, Cattaneo E, and MacDonald ME. Dominant phenotypes produced by the HD mutation in STHdh(q111) striatal cells. *Hum Mol Genet* (2000) 9: pp. 2799–2809.
16. Faber PW, Alter JR, MacDonald ME, and Hart AC. Polyglutamine-mediated dysfunction and apoptotic death of a Caenorhabditis elegans sensory neuron. *Proc Natl Acad Sci U S A* (1999) 96: pp. 179–184.
17. Marsh JL, Pallos J, and Thompson LM. Fly models of Huntington's disease. *Hum Mol Genet* (2003) 12 Spec No 2: p. R187–R193.
18. Hickey MA and Chesselet M. The use of transgenic and knock-in mice to study Huntington's disease. *Cytogenet Genome Res* (2003) 100: pp. 276–286.
19. von Hörsten S, Schmitt I, Nguyen HP, Holzmann C, Schmidt T, Walther T, Bader M, Pabst R, Kobbe P, Krotova J, et al. Transgenic rat model of Huntington's disease. *Hum Mol Genet* (2003) 12: pp. 617–624.
20. Mangiarini L, Sathasivam K, Seller M, Cozens B, Harper A, Hetherington C, Lawton M, Trottier Y, Lehrach H, Davies SW, et al. Exon 1 of the HD gene with an expanded CAG repeat is sufficient to cause a progressive neurological phenotype in transgenic mice. *Cell* (1996) 87: pp. 493–506.
21. Li JY, Popovic N, and Brundin P. The use of the R6 transgenic mouse models of Huntington's disease in attempts to develop novel therapeutic strategies. *NeuroRx* (2005) 2: pp. 447–464.
22. Reddy PH, Williams M, Charles V, Garrett L, Pike-Buchanan L, Whetsell WOJ, Miller G, and Tagle DA. Behavioural abnormalities and selective neuronal loss in HD transgenic mice expressing mutated full-length HD CDNA. *Nat Genet* (1998) 20: pp. 198–202.
23. Hodgson JG, Agopyan N, Gutekunst CA, Leavitt BR, LePiane F, Singaraja R, Smith DJ, Bissada N, McCutcheon K, Nasir J, et al. A YAC mouse model for Huntington's disease with full-length mutant huntingtin, cytoplasmic toxicity, and selective striatal neurodegeneration. *Neuron* (1999) 23: pp. 181–192.
24. Slow EJ, van Raamsdonk J, Rogers D, Coleman SH, Graham RK, Deng Y, Oh R, Bissada N, Hossain SM, Yang Y, et al. Selective striatal neuronal loss in a YAC128 mouse model of Huntington disease. *Hum Mol Genet* (2003) 12: pp. 1555–1567.
25. Yang X and Gong S. An overview on the generation of BAC transgenic mice for neuroscience research. *Curr Protoc Neurosci* (2005) Unit 5.20.
26. Griffioen G. Targeting disease, not disease targets: innovative approaches in tackling neurodegenerative disorders. *IDrugs* (2007) 10: pp. 259–263.
27. Zhang X, Smith DL, Meriin AB, Engemann S, Russel DE, Roark M, Washington SL, Maxwell MM, Marsh JL, Thompson LM, et al. A potent small molecule inhibits polyglutamine aggregation in Huntington's disease neurons and suppresses neurodegeneration in vivo. *Proc Natl Acad Sci U S A* (2005) 102: pp. 892–897.

28. Bodner RA, Outeiro TF, Altmann S, Maxwell MM, Cho SH, Hyman BT, McLean PJ, Young AB, Housman DE, and Kazantsev AG. Pharmacological promotion of inclusion formation: a therapeutic approach for Huntington's and Parkinson's diseases. *Proc Natl Acad Sci U S A* (2006) 103: pp. 4246–4251.

29. Varma H, Voisine C, DeMarco CT, Cattaneo E, Lo DC, Hart AC, and Stockwell BR. Selective inhibitors of death in mutant huntingtin cells. *Nat Chem Biol* (2007) 3: pp. 99–100.

30. Rego AC and de Almeida LP. Molecular targets and therapeutic strategies in Huntington's disease. *Curr Drug Targets CNS Neurol Disord* (2005) 4: pp. 361–381.

31. Lazo JS, Brady LS, and Dingledine R. Building a pharmacological lexicon: small molecule discovery in academia. *Mol Pharmacol* (2007) 72: pp. 1–7.

32. Zeitlin S, Liu JP, Chapman DL, Papaioannou VE, and Efstratiadis A. Increased apoptosis and early embryonic lethality in mice nullizygous for the Huntington's disease gene homologue. *Nat Genet* (1995) 11: pp. 155–163.

33. Zuccato C, Ciammola A, Rigamonti D, Leavitt BR, Goffredo D, Conti L, MacDonald ME, Friedlander RM, Silani V, Hayden MR, et al. Loss of huntingtin-mediated BDNF gene transcription in Huntington's disease. *Science* (2001) 293: pp. 493–498.

34. Gunawardena S, Her L, Brusch RG, Laymon RA, Niesman IR, Gordesky-Gold B, Sintasath L, Bonini NM, and Goldstein LSB. Disruption of axonal transport by loss of huntingtin or expression of pathogenic PolyQ proteins in Drosophila. *Neuron* (2003) 40: pp. 25–40.

35. Szebenyi G, Morfini GA, Babcock A, Gould M, Selkoe K, Stenoien DL, Young M, Faber PW, MacDonald ME, McPhaul MJ, et al. Neuropathogenic forms of huntingtin and androgen receptor inhibit fast axonal transport. *Neuron* (2003) 40: pp. 41–52.

36. Morfini G, Pigino G, and Brady ST. Polyglutamine expansion diseases: failing to deliver. *Trends Mol Med* (2005) 11: pp. 64–70.

37. Truant R, Atwal R, and Burtnik A. Hypothesis: huntingtin may function in membrane association and vesicular trafficking. *Biochem Cell Biol* (2006) 84: pp. 912–917.

38. Cooper JK, Schilling G, Peters MF, Herring WJ, Sharp AH, Kaminsky Z, Masone J, Khan FA, Delanoy M, Borchelt DR, et al. Truncated n-terminal fragments of huntingtin with expanded glutamine repeats form nuclear and cytoplasmic aggregates in cell culture. *Hum Mol Genet* (1998) 7: pp. 783–790.

39. Butler R, Leigh PN, McPhaul MJ, and Gallo JM. Truncated forms of the androgen receptor are associated with polyglutamine expansion in x-linked spinal and bulbar muscular atrophy. *Hum Mol Genet* (1998) 7: pp. 121–127.

40. Tai YF, Pavese N, Gerhard A, Tabrizi SJ, Barker RA, Brooks DJ, and Piccini P. Microglial activation in presymptomatic Huntington's disease gene carriers. *Brain* (2007) 130: pp. 1759–1766.

41. Fan MMY and Raymond LA. N-Methyl-d-aspartate (NMDA) receptor function and excitotoxicity in Huntington's disease. *Prog Neurobiol* (2007) 81: pp. 272–293.

42. Hickey MA and Chesselet MF. Apoptosis in Huntington's disease. *Prog Neuropsychopharmacol Biol Psychiatry* (2003) 27: pp. 255–265.

43. Pattison LR, Kotter MR, Fraga D, and Bonelli RM. Apoptotic cascades as possible targets for inhibiting cell death in Huntington's disease. *J Neurol* (2006) 253: pp. 1137–1142.

44. Browne SE and Beal MF. Oxidative damage in Huntington's disease pathogenesis. *Antioxid Redox Signal* (2006) 8: pp. 2061–2073.

45. Ramaswamy S, Shannon KM, and Kordower JH. Huntington's disease: pathological mechanisms and therapeutic strategies. *Cell Transplant* (2007) 16: pp. 301–312.

46. Vonsattel JP, Myers RH, Stevens TJ, Ferrante RJ, Bird ED, and Richardson EPJ. Neuropathological classification of Huntington's disease. *J Neuropathol Exp Neurol* (1985) 44: pp. 559–577.

47. Albin RL, Young AB, Penney JB, Handelin B, Balfour R, Anderson KD, Markel DS, Tourtellotte WW, and Reiner A. Abnormalities of striatal projection neurons and n-methyl-d-aspartate receptors in presymptomatic Huntington's disease. *N Engl J Med* (1990) 322: pp. 1293–1298.

48. Zuccato C and Cattaneo E. Role of brain-derived neurotrophic factor in Huntington's disease. *Prog Neurobiol* (2007) 81: pp. 294–330.

49. Strand AD, Baquet ZC, Aragaki AK, Holmans P, Yang L, Cleren C, Beal MF, Jones L, Kooperberg C, Olson JM, et al. Expression profiling of Huntington's disease models suggests that brain-derived neurotrophic factor depletion plays a major role in striatal degeneration. *J Neurosci* (2007) 27: pp. 11758–11768.

50. Guidetti P, Luthi-Carter RE, Augood SJ, and Schwarcz R. Neostriatal and cortical quinolinate levels are increased in early grade Huntington's disease. *Neurobiol Dis* (2004) 17: pp. 455–461.

51. Guidetti P, Bates GP, Graham RK, Hayden MR, Leavitt BR, MacDonald ME, Slow EJ, Wheeler VC, Woodman B, and Schwarcz R. Elevated brain 3-hydroxykynurenine and quinolinate levels in Huntington disease mice. *Neurobiol Dis* (2006) 23: pp. 190–197.

52. Giorgini F, Guidetti P, Nguyen Q, Bennett SC, and Muchowski PJ. A genomic screen in yeast implicates kynurenine 3-monooxygenase as a therapeutic target for Huntington disease. *Nat Genet* (2005) 37: pp. 526–531.

53. Aiken CT, Tobin AJ, and Schweitzer ES. A cell-based screen for drugs to treat Huntington's disease. *Neurobiol Dis* (2004) 16: pp. 546–555.

54. Miyashita T, Matsui J, Ohtsuka Y, Mami U, Fujishima S, Okamura-Oho Y, Inoue T, and Yamada M. Expression of extended polyglutamine sequentially activates initiator and effector caspases. *Biochem Biophys Res Commun* (1999) 257: pp. 724–730.

55. Wyttenbach A, Swartz J, Kita H, Thykjaer T, Carmichael J, Bradley J, Brown R, Maxwell M, Schapira A, Orntoft TF, et al. Polyglutamine expansions cause decreased Cre-mediated transcription and early gene expression changes prior to cell death in an inducible cell model of Huntington's disease. *Hum Mol Genet* (2001) 10: pp. 1829–1845.

56. MacDonald ME, Gines S, Gusella JF, and Wheeler VC. Huntington's disease. *Neuromolecular Med* (2003) 4: pp. 7–20.

57. Wang J, Gines S, MacDonald ME, and Gusella JF. Reversal of a full-length mutant huntingtin neuronal cell phenotype by chemical inhibitors of polyglutamine-mediated aggregation. *BMC Neurosci* (2005) 6: p. 1.

58. Sipione S, Rigamonti D, Valenza M, Zuccato C, Conti L, Pritchard J, Kooperberg C, Olson JM, and Cattaneo E. Early transcriptional profiles in huntingtin-inducible striatal cells by microarray analyses. *Hum Mol Genet* (2002) 11: pp. 1953–1965.

59. Gines S, Ivanova E, Seong I, Saura CA, and MacDonald ME. Enhanced Akt signaling is an early pro-survival response that reflects N-methyl-D-aspartate receptor activation in Huntington's disease knock-in striatal cells. *J Biol Chem* (2003) 278: pp. 50514–50522.

60. CHDI Inc. 3rd Annual HD Therapeutics Conference: a forum for drug discovery and development. February 4–7, 2008. Palm Springs, CA.

61. Varma H, Cheng R, Voisine C, Hart AC, and Stockwell BR. Inhibitors of metabolism rescue cell death in Huntington's disease models. *Proc Natl Acad Sci U S A* (2007) 104: pp. 14525–14530.

62. Carter RJ, Hunt MJ, and Morton AJ. Environmental stimulation increases survival in mice transgenic for exon 1 of the Huntington's disease gene. *Mov Disord* (2000) 15: pp. 925–937.

63. Woodman B, Butler R, Landles C, Lupton MK, Tse J, Hockly E, Moffitt H, Sathasivam K, and Bates GP. The Hdh(q150/q150) knock-in mouse model of HD and the R6/2 exon 1 model develop comparable and widespread molecular phenotypes. *Brain Res Bull* (2007) 72: pp. 83–97.

64. Levine MS, Cepeda C, Hickey MA, Fleming SM, and Chesselet M. Genetic mouse models of Huntington's and Parkinson's diseases: illuminating but imperfect. *Trends Neurosci* (2004) 27: pp. 691–697.

65. Loutfy MR, Harris M, Raboud JM, Antoniou T, Kovacs C, Shen S, Dufresne S, Smaill F, Rouleau D, Rachlis A, et al. A large prospective study assessing injection site reactions, quality of life and preference in patients using the biojector vs standard needles for enfuvirtide administration. *HIV Med* (2007) 8: pp. 427–432.

66. Eckstein J. *ISOA/ARF drug development tutorial.* Available online at http://www.alzforum.org/drg/tut/ISOATutorial.pdf (2005): accessed October 5, 2007.

67. Eli Lilly and Company and NIH Chemical Genomics Center. *Assay guidance manual version 4.1.* Available online at http://www.ncgc.nih.gov/guidance/manual_toc.html (2005): accessed October 5, 2007.

68. Ravina B, DiProspero N, Fagan SC, Fischbeck K, Harrison M, Hart R, Heemskerk J, Murphy D, Profenno L, Blumenstein R, Signer ER, et al. (for Huntington Study Group). Systematic Evaluation of Treatments for Huntington's Disease (SET-HD): An interim report. Presented at the Hereditary Disease Foundation CAG Symposium, August 2004, Cambridge, MA. Available online at http://www.huntingtonproject.org: accessed December 16, 2007.

69. Hay DG, Sathasivam K, Tobaben S, Stahl B, Marber M, Mestril R, Mahal A, Smith DL, Woodman B, and Bates GP. Progressive decrease in chaperone protein levels in a mouse model of Huntington's disease and induction of stress proteins as a therapeutic approach. *Hum Mol Genet* (2004) 13: pp. 1389–1405.

70. Kase H, Aoyama S, Ichimura M, Ikeda K, Ishii A, Kanda T, Koga K, Koike N, Kurokawa M, Kuwana Y, et al. Progress in pursuit of therapeutic A2A antagonists: the adenosine A2A receptor selective antagonist KW6002: research and development toward a novel nondopaminergic therapy for Parkinson's disease. *Neurology* (2003) 61: p. S97–100.

71. Qikprop, version 3.0 (Schrodinger, Inc., New York, 2005). See Jorgensen WL. The many roles of computation in drug discovery. *Science* (2004) 303: pp. 1813–1818.

72. Kamzalov S, Sumien N, Forster MJ, and Sohal RS. Coenzyme Q intake elevates the mitochondrial and tissue levels of coenzyme Q and alpha-tocopherol in young mice. *J Nutr* (2003) 133: pp. 3175–3180.

73. Miles MV. The uptake and distribution of coenzyme q10. *Mitochondrion* (2007) 7 Suppl: p. S72–S77.

74. Ohtsuki S, Tachikawa M, Takanaga H, Shimizu H, Watanabe M, Hosoya K, and Terasaki T. The blood-brain barrier creatine transporter is a major pathway for supplying creatine to the brain. *J Cereb Blood Flow Metab* (2002) 22: pp. 1327–1335.

75. Prass K, Royl G, Lindauer U, Freyer D, Megow D, Dirnagl U, Stöckler-Ipsiroglu G, Wallimann T, and Priller J. Improved reperfusion and neuroprotection by creatine in a mouse model of stroke. *J Cereb Blood Flow Metab* (2007) 27: pp. 452–459.

76. Janzen WP and Hodge CN. A chemogenomic approach to discovering target-selective drugs. *Chem Biol Drug Des* (2006) 67: pp. 85–86.

77. Keith CT, Borisy AA, and Stockwell BR. Multicomponent therapeutics for networked systems. *Nat Rev Drug Discov* (2005) 4: pp. 71–78.

78. Lehár J, Zimmermann GR, Krueger AS, Molnar RA, Ledell JT, Heilbut AM, Short GF3, Giusti LC, Nolan GP, Magid OA, et al. Chemical combination effects predict connectivity in biological systems. *Mol Syst Biol* (2007) 3: p. 80.

79. Boado RJ, Kazantsev A, Apostol BL, Thompson LM, and Pardridge WM. Antisense-mediated down-regulation of the human huntingtin gene. *J Pharmacol Exp Ther* (2000) 295: pp. 239–243.

80. Harper SQ, Staber PD, He X, Eliason SL, Martins IH, Mao Q, Yang L, Kotin RM, Paulson HL, and Davidson BL. RNA interference improves motor and neuropathological abnormalities in a Huntington's disease mouse model. *Proc Natl Acad Sci U S A* (2005) 102: pp. 5820–5825.

81. Pardridge WM. The blood-brain barrier: bottleneck in brain drug development. *NeuroRx* (2005) 2: pp. 3–14.
82. Skarlatos S, Yoshikawa T, and Pardridge WM. Transport of [125i]transferrin through the rat blood-brain barrier. *Brain Res* (1995) 683: pp. 164–171.
83. Kissel K, Hamm S, Schulz M, Vecchi A, Garlanda C, and Engelhardt B. Immuno-histochemical localization of the murine transferrin receptor (tfr) on blood-tissue barriers using a novel anti-tfr monoclonal antibody. *Histochem Cell Biol* (1998) 110: pp. 63–72.
84. Lee HJ, Engelhardt B, Lesley J, Bickel U, and Pardridge WM. Targeting rat anti-mouse transferrin receptor monoclonal antibodies through blood-brain barrier in mouse. *J Pharmacol Exp Ther* (2000) 292: pp. 1048–1052.
85. Friden PM, Walus LR, Musso GF, Taylor MA, Malfroy B, and Starzyk RM. Anti-transferrin receptor antibody and antibody-drug conjugates cross the blood-brain barrier. *Proc Natl Acad Sci U S A* (1991) 88: pp. 4771–4775.
86. Wu D, Yang J, and Pardridge WM. Drug targeting of a peptide radiopharmaceutical through the primate blood-brain barrier in vivo with a monoclonal antibody to the human insulin receptor. *J Clin Invest* (1997) 100: pp. 1804–1812.
87. Wu D. Neuroprotection in experimental stroke with targeted neurotrophins. *NeuroRx* (2005) 2: pp. 120–128.
88. Pardridge WM. Tyrosine hydroxylase replacement in experimental Parkinson's disease with transvascular gene therapy. *NeuroRx* (2005) 2: pp. 129–138.
89. Nellis DF, Giardina SL, Janini GM, Shenoy SR, Marks JD, Tsai R, Drummond DC, Hong K, Park JW, Ouellette TF, et al. Preclinical manufacture of anti-her2 liposome-inserting, scfv-peg-lipid conjugate. 2. Conjugate micelle identity, purity, stability, and potency analysis. *Biotechnol Prog* (2005) 21: pp. 221–232.
90. Nellis DF, Ekstrom DL, Kirpotin DB, Zhu J, Andersson R, Broadt TL, Ouellette TF, Perkins SC, Roach JM, Drummond DC, et al. Preclinical manufacture of an anti-her2 scfv-peg-dspe, liposome-inserting conjugate. 1. Gram-scale production and purification. *Biotechnol Prog* (2005) 21: pp. 205–220.
91. Hayes ME, Drummond DC, Kirpotin DB, Zheng WW, Noble CO, Park JW, Marks JD, Benz CC, and Hong K. Genospheres: self-assembling nucleic acid-lipid nanoparticles suitable for targeted gene delivery. *Gene Ther* (2006) 13: pp. 646–651.
92. Pangalos MN, Schechter LE, and Hurko O. Drug development for CNS disorders: strategies for balancing risk and reducing attrition. *Nat Rev Drug Discov* (2007) 6: pp. 521–532.
93. Davies B and Morris T. Physiological parameters in laboratory animals and humans. *Pharm Res* (1993) 10: pp. 1093–1095.
94. Hopkins SJ. Central nervous system recognition of peripheral inflammation: a neural, hormonal collaboration. *Acta Biomed* (2007) 78 Suppl 1: pp. 231–247.
95. Mehvar R and Jamali F. Concentration-effect relationships of tetrabenazine and dihydrotetrabenazine in the rat. *J Pharm Sci* (1987) 76: pp. 461–465.
96. Irwin S. Comprehensive observational assessment: Ia. A systematic, quantitative procedure for assessing the behavioral and physiologic state of the mouse. *Psychopharmacologia* (1968) 13: pp. 222–257.
97. Mievis S, Levivier M, Communi D, Vassart G, Brotchi J, Ledent C, and Blum D. Lack of minocycline efficiency in genetic models of Huntington's disease. *Neuromolecular Med* (2007) 9: pp. 47–54.
98. Smith DL, Woodman B, Mahal A, Sathasivam K, Ghazi-Noori S, Lowden PAS, Bates GP, and Hockly E. Minocycline and doxycycline are not beneficial in a model of Huntington's disease. *Ann Neurol* (2003) 54: pp. 186–196.
99. Wang X, Zhu S, Drozda M, Zhang W, Stavrovskaya IG, Cattaneo E, Ferrante RJ, Kristal BS, and Friedlander RM. Minocycline inhibits caspase-independent and -dependent mitochondrial cell death pathways in models of Huntington's disease. *Proc Natl Acad Sci U S A* (2003) 100: pp. 10483–10487.

100. Hoefnagel JJ, van Leeuwen RL, Mattie H, and Bastiaens MT. [Side effects of minocycline in the treatment of acne vulgaris]. *Ned Tijdschr Geneeskd* (1997) 141: pp. 1424–1427.

101. FDA. *ICH guidance for industry S7A safety pharmacology studies for human pharmaceuticals.* Available online at http://www.fda.gov/RegulatoryInformation/Guidances/ucm129150.htm: accessed February 23, 2010.

102. Biomarkers Definition Workgroup. Biomarkers and surrogate endpoints: preferred definitions and conceptual framework. *Clin Pharmacol Ther* (2001) 69: pp. 89–95.

103. Duff K, Paulsen JS, Beglinger LJ, Langbehn DR, Stout JC. Psychiatric symptoms in Huntington's disease before diagnosis: the predict-HD study. *Biol Psychiatry* (2007) 62: pp. 1341–1346.

104. Paulsen JS, Hayden M, Stout JC, Langbehn DR, Aylward E, Ross CA, Guttman M, Nance M, Kieburtz K, Oakes D, et al. Preparing for preventive clinical trials: the predict-HD study. *Arch Neurol* (2006) 63: pp. 883–890.

105. Huntington Study Group PHAROS Investigators. At risk for Huntington disease: the PHAROS (Prospective Huntington At Risk Observational Study) cohort enrolled. *Arch Neurol* (2006) 63: pp. 991–996.

106. Huntington Study Group. Unified Huntington's disease rating scale: reliability and consistency. Huntington Study Group. *Mov Disord* (1996) 11: pp. 136–142.

107. Aylward EH. Change in MRI striatal volumes as a biomarker in preclinical Huntington's disease. *Brain Res Bull* (2007) 72: pp. 152–158.

108. Rosas HD, Feigin AS, and Hersch SM. Using advances in neuroimaging to detect, understand, and monitor disease progression in Huntington's disease. *NeuroRx* (2004) 1: pp. 263–272.

109. Dunbar-Jacob J, Erlen JA, Schlenk EA, Ryan CM, Sereika SM, and Doswell WM. Adherence in chronic disease. *Annu Rev Nurs Res* (2000) 18: pp. 48–90.

9 RNA- and DNA-Based Therapies for Huntington's Disease

Meghan Sass and Neil Aronin

CONTENTS

INTRODUCTION

The goal of this chapter is to describe strategies for the development of RNA- and DNA-based therapies for Huntington's disease (HD). RNA-based therapies use endogenous cellular mechanisms for the suppression of gene expression and have the potential for exquisite sequence selectivity of target genes. DNA-based therapies, notably those based on antisense technologies, are not as flexible in target selectivity but can be more stable than those derived from RNA-based silencing. Both approaches are currently being aggressively pursued with promising results in preclinical studies.

HD is a particularly attractive target for DNA- and RNA-based therapies as it is an autosomal dominant disease resulting from mutation on one allele. Thus, in concept, eliminating expression of the mutant *huntingtin* allele would entirely prevent the neuronal pathology that it would otherwise cause. Because the mutant *huntingtin* messenger RNA (mRNA) transcript would be selectively targeted, the normal transcript would remain unaffected and able to mediate the normal functions of *huntingtin* that may be critical for neural development and function.

HD THERAPEUTICS BASED ON ANTISENSE TECHNOLOGIES

WHAT IS ANTISENSE?

The concept of antisense oligonucleotide gene silencing was first introduced in 1978 when Stephenson and Zamecnik (1978) used an antisense oligonucleotide to stop viral replication in cell culture. An antisense oligonucleotide is a single strand of nucleic acid or nucleic acid analogs, most often an oligodeoxyribonucleotide, usually 15–20 nucleotides in length with sequence complementary to a specific target mRNA. The antisense oligonucleotide and target mRNA bind together via Watson–Crick base pairing, and this hybridization leads to reduced levels of translation of the target transcript.

Antisense oligonucleotide-induced mechanisms of gene silencing include steric hindrance and interference with ribosomal function, as well as inhibition of mRNA splicing, which prevents mRNA maturation and destabilizes the pre-mRNA in the nucleus (Kurreck, 2003). However, the most prominent mechanism of antisense oligonucleotide-induced gene silencing is induction of RNase H endonuclease activity resulting in hydrolysis of the mRNA in the antisense oligonucleotide-target transcript duplex (Wu et al., 2004). The target mRNA is degraded whereas the exogenously introduced antisense oligonucleotide remains intact, allowing it to move on to the inhibition of another transcript (see Figure 9.1).

RNase H appears to be a ubiquitous enzyme in eukaryotes and bacteria. There are two human RNase enzymes, RNase H1 and H2. Although their endogenous cellular roles are not fully understood, RNase H1 has been shown to play the dominant role in the antisense silencing (Wu et al., 2004; Bennett and Swayze, 2010). RNase H1 is a 286-amino acid protein that is expressed in all human cells and tissues. The structure of RNase H1 contains binding, spacer, and catalytic domains. The spacer and catalytic domains are required for RNase H-mediated target cleavage; the RNA binding domain is not required but is responsible for increased binding affinity to target RNA and positional preference for cleavage by the enzyme (Wu et al., 2001). RNase H1 has been shown to exhibit a strong positional preference for cleavage at a site 7–12 nucleotides from the 5′-RNA/3′-DNA terminus of the heteroduplex (Lima et al., 2003).

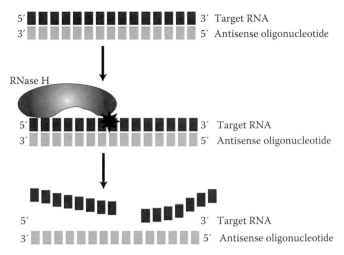

FIGURE 9.1   (See color insert following page 172.) The most prominent mechanism of antisense oligonucleotide gene silencing—induction of RNase H endonuclease activity. The antisense oligonucleotide (typically a deoxyribonucleotide) binds the target RNA to form the heteroduplex substrate. RNase H binds via its binding domain at the 3′ antisense oligonucleotide/5′ RNA pole and cleaves the target RNA approximately 7 base pairs from its binding site. The target mRNA is degraded, whereas the antisense oligonucleotide remains intact, allowing it to form another heteroduplex substrate for induction of RNase H cleavage.

ANTISENSE OLIGONUCLEOTIDE DESIGN

Critical characteristics of antisense oligonucleotide design include (1) the portion of the mRNA transcript to be targeted; (2) the nucleotide makeup of the antisense oligonucleotide; and (3) the binding energy between the antisense oligonucleotide and target. The secondary structure of the mRNA target is also an important consideration for antisense potency, as the target region on the mRNA must be physically accessible for hybridization by the antisense oligonucleotide (Andronescu et al., 2005; Vickers et al., 2000). Computer algorithms are now available that allow determination of mRNA secondary structure folding energy for selection of target sites with minimal overall free energy (Chan et al., 2006). Regions of mRNA transcripts amenable to oligonucleotide binding include their 5′ and 3′ terminal ends, internal loops, joint sequences, hairpins, and bulges (Far et al., 2001; Yang et al., 2003).

The antisense oligonucleotide sequence itself also plays an integral role in its efficacy. Antisense oligonucleotides are designed to span one of the known RNase H cleavage sites. Statistical analysis of more than 1,000 experiments that used antisense oligonucleotides determined that those containing CCAC, TCCC, ACTC, GCCA, and CTCT correlate with potent mRNA knockdown; conversely, GGGG, ACTG, AAA, and TAA motifs weaken the antisense oligonucleotide silencing effect (Matveeva et al., 2000). Although the exact reason the cytosine-rich antisense oligonucleotides are more effective at gene silencing is not known, the authors proposed that these sequences may represent RNase H preferred sequences, form more stable DNA-RNA duplexes, promote RNase H binding, or may encourage more efficient cellular uptake of the oligonucleotides. In addition to the specific sequences above, antisense oligonucleotides with high GC content correlate with high antisense oligonucleotide-mRNA binding stability and RNase H activity (Ho et al., 1996). A binding energy between the antisense oligonucleotide and target mRNA should be ≥ −8 kcal/mol, and binding energy between antisense oligonucleotides should be ≥ −1.1 kcal/mol to favor antisense oligonucleotide potency (Matveeva et al., 2003).

Finally, there are various chemical modifications of synthetic antisense molecules that improve binding efficiency to target mRNA, which translates into increased antisense potency. Because extracellular nucleases can rapidly degrade unmodified antisense oligonucleotides, these chemical modifications enhance nuclease resistance. In fact, the poor intrinsic stability of unmodified antisense oligonucleotides has been a major barrier to using antisense molecules in a clinical setting for drug delivery. The most commonly applied chemical modification is synthesis of the antisense molecule with a phosphorothioate (PS)-modified backbone, in which one of the nonbridging oxygen atoms in the phosphodiester bond is replaced by a sulfur atom (Eckstein, 2000). The PS backbone may decrease binding affinity of the antisense oligonucleotide to mRNA but prevents degradation by nuclease and thereby promotes RNase H-mediated cleavage of target.

The 2'-alkyl modifications, including the addition of 2'-O-methyl (2'-OMe) and 2'-O-methoxyethyl (2'-MOE) groups, are often added to a PS-modified antisense oligonucleotide. These modifications provide the DNA oligonucleotide with further resistance against nuclease activity and improve its binding to targets. Chimeric "gapmer" antisense oligonucleotides have been designed where 10 PS-modified nucleotides are flanked by five 2'OMe- or 2'-MOE-nucleotides, greatly promoting RNase H-mediated target destruction (McKay et al., 1999). Other chemical modifications of antisense oligonucleotides include modifications of the furanose ring—peptide nucleic acids, locked nucleic acids (LNAs), and phosphoroamidate morpholino oligomers (PMOs). These modifications do not promote RNase H-mediated cleavage *per se* but do promote further gene silencing through steric hindrance against the protein translational apparatus. LNA-gapmers, combined, have been designed similarly to improve oligonucleotide-mediated gene silencing (Chan et al., 2006).

ANTISENSE *IN VITRO* STUDIES FOR HD

Thoughtful antisense oligonucleotide design to ensure potency while avoiding toxicities has yielded effective antisense therapies in the clinic. The first Food and Drug Administration-approved antisense oligonucleotide drug was fomivirsen. Delivered by local intravitreous injection, fomivirsen is indicated in patients with AIDS to treat cytomegalovirus retinitis (Jabs and Griffiths, 2002).

Will antisense therapies be viable options for the treatment of neurodegenerative diseases, particularly for HD? Antisense oligonucleotides targeting huntingtin mRNA have already been shown to impair translation *in vitro* (Boado et al., 2000), and PS-modified antisense oligonucleotides decreased expression and aggregation of huntingtin protein, characteristic pathology caused by mutant huntingtin, as assessed by the fluorescence of fusion reporter construct in a cell-based model (Hasholt et al.,

2003). In another *in vitro* study, a PS-antisense oligonucleotide against *huntingtin* resulted in decreased endogenous huntingtin mRNA (Nellemann et al., 2000).

Do antisense oligonucleotides have enough sequence selectivity to target single nucleotide polymorphisms (SNPs)? Discrimination between mutant and normal alleles would, in theory, be important if not critical for small oligonucleotide-based therapy in autosomal dominant diseases such as HD—an SNP heterozygosity on the mutant allele could be selectively targeted by an antisense oligonucleotide to cause RNase H-degradation of the mutant allele, leaving expression of the normal allele intact. Although this concept has not been demonstrated yet in models of neurodegenerative disease using antisense oligonucleotides, allele-specific antisense oligonucleotides targeting an SNP on the RNA polymerase II gene have been demonstrated *in vitro* (ten Asbroek et al., 2000), and systemic administration of these antisense oligonucleotides has been shown to inhibit tumor growth in mice (Fluiter et al., 2002).

A direct silencing approach would target the CAG expansion itself. Use of standard RNA interference (RNAi) cannot distinguish between wild-type and expanded CAG repeats (Caplen, 2002). In cell culture, modifying antisense oligonucleotides that span part of the CAG in series and adjoining nucleotides in exon 1 of *huntingtin* initiated selective knockdown of the mutant *huntingtin* (Hu et al., 2009). The modifications include use of peptide nucleic acid or locked nucleic acid. The mechanism of the selective knockdown is not established, although elements of translational repression and mRNA degradation could be involved (Aronin, 2009). This novel strategy is more effective with longer CAG repeats (example 69 in series) compared with CAG repeats nearer the average for adults with HD (Snell et al., 1993). *In vivo* safety of peptide nucleic acids and locked nucleic acids needs to be established. With caveats, these creative approaches to reduce the mutant allele of *huntingtin* lie within the purview of antisense technology.

In Vivo Studies for Antisense Therapy

Antisense therapy has not yet been tested extensively for HD in *in vivo* models. Mice have been injected with a fluorescein-labeled antisense oligonucleotide targeting huntingtin mRNA, with this antisense oligonucleotide, 500 nM, injected daily into the striatum of rodents for 4 days. Subsequent immunocytochemistry revealed that the labeled antisense oligonucleotide gained access to the neurons within the striatum; however, Western blot analysis showed no decrease in huntingtin protein when compared with control mice (Haque and Isacson, 1997).

This study emphasized the additional challenges that antisense strategies face *in vivo* and questions that must be addressed when testing a new candidate antisense therapeutic agent. Can antisense oligonucleotides administered in brain block translation and have phenotypic effects? In another study, PS-modified gapmer antisense oligonucleotides with 5-nucleotide 2′-MOE-nucleotide substitutions flanking both 3′ and 5′ ends were pumped into the right lateral ventricle of rats and nonhuman primates. After 14 days of treatment, significant oligonucleotide concentrations (3.5–7.0 µM) were demonstrable in the brain and spinal cord in both species. Furthermore, with the same dosing regimen, antisense oligonucleotides were distributed in anatomical regions of the brain relevant to amyotrophic lateral

sclerosis (ALS) and other neurodegenerative diseases. Finally, antisense oligonucleotide against superoxide dismutase (SOD-1; mutations in SOD-1 are responsible for
20% of familial ALS) administered to the ventricle resulted in a 40% decrease in
SOD-1 mRNA after 2 weeks and a 50% decrease in SOD-1 protein in the frontal
cortex after 4 weeks of treatment. Importantly, such intraventricular administration
of SOD-1 antisense oligonucleotides could be shown to slow disease progression in
a rat model of ALS (Smith et al., 2006).

Additional antisense oligonucleotide studies in a mutant SOD-1 mouse model of
ALS have also been successful. In one, an antisense oligonucleotide was used to target
the p75 neurotrophin receptor, which is known to have increased expression in ALS.
In this case, the mice received the antisense oligonucleotide via intraperitoneal administration three times weekly. Treated mice showed uptake of fluorescence-labeled
antisense oligonucleotide and decreased p75 receptor expression in the spinal cord, as
well as improvement in motor deficits and decreased mortality compared with control
mice (Turner et al., 2003). In a separate study, mutant SOD-1 mice received intraperitoneal injections of an antisense oligonucleotide against the glutamate receptor
subunit 3 (GluR3), activation of which is implicated in several neurological diseases.
The treated mice survived longer than did controls, although there was no detectable
decrease in GluR3 protein levels in the spinal cord (Rembach et al., 2004).

Antisense Oligonucleotide Delivery Issues

Targeting the central nervous system (CNS) presents a unique set of obstacles. The
specialized and effective characteristics of the blood–brain barrier (BBB) and its
sophisticated vascular epithelial tight junctions in the CNS reduce accessibility
of most systemically administered molecules from the vasculature into the CNS.
CNS compounds that are hydrophilic ionized and greater than 18 Å in diameter or
500 Da in molecular mass do not generally penetrate the BBB. Molecules that pass
through the BBB do so by either lipid-soluble diffusion or catalyzed transport of
water-soluble molecules (Sanovich et al., 1995). In addition, to treat HD, an antisense
oligonucleotide will need to be delivered to the appropriate part of the brain (e.g.,
the striatum and cortex exhibit pathology early in the course of HD) and will need to
gain intracellular access to neurons in affected individuals.

Getting Antisense Oligonucleotides into Cells

Intracellular uptake of antisense oligonucleotides presents a major barrier to antisense therapies—the lipophilic cell membrane presents an impediment for anionic
molecules to pass, and they must enter cells primarily by absorptive endocytosis
(Lysik and Wu-Pong, 2003). In some studies, cellular uptake has been limited to
<2% of the dose (Marti et al., 1992; Wickstrom et al., 1988). Moreover, unmodi-
fied antisense oligonucleotides are rapidly degraded by single-stranded nucleases,
mainly the 3′ exonuclease.

Chemical modifications of antisense oligonucleotides can significantly affect
their bioavailability. Phosphorothioate modification of the antisense molecule promotes adhesion to cell surface proteins, whereas peptide nucleic acid and PMO

modifications are noncharged and have poor interaction with the negatively charged cellular membrane. Conjugation of a positively charged arginine-rich peptide to PMO-modified antisense oligonucleotides could be used to improve cellular delivery (Nelson et al., 2005).

In addition to chemical modifications, mechanical techniques including electroporation and shockwave application are useful for delivery *in vitro* but would not be practical for delivery *in vivo* to treat HD. Other intracellular delivery systems with potential *in vivo* applications include antisense oligonucleotide conjugation with cationic lipid carriers, carrier molecules that bind with cell-specific receptors, cyclodextrins, dendrimers, microparticles, and macromolecules (Lysik and Wu-Pong, 2003). These delivery systems can enhance intracellular delivery either by protecting antisense oligonucleotide from nuclease degradation and/or by promoting absorptive endocytosis. Included in the macromolecule class are cell-penetrating peptides (CPPs), short peptide sequences with a net positive charge that are conjugated to the antisense oligonucleotide via a disulphide bridge. Commonly used CPPs include penetratin, HIV TAT peptide 48–60, and transportan (Fattal et al., 2004; Jarver and Langel, 2004; Kaihatsu et al., 2004; Khan et al., 2004; Lysik and Wu-Pong, 2003; Nelson et al., 2005; Yoo and Juliano, 2000).

Another obstacle preventing antisense oligonucleotides from gaining access to target mRNA is that antisense oligonucleotides remain in the endosome, making them susceptible to lysosomal degradation. Endosomal sequestration of the antisense oligonucleotide (Shoji et al., 1991; Yakubov et al., 1989) and escape of antisense oligonucleotides from endosomes have both been demonstrated (Farrell et al., 1995; Marti et al., 1992; Zamecnik et al., 1994). The addition of dioleylphosphatidylethanolamine to liposome delivery systems results in the destabilization of endosomal membranes and promotion of release of the antisense oligonucleotide after endocytosis (Farhood et al., 1994; Jaaskelainen et al., 1994).

GETTING ANTISENSE OLIGONUCLEOTIDES INTO THE BRAIN

To gain intracellular access, the antisense molecule must first reach the target tissue—specifically the striatum in people affected with HD. In the *in vivo* studies cited previously, investigators administered antisense oligonucleotides into the ventricles by continuous infusion as a means to "bypass" the BBB. Similarly, antisense oligonucleotides have been delivered directly by stereotactic injection into the brain or intracranial tumor with measurable antisense effects in glioblastoma and closed head injury animal models (Shohami et al., 2000; Tyler et al., 1999; Yazaki et al., 1996). In the two studies outlined on the ALS mouse model, an antisense effect was demonstrated more simply by systemic intraperitoneal administration of the antisense oligonucleotide; however, these antisense oligonucleotides were meant only to target mRNAs in the spinal cord, not in the brain. Another strategy has been to use the bradykinin analogue RMP-7 to transiently increase BBB permeability via systemic coadministration with the antisense oligonucleotide (Riley et al., 1998).

Other investigators have shown that intraperitoneal injection of antisense oligonucleotides can reach the brain. An antisense oligonucleotide targeting the rat neurotensin receptor so delivered showed presence in the brain detected by gel shift assay and decreased receptor activity and binding sites for neurotensin, suggesting an antisense silencing effect directly in the brain (Tyler et al., 1999). In a glioblastoma mouse model, mice with intracranial implanted glioblastoma cells showed decreased intracranial tumor growth and increased survival time after daily intraperitoneal injection with antisense oligonucleotides targeting protein kinase C α mRNA for 82 days (Yazaki et al., 1996). How antisense oligonucleotides crossed the BBB into the CNS has not been established.

ANTICIPATED SIDE EFFECTS OF ANTISENSE OLIGONUCLEOTIDE THERAPY

Side effects after *in vivo* administration of antisense oligonucleotides remain a concern. Antisense oligonucleotide drugs can cause dose-dependent, mild-to-moderate toxicities when administered to animals and humans (Chan et al., 2006). Mild toxicities observed at high plasma antisense oligonucleotide concentrations include thrombocytopenia, hyperglycemia, and elevation of liver enzymes (Jason et al., 2004). *In vivo* activation of the complement cascade and inhibition of the clotting cascade have also been observed. These effects are sequence independent and result from antisense oligonucleotide backbone chemistry, as PS-antisense oligonucleotides have been shown to bind nonspecifically to plasma proteins at high concentrations (Chan et al., 2006).

Immune stimulation by antisense oligonucleotides can cause splenomegaly, lymphoid hyperplasia, and diffuse mononuclear cell infiltrates (Crooke, 2008). A proposed mechanism of immune stimulation by the antisense oligonucleotide is that an unmethylated cytosine-phosphorous-guanine (CpG) motif on the antisense oligonucleotide is recognized by the Toll-like receptor in the immune system, resulting in release of cytokines, B-cell proliferation, antibody production, and activation of T lymphocytes and natural killer cells. To avoid this immunostimulatory effect, antisense oligonucleotides can be designed lacking the CpG motif (Vollmer et al., 2004). In addition, inclusion of locked nucleic acids in the antisense oligonucleotides may also reduce immune activation (Uhlmann and Vollmer, 2003).

HD THERAPEUTICS BASED ON RNA INTERFERENCE

THE DISCOVERY OF RNA INTERFERENCE

In 1990, plant biologists attempted to deepen the color of pigmented petunias by introducing genes encoding enzymes involved in reactions that produce flavonoids, the most common flower pigment. The investigators observed an unexpected phenotype—up to 42% of the plants produced nonpigmented flowers. In addition, mRNA levels of the introduced gene were significantly reduced in the unpigmented part of the petal (Napoli et al., 1990). This phenomenon was termed *cosuppression* because the endogenous and transgenes were both repressed.

During the following decade, major advances were made in understanding the mechanism of this endogenous cellular process of RNA silencing, since renamed RNA interference (RNAi). RNAi was demonstrated to occur in many eukaryotic cells, including mammalian cells. Many of the initial RNAi studies used *Caenorhabditis elegans*; researchers showed RNA silencing with both antisense RNA and a "control" sense RNA targeting the transcript (Guo and Kemphues, 1995). Later it was realized that this silencing was the result of RNAi. In 1998, a major breakthrough in the understanding of RNAi was the demonstration that double-stranded RNA (dsRNA) silenced the gene *unc-22* in *C. elegans*, whereas single-stranded RNA introduced into the cell had only a marginal effect on silencing. In addition, a small amount of dsRNA produced potent gene silencing—only a few molecules of dsRNA were required for gene silencing, arguing against a stoichiometric process (Fire et al., 1998).

Soon it was discovered that RNAi-mediated gene silencing was a post-transcriptional event. Endogenous mRNA was the target for the interference. After introduction of dsRNA into *C. elegans*, mRNA was degraded before translation could occur, making RNAi a type of post-transcriptional gene silencing (Montgomery et al., 1998). During the following few years, investigators showed that within the cell longer dsRNA molecules are processed into 21–23 nucleotide segments that were the "triggers" or mediators of RNAi. They appeared to guide sequence-specific mRNA cleavage of a homologous target RNA with the assistance of an endonuclease (Hamilton and Baulcombe, 1999; Hammond et al., 2000; Zamore et al., 2000). The target RNA cleavage sites occurred 21–23 nucleotides apart, only in areas homologous to the small dsRNA, further evidence that the small dsRNA was guiding cleavage of the homologous target RNA (Hammond et al., 2000). The 21–23 nucleotide dsRNAs are now known as small interfering RNAs (siRNAs).

OTHER GENE-SILENCING TRIGGERS: THE MICRORNAS

Another class of small RNAs is the microRNAs (miRNAs). Unlike siRNAs, which are derived from long dsRNAs up to thousands of base pairs in length, miRNAs precursors are encoded by the genome. miRNA precursor transcripts contain a stem and a loop, which are subsequently processed into mature, functional miRNAs by proteins common to the siRNA-mediated gene silencing pathway. miRNAs often have nucleotide mismatches with their targets and effect gene silencing through translational block (repression) of the target gene. Since the discovery of the first miRNA, *lin-4*, in 1993 (Lee et al., 1993), >1,000 miRNAs have been discovered in plants, animals, and viruses. miRNAs perform an integral role in regulation of gene expression and in humans are estimated to be nearly 1,000 in number (Berezikov et al., 2005; Lewis et al., 2005).

MECHANISM OF SIRNA-MEDIATED RNAI

siRNA-mediated RNAi has been a major focus of RNAi therapeutic development for HD. The RNAi mechanism makes it a promising modality to treat disease as siRNA-mediated RNAi therapies have the potential to be allele specific—the mutant transcript can be targeted selectively without affecting nonmutant transcripts. siRNA therapies can also be potent—delivery of a relatively small amount of trigger or "drug" could have a persistent effect on gene silencing because RNAi is reiterative.

In siRNA-mediated RNAi, dsRNAs (about 21 nucleotides in each strand) are used to initiate the assembly of RNA/protein into the RNAi silencing complex (RISC). RISC formation leads to a cascade of intracellular events that result in cleavage of the target mRNA, which has complementarity to the guide (antisense) strand of the siRNA duplex (Hammond et al., 2000). siRNAs can be made from longer dsRNA or alternatively introduced into cells as transfected DNA sequences via gene transfer using a range of methods, including viral delivery. In the latter case, the dsRNA is encoded as a "hairpin sequence" that is produced on the transcription of the transfected DNA sequence.

The dsRNA is substrate for Dicer, a dsRNA-specific endonuclease that converts the long dsRNA into 21–25 nucleotide dsRNA with 2-nucleotide 3′ overhangs— the siRNAs. Dicer contains both a dsRNA-binding domain and a catalytic component, an RNase III that processes long dsRNA into siRNAs (Bernstein et al., 2001; Elbashir et al., 2001a; Ketting et al., 2001). The siRNAs serve as the guides for the remainder of the RNAi pathway.

A critical moment in establishing RISC activity is selection of the RNA strand in the siRNA. Either the passenger strand (sense) or the guide strand (antisense) can be loaded into RISC. Assembly of the desired guide strand favors RNAi, whereas loading the passenger strand dampens gene silencing. In the favorable condition, the guide strand has relative 5′ thermodynamic instability and is selected to enter the RISC complex by R2D2 (Liu et al., 2003; Tomari et al., 2004). Within the RISC, the guide strand binds to the Argonaute 2 (Ago2) protein, a component of RISC that contains the PAZ domain (that binds the guide strand at the 3′ end) and PIWI domain (that binds the guide strand at the 5′ end). The PIWI domain also contains the endonuclease that cleaves target RNAs. Although associated with the Ago2 protein, the guide strand binds its target mRNA by Watson–Crick base pairing. However, not all ~20 nucleotides of the guide strand contribute equally to target RNA binding. Bases 2–8, known as the seed sequence at the 5′ end of the guide strand, contribute disproportionately to target binding by the guide strand. After binding, the Ago2-contained endonuclease cleaves the target RNA at the phosphodiester bond across from nucleotides 10 and 11 of the guide. This cleavage site is known as the "scissile (cutting) phosphate" (Elbashir et al., 2001a, 2001b) (see Figure 9.2).

The Ago2 protein is a multiple-turnover enzyme such that each small RNA directs cleavage of hundreds of target mRNA molecules (Elbashir et al., 2001b; Haley and Zamore, 2004). This reiterative process is an essential property of RNAi. Release of the cleaved RNA fragments requires ATP, and the fragments are subsequently degraded; the 3′ end is destroyed in the cytoplasm by exonuclease Xrn1 and the 5′ end by the exosome, a collection of exonucleases dedicated to 3′ to 5′ degradation (Orban and Izaurralde, 2005). What happens to the passenger strand of the siRNA duplex? It is destroyed by Ago2—after cleavage, the strand remnants leave RISC, and then the RISC complex/guide strand continues the RNAi process through another cycle (Matranga et al., 2005; Rand et al., 2005).

RNAi Is Not Transitive in Mammalian Cells

The RNAi pathway is a part of a group of post-transcriptional silencing events that require a set of related proteins common to almost all eukaryotes (Aravind et al., 2000). In some, the silencing serves likely as a defense mechanism that protects the cell from viral invasion or genome disruption resulting from transposon jumping. It

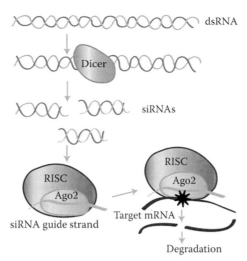

FIGURE 9.2   (See color insert following page 172.) siRNA-mediated RNAi. dsRNAs are bound by the enzyme Dicer and cleaved by Dicer's catalytic component (RNase III) into siRNAs—21-nucleotide dsRNA with 2-nucleotide 3′ overhangs. With assistance of R2D2, the guide strand is loaded into the RISC. Within RISC, the guide strand binds to Ago2. After complementary base pairing between siRNA guide strand and target mRNA, the endonuclease Ago2 cleaves the target mRNA at the scissile phosphate. The 5′ end of the target is degraded in the exosome, and the 3′ end is degraded in the cytoplasm. Ago2 is a multiple turnover enzyme, and one Ago-bound guide strand directs cleavage of hundreds of target mRNAs.

is not clear whether siRNA-mediated RNAi serves this function in mammals, but RISC proteins do process siRNAs to effect RNAi in mammalian cells. The availability of this innate cellular process holds promise for the use of RNAi to treat human disease; as described previously, synthetic siRNAs or virally delivered short-hairpin RNAs (shRNAs) can be introduced into cells or tissues, with siRNA sequences designed to have complementarity to a disease causing mRNA to be targeted for destruction by RISC.

Fortunately, RNAi is not "transitive" in mammalian cells: in the original descriptions of the RNAi, as realized in *C. elegans*, the RNAi mechanism included synthesis of dsRNA by RNA-dependent RNA polymerase, with templates emanating from degradation products of the pathway. These dsRNAs subsequently guide more homologous RNA cleavage. Mammals lack this enzyme, therefore RNAi therapies can be sequence selective.

To emphasize, nonspecific amplification of silencing that occurs in worms does not occur in mammalian cells or in *Drosophila*, a model system applicable to RNAi mechanisms in mammals (Celotto and Graveley, 2002; Chiu and Rana, 2002; Schwarz et al., 2002). This serendipitous characteristic of mammalian RNAi for therapeutics allows the introduction of a synthetic siRNA to be site selective and remain site selective.

Critically for autosomal dominant diseases such as HD, this difference between the RNAi mechanisms in nematodes and humans allows allelic discrimination in humans. In the nematode, the entire target mRNA could be used as a template for the production of new siRNAs. Single nucleotide allelic differences, as might occur in SNP heterozygosities or oncogene changes, would be lost if cleaved mRNA target generated a host of siRNAs along the target's sequence. Another fortunate characteristic difference in RNAi mechanisms is that, unlike in worms, in mammalian and *Drosophila* cells (Roignant et al., 2003) the initial silencing trigger does not spread to silence genes in other tissues.

HD Is Conducive for RNAi Therapy

HD is an autosomal dominant disease that is caused by an expanded CAG repeat on one mutant allele. Thus, in theory, RNAi therapies can allow allele-specific targeting such that the mutant mRNA is degraded, leaving the nonmutant *huntingtin* transcript unaffected. The autosomal dominant mode of inheritance can be used to identify affected patients before onset of symptoms because those affected with HD almost always have inherited the disease. RNAi therapies could be administered early to an HD patient in an attempt to delay onset of symptoms or treat the disease before excessive neuropathology occurs. Because the clinical manifestations of HD occur over decades, there is a significant therapeutic window to institute treatment.

Synthetic Small RNAs—siRNAs and shRNAs

siRNAs can be synthesized with chemical structure characteristic of endogenous Dicer cleavage products: 21-nucleotide synthetic dsRNA with 5′-phosphorylated ends and 2-nucleotide 3′-unphosphorylated overhangs. As described previously, another means of introducing synthetic small RNAs into the cell is by delivery of an shRNA. shRNAs are commonly expressed in mammalian cells using expression vectors with a pol III promoter, with sense and antisense sequences from a target gene connected by a loop. The shRNA expressed from the vector is subsequently processed in the cytoplasm by Dicer into siRNAs (Brummelkamp et al., 2002; Paddison et al., 2002a, 2002b; Paul et al., 2002; Yu et al., 2002) (see Figure 9.3).

Synthetic siRNA Design Principles

In the optimal circumstance, an siRNA is designed such that the guide strand is preferentially loaded into RISC and the passenger strand is destroyed. The unequal thermodynamic stability at selected sites of the functional siRNA duplex (Aza-Blanc et al., 2003; Khvorova et al., 2003; Reynolds et al., 2004; Schwarz et al., 2003; Ui-Tei et al., 2004) favors the guide strand for RISC processing—the more RISC assemblies that contain guide strands, the more effective gene silencing will be. Lowering thermodynamic stability at one end of the duplex improves RISC loading of the siRNA, and, importantly, the strand with its 5′ end at the unstable site preferably enters RISC (Khvorova et al., 2003; Schwarz et al., 2003). Thus, when designing a synthetic siRNA, a mismatch is purposefully placed at the 5′ end of the antisense strand to promote antisense strand-guided silencing. This thermodynamic principle of asymmetry is used to design siRNAs to target the HD gene (see below).

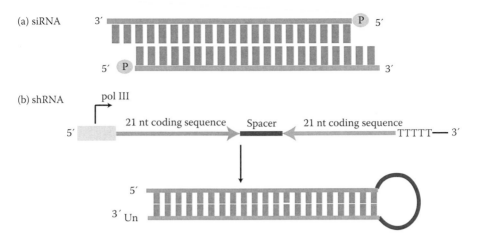

FIGURE 9.3 **(See color insert following page 172.)** (a) siRNAs: siRNAs can be synthe-sized to mimic endogenous RNase III Dicer cleavage products—21-nucleotide dsRNA with 5′ phosphates and 2-nucleotide 3′ unphosphorylated overhangs. (b) shRNAs: vectors introduced into the cell contain the DNA template with the pol III promoter, which drives expression of the ssRNA transcript through the polyT termination sequence. The sense and antisense strand on the ssRNA associate in cis to form the shRNA—a dsRNA separated by a loop (Paul et al., 2002; Sui et al., 2002; Yu et al., 2002). The shRNA is processed by Dicer into an siRNA (Dykxhoorn et al., 2003).

The 5′ phosphorylation of the guide strand is one of the ATP-dependent steps in RNAi (Du et al., 2005) and is required in synthetic siRNAs for the guide strand to trigger RNAi (Schwarz et al., 2002, 2003). Thus, modifications at the 5′ end that block phosphorylation (including 2′-*O*-methylation) will block RNAi, whereas the same modifications at the 3′ end of the guide strand will not affect gene silencing. Interestingly, although 5′ phosphorylation is necessary for a functional siRNA, the initial 5′ nucleotide does not need to be matched to target RNA for effective RNAi (Schwarz et al., 2003).

A major contributor to target specificity is encoded in nucleotides 2–8 at the 5′ end— the seed sequence that binds to the complementary site on the target mRNA (Doench and Sharp, 2004; Haley and Zamore, 2004). Furthermore, mismatched nucleotides between the guide strand and the target mRNA at the scissile phosphate, between nucleotides 10 and 11 of the guide strand, impair RNAi (Elbashir et al., 2001b). The 3′ end does not play an integral role in target recognition but does orient RISC and stabilizes the guide strand-target duplex for guided cleavage. Therefore, mismatches at the 3′ end of the guide strand are tolerated, although mismatched 3′ siRNAs have diminished target cleavage (Haley and Zamore, 2004). Thus, the seed sequence serves to identify the mRNA target, whereas the 3′ sequence contributes to catalysis.

ALLELIC DISCRIMINATION IN HD

A challenge with using RNAi to treat autosomal dominant disease is discrimination between the mutant and wild-type alleles. Fortunately, a single normal *huntingtin* allele suffices to retain normal neuronal development and function (Dixon, 2004); therefore, an objective of the RNAi is to silence selectively the mutant huntingtin mRNA. However, although HD is a trinucleotide repeat disease arising from expansions in its CAG repeat domain, a CAG repeat ~70 base pairs in length is also present on the normal allele, and many other mRNAs in the genome encode protein with a polyglutamine stretch (Perutz, 1999). Thus, designing an siRNA targeting the expanded CAG repeat would result in degradation of normal huntingtin mRNA and numerous other gene products containing CAG repeats. siRNAs targeting CAG repeats are unable to distinguish between wild-type and mutant *huntingin* transcripts (Caplen et al., 2002).

An alternative approach to achieve allele-specific silencing is to target an SNP in the mutant *huntingtin* allele—a DNA sequence variation elsewhere in the *huntingtin* gene sequence coinherited along with the CAG expansion that involves a change in a single nucleotide (Schwarz et al., 2006). The idea is that siRNAs would distinguish between SNP heterozygosities—sites that have different nucleotides at the same position when comparing the patients' two *huntingtin* alleles. The goal would be to preferentially target the site of the SNP heterozygosity on the mutant *huntingtin* allele. Synthetic small RNAs can be designed with complementarity to the mutant allele SNP nucleotide but mismatched with the wild-type allele. In fact, synthetic siRNAs have been designed to target point mutations on the *SOD-1* gene and *Tau* gene (cause of frontotemporal dementia), allowing for selective mutant allele silencing (Ding et al., 2003; Miller et al., 2003). Furthermore, siRNAs have been successfully designed against an SNP for allelic discrimination on the mutant Machado-Joseph disease/spinocerebellar ataxia (SCA) type 3 gene, the cause of another polyglutamine neurodegenerative disorder (Miller et al., 2003).

Two considerations are important when selecting an SNP heterozygosity to target and when designing an siRNA to target this site: the location of SNP position relative to the siRNA antisense strand and the type of nucleotide mismatch at the SNP site between siRNA nonmutant RNA. To target SNP heterozygosity, the SNP site on the guide strand should be complementary to the mutant allele SNP and will therefore be noncomplementary to the corresponding sequence on the nonmutant allele. Certain guide strand/nonmutant mismatches are conducive to preventing RNAi (A/A, A/G mismatches). On the other hand, A/C and G/U mismatches between guide and target appear to be compatible with RNAi (Du et al., 2005). Designing the siRNA such that the nucleotide complementary to the target SNP site is at position 10 (at the scissile phosphate, as described previously) or at position 16 of the guide strand has been shown to promote allele-specific silencing (Schwarz et al., 2006). The mechanism for discrimination at position 16 is not established, but a mismatch there may impede RISC orientation for mRNA catalysis (see Figure 9.4).

RNAi *In Vitro* Studies

In vitro studies have shown that RNAi can be effectively used to reduce *huntingtin* gene expression, although not specifically addressing the issue of allelic discrimination. Early *in vitro* studies using RNAi to target expanded polyglutamine transcripts resulted in reduction of cellular toxicity and death in a cellular model of spinobulbar

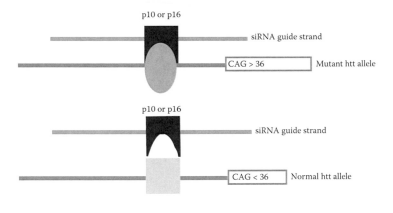

 Represents complementarity basepairing between siRNA guide and mutant huntingtin SNP site to promote siRNA targeting of mutant huntingtin allele for degradation

 Represents noncomplementarity basepairing between siRNA guide and normal huntingtin SNP site; A/A or A/G mismatches at this site promote preservation of normal allele, whereas A/C and G/U mismatches are "tolerated" and should be avoided because they may promote silencing of normal allele

FIGURE 9.4 Small RNA design principles to selectively target a mutant allele at an SNP heterozygosity. The SNP should be targeted at position 10 or position 16 (p10 or p16) of the siRNA guide.

muscular atrophy, another CAG repeat neurogenic disease (Caplen et al., 2002). When compared with controls, siRNA expression vectors targeting various positions of human huntingtin mRNA resulted in decreased huntingtin protein and huntingtin mRNA in DAOY cerebellar meduloblastoma and HeLa cells (Chen et al., 2005). Similarly, adeno-associated virus (AAV)-delivered shRNAs against *huntingtin* produced a decrease in both huntingtin protein and mRNA (Harper et al., 2005).

Principles of nucleotide-specific RNAi have guided a therapeutic approach for selective knockdown of the mutant *huntingtin* allele while preserving the wild-type allele. Multiple SNPs have been confirmed (Lombardi et al., 2009; Pfister et al., 2009; Warby et al., 2009). An SNP heterozygosity in the 3′ untranslated region of *huntingtin* at site *rs*362307 has a greater frequency associated with the mutant *huntingtin* allele, and in combination with other frequent SNP heterozygosities, as few as five siRNAs can be used to knock down the mutant *huntingtin* allele in most patients with HD (Pfister et al., 2009). To accomplish the selective silencing, one or two mismatched nucleotides need to be incorporated in the guide strand of siRNA, thereby avoiding knockdown of the wild-type *huntingtin* allele while destroying the mutant *huntingtin* allele (Pfister et al., 2009).

RNAi *IN VIVO* STUDIES

Critical for RNAi to treat HD is delivery of siRNA into the neuron. To overcome this hurdle, intracerebral delivery of modified siRNA viral vectors coding for shRNAs have been developed and shown to be effective in transgenic mouse models of neurodegenerative diseases, including CAG repeat diseases (Farah, 2007). Intracerebellar injection of recombinant AAV vectors expressing shRNAs against the mutant *ataxin-1* transcript, the cause of the expanded polyglutamine disease SCA1, resulted in restored cerebellar morphology, reduced ataxin-1 inclusions in Purkinje cells (both characteristic of the disease), and improved motor coordination in treated mice (Xia et al., 2004).

A similar approach has been used in human *huntingtin* transgene mouse models. Two groups used AAV vectors to deliver shRNAs into the striatum by stereotactic injection. The shRNAs targeted the human *huntingtin* transgene and resulted in decreased human huntingtin mRNA, protein, and neuronal aggregates (inclusions) and improved motor deficits (Harper et al., 2005; Rodriguez-Lebron et al., 2005). Improved motor behaviors were noted by 14 weeks after treatment (Harper et al., 2005). Delivered after the onset of motor changes, *huntingtin*-targeting AAV-shRNAs decreased neuronal aggregates and ameliorated down-regulation of DARPP-32 expression (characteristic of mutant *huntingtin*-affected neurons) (Machida et al., 2006). The implication of this study is that RNAi therapies could benefit HD patients even after the onset of symptoms.

Viral delivery systems in humans raise concerns of host immune activation, potential for sustained small RNA delivery, viral vector spread (Davidson et al., 2000; Liu et al., 2005; Passini et al., 2003), and inability to turn off small RNA delivery. AAV and adenovirus are nonintegrating vectors that remain episomal and deliver shRNAs to dividing and nondividing cells (Grimm and Kay, 2006; Kim and Rossi, 2007), whereas lentiviral vectors also target both dividing and nondividing cells but express shRNA only after transgene integration into the host cell genome, with potential for long-term expression of the shRNA (Kim and Rossi, 2007). In studies in murine models of HD and SCA1 and in wild-type mice, expression of the shRNAs for several months had no observable effects on striatal or cerebellar morphology and behavioral phenotype (Harper et al., 2005; Rodriguez-Lebron et al., 2005; Xia et al., 2004). The ability to stop shRNA production is paramount for safety in the event of unanticipated side effects, and shRNA gene expression that can be securely switched off may be most suitable for human therapy. Nevertheless, as HD is chronic and progressive, long-term RNAi offers an attractive and effective therapeutic strategy.

Recent data have also raised a concern for toxicity from shRNA after delivery into the cell. Both control and huntingtin-targeting AAV-delivered shRNAs injected into the striatum resulted in neuronal cell death associated with increased inflammation evidenced by microglia in the striatum 4 months after initial delivery. These toxicities are thought to be secondary to the specific shRNA introduced and might be related to the absolute amount of siRNA generated, as some AAV-delivered shRNAs did not produce toxicity. The nontoxic shRNAs produced lower levels of antisense strand RNA compared with the toxic shRNAs. Furthermore, AAV delivery of microRNA was found to be safer than shRNA. A possible explanation is that the microRNA was expressed at lower amounts than the siRNA, with fewer side effects. To emphasize, regulated expression of virally delivered microRNA or shRNA might be warranted. A caveat of pharmacological therapy is that the drug can be stopped if side effects occur (an exit strategy) (McBride et al., 2008).

Use of synthetic siRNAs offers an alternative approach for gene silencing that avoids the problem of switching off a virally delivered shRNA and toxicities associated with specific shRNAs. However, siRNA delivery to the brain remains a challenge. Chemical modifications to the synthetic siRNA that may improve *in vivo* bioavailability are locked nucleic acid substitutions and modified RNA nucleotides that improve dsRNA stability (Elmen et al., 2005; Mook et al., 2007). The 2′ alkyl substitutions also protect against *in vivo* siRNA degradation and improve dsRNA stability (Morrissey et al., 2005), including 2′-F, 2′-OMe, or 2′-H substitutions for the 2′-OH residues. In two transgenic mouse lines expressing mutant *huntingtin*, intraventricular injection of 0.2 μg of siRNA against *huntingtin* with a transfection reagent resulted in improved motor deficits and prolonged survival associated with decreased huntingtin mRNA and protein (Wang et al., 2005). However, delivery of siRNA with a transfection reagent is not a practical means to treat humans because of the neuronal toxicity of most lipid-based transfection agents.

In another study, a cholesterol-conjugated siRNA against *huntingtin* was injected directly into the striatum and cortex that also received mutant *huntingtin* in an AAV vector. After intrastriatal injection of AAV expressing a mutant huntingtin cDNA including amino acids 1–400 with the 100 CAG repeat, mice developed rapid-onset characteristic Huntington's neuropathology and motor deficits within 2 weeks. Coadministration of a cholesterol-conjugated siRNA improved neuronal survival, reduced aggregate formation, significantly delayed onset of behavioral abnormalities, and decreased expression of mutant huntingtin protein (DiFiglia et al., 2007).

Thus, direct administration of siRNAs directed against *huntingtin* has therapeutic potential, with the caveat that the siRNAs might need to be conjugated to improve neuronal entry. Other approaches have shown promise:

- Small RNAs delivered into the brain via an infusion pump into the third ventricle (400 μg siRNA/day for up to 2 weeks) decreased mRNA and protein for a dopamine transporter in mice (Thakker et al., 2004).
- Intrathecal pump administration of siRNA at the same dose (400 μg siRNA/day for 6 days) targeting a pain-related cation channel decreased mRNA levels and acute and chronic pain responses in a rat model (Dorn et al., 2004).
- Small RNAs conjugated to carrier peptides have been introduced into venous circulation to facilitate siRNA entry into the CNS (Kumar et al., 2007).

Additionally, encapsulation of shRNAs in liposomes, to protect the plasmid DNA from degradation by ubiquitous endonucleases *in vivo*, has been used as a delivery strategy to the CNS. shRNA against luciferase encoded in DNA plasmids was encapsulated in the interior of PEGylated immunoliposomes, targeted toward implanted brain glioma cells expressing luciferase. The PEGylated immunoliposomes, tethered to a rat transferrin receptor monoclonal antibody, were injected intravenously into rats weekly for 4 weeks and resulted in a 90% decrease in luciferase production in the glioma cells. The authors concluded that the rat transferrin receptor monoclonal antibody allowed the PEGylated immunoliposomes to cross the epithelial tight junctions of the BBB and allowed for entry into the glioma cells (Zhang et al., 2003).

A similar protocol was used to target human epidermal growth factor receptor in human gliomas implanted into the brains of mice; epidermal growth factor receptor function was decreased by 95%, and mice with advanced intracranial brain cancer had an 88% increase in survival time (Zhang et al., 2003).

Ligands or Fabs designed to interact specifically with a cell surface receptor on the target cell have also been conjugated directly to siRNA molecules to mediate cell-specific delivery. Such strategies have been used to target HIV-infected cells, prostate cancer cells, and Ewing sarcoma cells *in vivo* (Hu-Lieskovan et al., 2005; McNamara et al., 2006; Song et al., 2005). siRNA has also been associated with a rabies virus glycoprotein (RVG) peptide, the portion of the RVG responsible for binding to the acetylcholine (ACh) receptor. The synthetic RVG peptide has nine positively charged amino acids that allow it to bond ionically with the negatively charged siRNA. The RVG peptide was shown to bind ACh receptors on neuronal cells, and RVG associated with green fluorescent protein (GFP) targeting siRNA was able to silence GFP expression *in vitro*. Furthermore, after systemic administration of fluorescein isothiocyanate-labeled siRNA with RVG, the labeled siRNA was evident in the brain but not the liver or spleen. siRNA against *SOD-1* assembled with RVG peptide injected into the peripheral venous circulation resulted in decreased SOD-1 mRNA and protein production in the brain, and the presence of siRNA in the brain was confirmed directly by Northern blot analysis. Similarly using RVG delivery, siRNAs targeting flavivirus improved survival by 80% in a mouse model of flaviviral encephalitis (Kumar et al., 2007). Taking all the data from this study into consideration, it appears that RVG peptide associated with siRNA promotes intraneuronal delivery of siRNA after systemic administration either by providing a means to bypass the BBB and/or promoting cellular uptake into the neuron via the ACh receptor. Although the exact mechanism by which the RVG peptide bypasses the BBB was not addressed in this study, the rabies virus uses this essential glycoprotein to enter peripheral neurons and enter CNS via retrograde axonal transport (Faber et al., 2004). Confirmation of the RVG peptide approach will be critical to predict its widespread use.

Finally, a hydrodynamic approach to deliver siRNA should be noted. In this approach, injection of 20%–40% blood volume equivalent into the mouse circulation causes massive endocytosis of siRNA into cells (Lewis and Wolff, 2005). Although this method has been shown in the experimental setting to be effective in delivering siRNAs to protect mice from severe hepatitis, it not a practical mode of delivery for humans because it causes heart failure in mice (Song et al., 2003).

Anticipated Side Effects of RNAi Therapies

Unwanted effects of RNAi therapy include "off-target effects" (recognition of mRNAs other than that of the desired target by an siRNA), saturation of endogenous RNAi machinery, and induction of immune responses. Off-target effects of siRNAs can be widespread and may change expression profiles of numerous nontarget mRNAs (Jackson et al., 2003; Persengiev et al., 2004; Saxena et al., 2003; Scacheri et al., 2004; Sledz et al., 2003) with consequent effects on protein expression levels, phenotype, and clinical manifestations not yet known. mRNA array assays have been used to study such off-target effects, although the array data should be interpreted with caution because of many potentially confounding issues: (1) array analysis requires the setting of somewhat arbitrary limits to "decide" whether there is a significant change in the level of a particular mRNA; (2) the causes of changes in mRNA levels are not known, including whether these may be direct effects of siRNA or indirect effects from other cellular influences; (3) changes in mRNA levels may or may not predict changes in protein levels; and (4) array studies are often done in short-term experiments, whereas clinical therapies for chronic conditions such as HD are likely to be long term.

As synthetic siRNAs can cause translational repression when they match to untranslated regions in mRNAs, off-target translational repression in principle can be reduced by designing siRNAs homologous to target transcript open reading frames. However, it should be noted that mismatches up to four nucleotides and G/U wobbles can lead to translational repression of nontarget mRNAs that are difficult to predict.

Introduction of synthetic siRNAs into cells also raises the concern that the exogenously introduced small RNAs could compete with and interfere with the endogenous miRNA silencing system. Mammalian cells are known to have many miRNAs that share proteins with the RNAi pathway for gene silencing. It has been shown that AAV-introduced shRNA into the liver of mice resulted in dose-related liver injury and death, partially explicable by down-regulation of liver-derived miRNAs (Grimm et al., 2006).

Certain characteristics of synthetic siRNAs may also cause activation of the immune system, including inducing an interferon response that causes global, non-specific inhibition of protein translation. RNA molecules less than 30 base pairs in length have been thought to avoid this response, and 21-base pair siRNAs injected into the tail vein of mice do not elicit an interferon or interleukin 12 response (Heidel et al., 2004). Contrary to this finding, however, siRNAs have been shown to cause an interferon response in cell culture (Sledz et al., 2003). Moreover, specific immunostimulatory motifs on synthetic siRNAs have been shown to cause plasmacytoid dendritic cells to produce type 1 interferon; Toll-3 receptors have been shown to recognize dsRNA leading to activation of nuclear factor κB and regulation of interleukin response (Alexopoulou et al., 2001). Additionally, GU-rich sequences, blunt-ended duplexes, and 5′ triphosphates recruit immunostimulatory responses and should be avoided in siRNAs (Crooke, 2008). Incorporation of 2′-OMe uridine or guanosine nucleosides into one strand of the synthetic siRNA has been shown to abrogate immune activation after systemic administration in mice (Judge et al., 2006).

Finally, how the siRNA is packaged and its mode of delivery may play a role in eliciting an immune response. For example, siRNAs administered systemically in liposomes were reported to increase interleukin 6 and tumor necrosis factor α (Judge et al., 2005), although intrastriatal injection of cholesterol-conjugated siRNAs showed no increase in reactive microglia or astrocyte number by immunocytochemistry of the striatum compared with phosphate-buffered saline injection (DiFiglia et al., 2007). Further investigations will clearly be needed to determine factors that lead to immune responses against intracerebral administration of siRNAs. In particular, it will be important to determine whether repeated injection of small RNAs into the brain will cause greater immune activation, as it is anticipated that people affected with HD will require intermittent administration of small RNAs because of the chronic and progressive nature of the disease.

ANTISENSE VERSUS RNAi

RNAi and antisense strategies offer promising strategies for the treatment of HD. They share the therapeutic goal to eliminate, or reduce, mutant huntingtin mRNA with the expectation that the mutant huntingtin protein is eliminated or reduced. Mechanistic differences between RNAi and antisense technology might sway preference of one over the other. RNAi uses available cellular processes, proteins that are conserved among many life forms to regulated gene expression. Small RNAs serve in a protein–nucleotide assembly to recognize mRNA sites and either cleave mRNA or repress translation. Antisense oligonucleotides recruit RNase H; rather than RNA/RNA interactions in RNAi, antisense involves DNA/RNA duplexes that cleave mRNA. Possibly, single-stranded, isolated DNA is more stable than RNA duplex, but stability of the guide strand of siRNA in RISC is not known.

It would seem that RNAi might be a more selective therapy than antisense. The RISC assemblies can be used to cleave many mRNA molecules *in vivo* for at least a few days (DiFiglia et al., 2007). It is not clear whether antisense oligonucleotides form stable complexes. However, modifications of antisense oligonucleotides might improve stability by impairing their cellular clearance. Differences in nucleotide stability could account for differences in potency results between RNAi and antisense. On a molar basis at the same target site, some studies found that siRNAs are more potent than antisense oligonucleotides (Dorn et al., 2001; Hemmings-Mieszcak et al., 2003). Compared with antisense oligonucleotides, the increased potency for siRNA has reached 100-fold (Miyagishi et al., 2003). Similar dose potencies have been reported in *in vitro* experiments (Vickers et al., 2003). Variables of delivery and nucleotide stability confound a simplistic choice of RNAi over antisense. Furthermore, although principles of allele selectivity have not been formulated for antisense oligonucleotides as has been established for RNAi, antisense is capable of allele-selective knockdown (ten Asbroek et al., 2000).

We believe that knockdown of the mutant *huntingtin* gene product has a dominant conceptual advantage because mutant huntingtin initiates many pathophysiological events. Wild-type huntingtin has salubrious effects to counter harmful effects of mutant huntingtin (Leavitt et al., 2001). Caspase-3 activation is thought to mediate some of the pathogenic effects of mutant huntingtin, and wild-type huntingtin inhibits caspase-3 activity (Zhang et al., 2006). Therefore, indiscriminant knockdown of wild-type huntingtin might be risky. How much wild-type knockdown can be tolerated is unclear; up to 50% decrease could be tolerated in the brain (Drouet et al., 2009). Nonetheless, if practical, in a genetic disease, abrogation of the disease allele product makes sense. Target selectivity favors RNAi, in part because there are more compelling experimental data for RNAi compared with antisense. siRNAs can be designed to improve guide strand incorporation in an RISC assembly, thereby improving knockdown activity of siRNAs (Schwarz et al., 2003). Furthermore for synthetic siRNA or shRNA design, certain design rules have been established that may allow the small RNA to target the mutant *huntingtin* allele at an SNP site while preserving nonmutant huntingtin mRNA (Schwarz et al., 2006). Numerous SNPs have been catalogued (Lombardi et al., 2009; Pfister et al., 2009; Warby et al., 2009). We propose that SNP targeting will be a useful approach for selective RNAi silencing of the mutant *huntingtin* allele. Single nucleotide discrimination by antisense oligonucleotides for *huntingtin* might be possible and needs to be tested.

Safety of RNAi and antisense ought to be clarified. Both antisense oligonucleotides and small RNAs can result in a range of nonspecific side effects. Antisense oligonucleotides administered at a concentration of 200 nM or greater cause nonspecific effects on protein synthesis *in vitro* (Hu-Lieskovan et al., 2005). Off-target effects of siRNAs have been shown to be dose dependent (Persengiev et al., 2004; Saxena et al., 2003). Detailed analysis of immune responses to both types of therapy should be tested.

OPINIONS AND SPECULATIONS

The press release for the 2006 Nobel Prize in Physiology or Medicine emphasized a potential therapeutic role of RNAi in human disease (Nobelprize.org, 2006). We think the Nobel press release to be prescient. The essential idea is to knock down the mutant *huntingtin* gene product, thereby preventing the many aberrant consequences of expression of the mutant huntingtin protein. Gene silencing in neurodegenerative disease remains a concept inchoate. It is full of promise. The primary obstacles are delivery and safety (unanticipated side effects). Delivery of siRNA could be a matter of increasing doses to favor uptake of small RNAs into brain cells. In the past few years, experimental data demonstrated that—once in neurons—siRNAs can knock down mutant huntingtin in HD animal models with salubrious effects. Antisense oligonucleotides have been given at doses manifold greater than siRNA doses; perhaps, similar doses of siRNAs would improve their neuronal uptake. We should consider an exit strategy for viral delivery of shRNA or miRNA. In devastating diseases such as HD, with few treatment options and unrelenting disease progression, the urge by physician and patient alike is to administer therapy. RNAi therapy (or antisense) would probably be started before substantial neuronal loss early in the disease. In late stages of HD, neuronal loss has yielded inexorable deficits in motor movement and cognition, unlikely to be ameliorated by changes in mutant huntingtin expression. Furthermore, because of disease severity, achieving statistical differences in late-stage treatment groups would be difficult. Clinical studies of RNAi would probably be performed in HD patients with early stages of disease. Therefore, safety becomes a prime issue; small RNA or antisense will need to be stopped if patients develop unexpected problems. Nonselective reduction of wild-type huntingtin might have untoward effects on HD patients. Wild-type huntingtin inhibits caspase-3 that mediates apoptosis (Zhang et al., 2006). Knocking down wild-type *huntingtin*, in theory, could potentiate neuronal loss, a possibility that warrants study and favors selective knockdown against mutant *huntingtin*, invoking use of RNAi. The dictum *primum non nocere* pertains (first, do no harm). We propose that selective allele silencing and regulatable viral delivery of small RNAs are testable ideas that should be included in preclinical and clinical trials for therapy in HD. That antisense and

RNAi have different mechanisms of action, and presumably different toxicities, might be used to a therapeutic advantage. A combination of antisense oligonucleotide and siRNA treatment might allow lower concentrations of either, to strengthen reduction of mutant *huntingtin* by selective and nonselective strategies, while avoiding overlapping toxicities (Aronin, 2009).

REFERENCES

Alexopoulou L, Holt AC, Medzhitov R, Flavell RA (2001) Recognition of double-stranded RNA and activation of NF-kappaB by Toll-like receptor 3. *Nature* 413:732–738.

Andronescu M, Zhang ZC, Condon A (2005) Secondary structure prediction of interacting RNA molecules. *J Mol Biol* 345:987–1001.

Aravind L, Watanabe H, Lipman DJ, Koonin EV (2000) Lineage-specific loss and divergence of functionally linked genes in eukaryotes. *Proc Natl Acad Sci U S A* 97:11319–11324.

Aronin N (2009) Expanded CAG repeats in the crosshairs. *Nat Biotechnol* 27:451–452.

Aza-Blanc P, Cooper CL, Wagner K, Batalov S, Deveraux QL, Cooke MP (2003) Identification of modulators of TRAIL-induced apoptosis via RNAi-based phenotypic screening. *Mol Cell* 12:627–637.

Bennett CF, Swayze EE (2010) RNA targeting therapeutics: molecular mechanisms of antisense oligonucleotides as a therapeutic platform. *Annu Rev Pharmacol Toxicol* 50: 259–293.

Berezikov E, Guryev V, van de Belt J, Wienholds E, Plasterk RH, Cuppen E (2005) Phylogenetic shadowing and computational identification of human microRNA genes. *Cell* 120:21–24.

Bernstein E, Caudy AA, Hammond SM, Hannon GJ (2001) Role for a bidentate ribonuclease in the initiation step of RNA interference. *Nature* 409:363–366.

Boado RJ, Kazantsev A, Apostol BL, Thompson LM, Pardridge WM (2000) Antisense-mediated down-regulation of the human huntingtin gene. *J Pharmacol Exp Ther* 295:239–243.

Brummelkamp TR, Bernards R, Agami R (2002) A system for stable expression of short interfering RNAs in mammalian cells. *Science* 296:550–553.

Caplen NJ, Taylor JP, Statham VS, Tanaka F, Fire A, Morgan RA (2002) Rescue of polyglutamine-mediated cytotoxicity by double-stranded RNA-mediated RNA interference. *Hum Mol Genet* 11:175–184.

Celotto AM, Graveley BR (2002) Exon-specific RNAi: a tool for dissecting the functional relevance of alternative splicing. *RNA* 8:718–724.

Chan JH, Lim S, Wong WS (2006) Antisense oligonucleotides: from design to therapeutic application. *Clin Exp Pharmacol Physiol* 33:533–540.

Chen ZJ, Kren BT, Wong PY, Low WC, Steer CJ (2005) Sleeping Beauty-mediated down-regulation of huntingtin expression by RNA interference. *Biochem Biophys Res Commun* 329:646–652.

Chi JT, Chang HY, Wang NN, Chang DS, Dunphy N, Brown PO (2003) Genomewide view of gene silencing by small interfering RNAs. *Proc Natl Acad Sci U S A* 100: 6343–6346.

Chiu YL, Rana TM (2002) RNAi in human cells: basic structural and functional features of small interfering RNA. *Mol Cell* 10:549–561.

Crooke ST (2008) *Antisense drug technology: principles, strategies, and applications,* 2nd edition. Boca Raton: CRC Press.

Davidson BL, Stein CS, Heth JA, Martins I, Kotin RM, Derksen TA, Zabner J, Ghodsi A, Chiorini JA (2000) Recombinant adeno-associated virus type 2, 4, and 5 vectors: transduction of variant cell types and regions in the mammalian central nervous system. *Proc Natl Acad Sci U S A* 97:3428–3432.

DiFiglia M, Sena-Esteves M, Chase K, Sapp E, Pfister E, Sass M, Yoder J, Reeves P, Pandey RK, Rajeev KG, et al. (2007) Therapeutic silencing of mutant huntingtin with siRNA attenuates striatal and cortical neuropathology and behavioral deficits. *Proc Natl Acad Sci U S A* 104:17204–17209.

Ding H, Schwarz DS, Keene A, Affar el B, Fenton L, Xia X, Shi Y, Zamore PD, Xu Z (2003) Selective silencing by RNAi of a dominant allele that causes amyotrophic lateral sclerosis. *Aging Cell* 2:209–217.

Dixon KT, Cearley JA, Hunter JM, Detloff PJ (2004) Mouse Huntington's disease homolog mRNA levels: variation and allele effects. *Gene Expr* 11:221–231.

Doench JG, Sharp PA (2004) Specificity of microRNA target selection in translational repression. *Genes Dev* 18:504–511.

Dorn G, Abdel'Al S, Natt FJ, Weiler J, Hall J, Meigel I, Mosbacher J, Wishart W (2001) Specific inhibition of the rat ligand-gated ion channel P2X3 function via methoxyethoxy-modified phosphorothioated antisense oligonucleotides. *Antisense Nucleic Acid Drug Dev* 11:165–174.

Dorn G, Patel S, Wotherspoon G, Hemmings-Mieszczak M, Barclay J, Natt FJ, Martin P, Bevan S, Fox A, Ganju P, et al. (2004) siRNA relieves chronic neuropathic pain. *Nucleic Acids Res* 32:e49.

Drouet V, Perrin V, Hassig R, Dufour MS, Auregan G, Alves S, Bonvento G, Brouillet E, Luthi-Carter R, Hantraye P, et al. (2009) Sustained effects of nonallele-specific *huntingtin* silencing. *Ann Neurol* 65:276–285.

Du Q, Thonberg H, Wang J, Wahlestedt C, Liang Z (2005) A systematic analysis of the silencing effects of an active siRNA at all single-nucleotide mismatched target sites. *Nucleic Acids Res* 33:1671–1677.

Dykxhoorn DM, Novina CD, Sharp PA (2003) Killing the messenger: short RNAs that silence gene expression. *Nat Rev Mol Cell Biol* 4:457–467.

Eckstein F (2000) Phosphorothioate oligodeoxynucleotides: what is their origin and what is unique about them? *Antisense Nucleic Acid Drug Dev* 10:117–121.

Elbashir SM, Lendeckel W, Tuschl T (2001a) RNA interference is mediated by 21- and 22-nucleotide RNAs. *Genes Dev* 15:188–200.

Elbashir SM, Martinez J, Patkaniowska A, Lendeckel W, Tuschl T (2001b) Functional anatomy of siRNAs for mediating efficient RNAi in Drosophila melanogaster embryo lysate. *EMBO J* 20:6877–6888.

Elmen J, Thonberg H, Ljungberg K, Frieden M, Westergaard M, Xu Y, Wahren B, Liang Z, Orum H, Koch T, et al. (2005) Locked nucleic acid (LNA) mediated improvements in siRNA stability and functionality. *Nucleic Acids Res* 33:439–447.

Faber M, Pulmanausahakul R, Nagao K, Prosniak M, Rice AB, Koprowski H, Schnell MJ, Dietzschold B (2004) Identification of viral genomic elements responsible for rabies virus neuroinvasiveness. *Proc Natl Acad Sci U S A* 101:16328–16332.

Far RK, Nedbal W, Sczakiel G (2001) Concepts to automate the theoretical design of effective antisense oligonucleotides. *Bioinformatics* 17:1058–1061.

Farah MH (2007) RNAi silencing in mouse models of neurodegenerative diseases. *Curr Drug Deliv* 4:161–167.

Farhood H, Gao X, Son K, Yang YY, Lazo JS, Huang L, Barsoum J, Bottega R, Epand RM (1994) Cationic liposomes for direct gene transfer in therapy of cancer and other diseases. *Ann N Y Acad Sci* 716:23–34; discussion 34–35.

Farrell CL, Bready JV, Kaufman SA, Qian YX, Burgess TL (1995) The uptake and distribution of phosphorothioate oligonucleotides into vascular smooth muscle cells in vitro and in rabbit arteries. *Antisense Res Dev* 5:175–183.

Fattal E, Couvreur P, Dubernet C (2004) "Smart" delivery of antisense oligonucleotides by anionic pH-sensitive liposomes. *Adv Drug Deliv Rev* 56:931–946.

Fire A, Xu S, Montgomery MK, Kostas SA, Driver SE, Mello CC (1998) Potent and specific genetic interference by double-stranded RNA in Caenorhabditis elegans. *Nature* 391:806–811.

Fluiter K, ten Asbroek AL, van Groenigen M, Nooij M, Aalders MC, Baas F (2002) Tumor genotype-specific growth inhibition in vivo by antisense oligonucleotides against a polymorphic site of the large subunit of human RNA polymerase II. *Cancer Res* 62:2024–2028.

Grimm D, Kay MA (2006) Therapeutic short hairpin RNA expression in the liver: viral targets and vectors. *Gene Ther* 13:563–575.

Grimm D, Streetz KL, Jopling CL, Storm TA, Pandey K, Davis CR, Marion P, Salazar F, Kay MA (2006) Fatality in mice due to oversaturation of cellular microRNA/short hairpin RNA pathways. *Nature* 441:537–541.

Group VS (2002) A randomized controlled clinical trial of intravitreous fomivirsen for treatment of newly diagnosed peripheral cytomegalovirus retinitis in patients with AIDS. *Am J Ophthalmol* 133:467–474.

Guo S, Kemphues KJ (1995) par-1, a gene required for establishing polarity in C. elegans embryos, encodes a putative Ser/Thr kinase that is asymmetrically distributed. *Cell* 81:611–620.

Haley B, Zamore PD (2004) Kinetic analysis of the RNAi enzyme complex. *Nat Struct Mol Biol* 11:599–606.

Hamilton AJ, Baulcombe DC (1999) A species of small antisense RNA in posttranscriptional gene silencing in plants. *Science* 286:950–952.

Hammond SM, Bernstein E, Beach D, Hannon GJ (2000) An RNA-directed nuclease mediates post-transcriptional gene silencing in Drosophila cells. *Nature* 404:293–296.

Haque N, Isacson O (1997) Antisense gene therapy for neurodegenerative disease? *Exp Neurol* 144:139–146.

Harper SQ, Staber PD, He X, Eliason SL, Martins IH, Mao Q, Yang L, Kotin RM, Paulson HL, Davidson BL (2005) RNA interference improves motor and neuropathological abnormalities in a Huntington's disease mouse model. *Proc Natl Acad Sci U S A* 102:5820–5825.

Hasholt L, Abell K, Norremolle A, Nellemann C, Fenger K, Sorensen SA (2003) Antisense downregulation of mutant huntingtin in a cell model. *J Gene Med* 5:528–538.

Heidel JD, Hu S, Liu XF, Triche TJ, Davis ME (2004) Lack of interferon response in animals to naked siRNAs. *Nat Biotechnol* 22:1579–1582.

Hemmings-Mieszczak M, Dorn G, Natt FJ, Hall J, Wishart WL (2003) Independent combinatorial effect of antisense oligonucleotides and RNAi-mediated specific inhibition of the recombinant rat P2X3 receptor. *Nucleic Acids Res* 31:2117–2126.

Ho SP, Britton DH, Stone BA, Behrens DL, Leffet LM, Hobbs FW, Miller JA, Trainor GL (1996) Potent antisense oligonucleotides to the human multidrug resistance-1 mRNA are rationally selected by mapping RNA-accessible sites with oligonucleotide libraries. *Nucleic Acids Res* 24:1901–1907.

Hu J, Matsui M, Gagnon KT, Schwartz JC, Gabillet S, Arar K, Wu J, Bezprozvanny I, Corey DR (2009) Allele-specific silencing of mutant huntingtin and ataxin-3 genes by targeting expanded CAG repeats in mRNAs. *Nat Biotechnol* 27:478–484.

Hu-Lieskovan S, Heidel JD, Bartlett DW, Davis ME, Triche TJ (2005) Sequence-specific knockdown of EWS-FLI1 by targeted, nonviral delivery of small interfering RNA inhibits tumor growth in a murine model of metastatic Ewing's sarcoma. *Cancer Res* 65:8984–8992.

Jaaskelainen I, Monkkonen J, Urtti A (1994) Oligonucleotide-cationic liposome interactions. A physicochemical study. *Biochim Biophys Acta* 1195:115–123.

Jabs DA, Griffiths PD (2002) Fomivirsen for the treatment of cytomegalovirus retinitis. *Am J Ophthalmol* 133:552–556.

Jackson AL, Bartz SR, Schelter J, Kobayashi SV, Burchard J, Mao M, Li B, Cavet G, Linsley PS (2003) Expression profiling reveals off-target gene regulation by RNAi. *Nat Biotechnol* 21:635–637.

Jarver P, Langel U (2004) The use of cell-penetrating peptides as a tool for gene regulation. *Drug Discov Today* 9:395–402.

Jason TL, Koropatnick J, Berg RW (2004) Toxicology of antisense therapeutics. *Toxicol Appl Pharmacol* 201:66–83.

Judge AD, Bola G, Lee AC, MacLachlan I (2006) Design of noninflammatory synthetic siRNA mediating potent gene silencing in vivo. *Mol Ther* 13:494–505.

Judge AD, Sood V, Shaw JR, Fang D, McClintock K, MacLachlan I (2005) Sequence-dependent stimulation of the mammalian innate immune response by synthetic siRNA. *Nat Biotechnol* 23:457–462.

Kaihatsu K, Huffman KE, Corey DR (2004) Intracellular uptake and inhibition of gene expression by PNAs and PNA-peptide conjugates. *Biochemistry* 43:14340–14347.

Ketting RF, Fischer SE, Bernstein E, Sijen T, Hannon GJ, Plasterk RH (2001) Dicer functions in RNA interference and in synthesis of small RNA involved in developmental timing in C. elegans. *Genes Dev* 15:2654–2659.

Khan A, Benboubetra M, Sayyed PZ, Ng KW, Fox S, Beck G, Benter IF, Akhtar S (2004) Sustained polymeric delivery of gene silencing antisense ODNs, siRNA, DNAzymes and ribozymes: in vitro and in vivo studies. *J Drug Target* 12:393–404.

Khvorova A, Reynolds A, Jayasena SD (2003) Functional siRNAs and miRNAs exhibit strand bias. *Cell* 115:209–216.

Kim DH, Rossi JJ (2007) Strategies for silencing human disease using RNA interference. *Nat Rev Genet* 8:173–184.

Kumar P, Wu H, McBride JL, Jung KE, Kim MH, Davidson BL, Lee SK, Shankar P, Manjunath N (2007) Transvascular delivery of small interfering RNA to the central nervous system. *Nature* 448:39–43.

Kurreck J (2003) Antisense technologies. Improvement through novel chemical modifications. *Eur J Biochem* 270:1628–1644.

Leavitt BR, Guttman JA, Hodgson JG, Kimel GH, Singaraja R, Vogl AW, Hayden MR (2001) Wild-type huntingtin reduces the cellular toxicity of mutant huntingtin in vivo. *Am J Hum Genet* 68:313–324.

Lee RC, Feinbaum RL, Ambros V (1993) The C. elegans heterochronic gene lin-4 encodes small RNAs with antisense complementarity to lin-14. *Cell* 75:843–854.

Lewis BP, Burge CB, Bartel DP (2005) Conserved seed pairing, often flanked by adenosines, indicates that thousands of human genes are microRNA targets. *Cell* 120: 15–20.

Lewis DL, Wolff JA (2005) Delivery of siRNA and siRNA expression constructs to adult mammals by hydrodynamic intravascular injection. *Methods Enzymol* 392: 336–350.

Lima WF, Wu H, Nichols JG, Prakash TP, Ravikumar V, Crooke ST (2003) Human RNase H1 uses one tryptophan and two lysines to position the enzyme at the 3'-DNA/5'-RNA terminus of the heteroduplex substrate. *J Biol Chem* 278:49860–49867.

Liu G, Martins IH, Chiorini JA, Davidson BL (2005) Adeno-associated virus type 4 (AAV4) targets ependyma and astrocytes in the subventricular zone and RMS. *Gene Ther* 12:1503–1508.

Liu Q, Rand TA, Kalidas S, Du F, Kim HE, Smith DP, Wang X (2003) R2D2, a bridge between the initiation and effector steps of the Drosophila RNAi pathway. *Science* 301:1921–1925.

Lombardi MS, Jaspers L, Spronkmans C, Gellera C, Taroni F, DiMaria E, DiDonato S, Kaemmerer WF (2009). A majority of Huntington's disease patients may be treatable by individualized allele-specific RNA interference. *Experimental Neurology* 217:312–319.

Lysik MA, Wu-Pong S (2003) Innovations in oligonucleotide drug delivery. *J Pharm Sci* 92:1559–1573.

Machida Y, Okada T, Kurosawa M, Oyama F, Ozawa K, Nukina N (2006) rAAV-mediated shRNA ameliorated neuropathology in Huntington disease model mouse. *Biochem Biophys Res Commun* 343:190–197.

Marti G, Egan W, Noguchi P, Zon G, Matsukura M, Broder S (1992) Oligodeoxyribonucleotide phosphorothioate fluxes and localization in hematopoietic cells. *Antisense Res Dev* 2:27–39.

Matranga C, Tomari Y, Shin C, Bartel DP, Zamore PD (2005) Passenger-strand cleavage facilitates assembly of siRNA into Ago2-containing RNAi enzyme complexes. *Cell* 123:607–620.

Matveeva OV, Mathews DH, Tsodikov AD, Shabalina SA, Gesteland RF, Atkins JF, Freier SM (2003) Thermodynamic criteria for high hit rate antisense oligonucleotide design. *Nucleic Acids Res* 31:4989–4994.

Matveeva OV, Tsodikov AD, Giddings M, Freier SM, Wyatt JR, Spiridonov AN, Shabalina SA, Gesteland RF, Atkins JF (2000) Identification of sequence motifs in oligonucleotides whose presence is correlated with antisense activity. *Nucleic Acids Res* 28:2862–2865.

McBride JL, Boudreau RL, Harper SQ, Staber PD, Monteys AM, Martins I, Gilmore BL, Burstein H, Peluso RW, Polisky B, et al. (2008) Artificial miRNAs mitigate shRNA-mediated toxicity in the brain: implications for the therapeutic development of RNAi. *Proc Natl Acad Sci U S A* 105:5868–5873.

McKay RA, Miraglia LJ, Cummins LL, Owens SR, Sasmor H, Dean NM (1999) Characterization of a potent and specific class of antisense oligonucleotide inhibitor of human protein kinase C-alpha expression. *J Biol Chem* 274:1715–1722.

McNamara JO 2nd, Andrechek ER, Wang Y, Viles KD, Rempel RE, Gilboa E, Sullenger BA, Giangrande PH (2006) Cell type-specific delivery of siRNAs with aptamer-siRNA chimeras. *Nat Biotechnol* 24:1005–1015.

Miller VM, Xia H, Marrs GL, Gouvion CM, Lee G, Davidson BL, Paulson HL (2003) Allele-specific silencing of dominant disease genes. *Proc Natl Acad Sci U S A* 100:7195–7200.

Miyagishi M, Hayashi M, Taira K (2003) Comparison of the suppressive effects of antisense oligonucleotides and siRNAs directed against the same targets in mammalian cells. *Antisense Nucleic Acid Drug Dev* 13:1–7.

Montgomery MK, Xu S, Fire A (1998) RNA as a target of double-stranded RNA-mediated genetic interference in Caenorhabditis elegans. *Proc Natl Acad Sci U S A* 95:15502–15507.

Mook OR, Baas F, de Wissel MB, Fluiter K (2007) Evaluation of locked nucleic acid-modified small interfering RNA in vitro and in vivo. *Mol Cancer Ther* 6:833–843.

Morrissey DV, Lockridge JA, Shaw L, Blanchard K, Jensen K, Breen W, Hartsough K, Machemer L, Radka S, Jadhav V, et al. (2005) Potent and persistent in vivo anti-HBV activity of chemically modified siRNAs. *Nat Biotechnol* 23:1002–1007.

Napoli C, Lemieux C, Jorgensen R (1990) Introduction of a chimeric chalcone synthase gene into petunia results in reversible co-suppression of homologous genes in trans. *Plant Cell* 2:279–289.

Nellemann C, Abell K, Norremolle A, Lokkegaard T, Naver B, Ropke C, Rygaard J, Sorensen SA, Hasholt L (2000) Inhibition of Huntington synthesis by antisense oligodeoxynucleotides. *Mol Cell Neurosci* 16:313–323.

Nelson MH, Stein DA, Kroeker AD, Hatlevig SA, Iversen PL, Moulton HM (2005) Arginine-rich peptide conjugation to morpholino oligomers: effects on antisense activity and specificity. *Bioconjug Chem* 16:959–966.

Nobelprize.org (2006) Press release: the Nobel Prize in Physiology or Medicine 2006. http://nobel prize.org/nobel_prizes/medicine/laureates/2006/press.html. Accessed March 18, 2010.

Orban TI, Izaurralde E (2005) Decay of mRNAs targeted by RISC requires XRN1, the Ski complex, and the exosome. *RNA* 11:459–469.

Paddison PJ, Caudy AA, Bernstein E, Hannon GJ, Conklin DS (2002a) Short hairpin RNAs (shRNAs) induce sequence-specific silencing in mammalian cells. *Genes Dev* 16:948–958.

Paddison PJ, Caudy AA, Hannon GJ (2002b) Stable suppression of gene expression by RNAi in mammalian cells. *Proc Natl Acad Sci U S A* 99:1443–1448.

Passini MA, Watson DJ, Vite CH, Landsburg DJ, Feigenbaum AL, Wolfe JH (2003) Intraventricular brain injection of adeno-associated virus type 1 (AAV1) in neonatal mice results in complementary patterns of neuronal transduction to AAV2 and total long-term correction of storage lesions in the brains of beta-glucuronidase-deficient mice. *J Virol* 77:7034–7040.

Paul CP, Good PD, Winer I, Engelke DR (2002) Effective expression of small interfering RNA in human cells. *Nat Biotechnol* 20:505–508.

Persengiev SP, Zhu X, Green MR (2004) Nonspecific, concentration-dependent stimulation and repression of mammalian gene expression by small interfering RNAs (siRNAs). *RNA* 10:12–18.

Perutz MF (1999) Glutamine repeats and neurodegenerative diseases. *Brain Res Bull* 50:467.

Pfister EL, Kennington L, Straubhaar J, Wagh S, Liu W, DiFiglia M, Landwehrmeyer B, Vonsattel J-P, Zamore PD, Aronin N (2009) Five siRNAs targeting three SNPs may provide therapy for three-quarters of Huntington's disease patients. *Curr Biol* 19:774–778.

Rand TA, Petersen S, Du F, Wang X (2005) Argonaute2 cleaves the anti-guide strand of siRNA during RISC activation. *Cell* 123:621–629.

Rembach A, Turner BJ, Bruce S, Cheah IK, Scott RL, Lopes EC, Zagami CJ, Beart PM, Cheung NS, Langford SJ, et al. (2004) Antisense peptide nucleic acid targeting GluR3 delays disease onset and progression in the SOD1 G93A mouse model of familial ALS. *J Neurosci Res* 77:573–582.

Reynolds A, Leake D, Boese Q, Scaringe S, Marshall WS, Khvorova A (2004) Rational siRNA design for RNA interference. *Nat Biotechnol* 22:326–330.

Riley MG, Kim NN, Watson VE, Gobin YP, LeBel CP, Black KL, Bartus RT (1998) Intra-arterial administration of carboplatin and the blood brain barrier permeabilizing agent, RMP-7: a toxicologic evaluation in swine. *J Neurooncol* 36:167–178.

Rodriguez-Lebron E, Denovan-Wright EM, Nash K, Lewin AS, Mandel RJ (2005) Intrastriatal rAAV-mediated delivery of anti-huntingtin shRNAs induces partial reversal of disease progression in R6/1 Huntington's disease transgenic mice. *Mol Ther* 12:618–633.

Roignant JY, Carre C, Mugat B, Szymczak D, Lepesant JA, Antoniewski C (2003) Absence of transitive and systemic pathways allows cell-specific and isoform-specific RNAi in Drosophila. *RNA* 9:299–308.

Sanovich E, Bartus RT, Friden PM, Dean RL, Le HQ, Brightman MW (1995) Pathway across blood-brain barrier opened by the bradykinin agonist, RMP-7. *Brain Res* 705:125–135.

Saxena S, Jonsson ZO, Dutta A (2003) Small RNAs with imperfect match to endogenous mRNA repress translation. Implications for off-target activity of small inhibitory RNA in mammalian cells. *J Biol Chem* 278:44312–44319.

Scacheri PC, Rozenblatt-Rosen O, Caplen NJ, Wolfsberg TG, Umayam L, Lee JC, Hughes CM, Shanmugam KS, Bhattacharjee A, Meyerson M, et al. (2004) Short interfering RNAs can induce unexpected and divergent changes in the levels of untargeted proteins in mammalian cells. *Proc Natl Acad Sci U S A* 101:1892–1897.

Schwarz DS, Ding H, Kennington L, Moore JT, Schelter J, Burchard J, Linsley PS, Aronin N, Xu Z, Zamore PD (2006) Designing siRNA that distinguish between genes that differ by a single nucleotide. *PLoS Genet* 2:e140.

Schwarz DS, Hutvagner G, Du T, Xu Z, Aronin N, Zamore PD (2003) Asymmetry in the assembly of the RNAi enzyme complex. *Cell* 115:199–208.

Schwarz DS, Hutvagner G, Haley B, Zamore PD (2002) Evidence that siRNAs function as guides, not primers, in the Drosophila and human RNAi pathways. *Mol Cell* 10:537–548.

Semizarov D, Frost L, Sarthy A, Kroeger P, Halbert DN, Fesik SW (2003) Specificity of short interfering RNA determined through gene expression signatures. *Proc Natl Acad Sci U S A* 100:6347–6352.

Shohami E, Kaufer D, Chen Y, Seidman S, Cohen O, Ginzberg D, Melamed-Book N, Yirmiya R, Soreq H (2000) Antisense prevention of neuronal damages following head injury in mice. *J Mol Med* 78:228–236.

Shoji Y, Akhtar S, Periasamy A, Herman B, Juliano RL (1991) Mechanism of cellular uptake of modified oligodeoxynucleotides containing methylphosphonate linkages. *Nucleic Acids Res* 19:5543–5550.

Sledz CA, Holko M, de Veer MJ, Silverman RH, Williams BR (2003) Activation of the interferon system by short-interfering RNAs. *Nat Cell Biol* 5:834–839.

Smith RA, Miller TM, Yamanaka K, Monia BP, Condon TP, Hung G, Lobsiger CS, Ward CM, McAlonis-Downes M, Wei H, et al. (2006) Antisense oligonucleotide therapy for neurodegenerative disease. *J Clin Invest* 116:2290–2296.

Snell RG, MacMillan JC, Cheadle JP, Fenton I, Lazarou LP, Davies P, MacDonald ME, Gusella JF, Harper PS, Shaw DJ (1993) Relationship between trinucleotide repeat expansion and phenotypic variation in Huntington's disease. *Nat Genet* 4:393–397.

Song E, Lee SK, Wang J, Ince N, Ouyang N, Min J, Chen J, Shankar P, Lieberman J (2003) RNA interference targeting Fas protects mice from fulminant hepatitis. *Nat Med* 9:347–351.

Song E, Zhu P, Lee SK, Chowdhury D, Kussman S, Dykxhoorn DM, Feng Y, Palliser D, Weiner DB, Shankar P, et al. (2005) Antibody mediated in vivo delivery of small interfering RNAs via cell-surface receptors. *Nat Biotechnol* 23:709–717.

Stephenson ML, Zamecnik PC (1978) Inhibition of Rous sarcoma viral RNA translation by a specific oligodeoxyribonucleotide. *Proc Natl Acad Sci U S A* 75:285–288.

Sui G, Soohoo C, Affar el B, Gay F, Shi Y, Forrester WC (2002) A DNA vector-based RNAi technology to suppress gene expression in mammalian cells. *Proc Natl Acad Sci U S A* 99:5515–5520.

ten Asbroek AL, Fluiter K, van Groenigen M, Nooij M, Baas F (2000) Polymorphisms in the large subunit of human RNA polymerase II as target for allele-specific inhibition. *Nucleic Acids Res* 28:1133–1138.

Thakker DR, Natt F, Husken D, Maier R, Muller M, van der Putten H, Hoyer D, Cryan JF (2004) Neurochemical and behavioral consequences of widespread gene knockdown in the adult mouse brain by using nonviral RNA interference. *Proc Natl Acad Sci U S A* 101:17270–17275.

Tomari Y, Matranga C, Haley B, Martinez N, Zamore PD (2004) A protein sensor for siRNA asymmetry. *Science* 306:1377–1380.

Turner BJ, Cheah IK, Macfarlane KJ, Lopes EC, Petratos S, Langford SJ, Cheema SS (2003) Antisense peptide nucleic acid-mediated knockdown of the p75 neurotrophin receptor delays motor neuron disease in mutant SOD1 transgenic mice. *J Neurochem* 87:752–763.

Tyler BM, Jansen K, McCormick DJ, Douglas CL, Boules M, Stewart JA, Zhao L, Lacy B, Cusack B, Fauq A, et al. (1999) Peptide nucleic acids targeted to the neurotensin receptor and administered i.p. cross the blood-brain barrier and specifically reduce gene expression. *Proc Natl Acad Sci U S A* 96:7053–7058.

Uhlmann E, Vollmer J (2003) Recent advances in the development of immunostimulatory oligonucleotides. *Curr Opin Drug Discov Devel* 6:204–217.

Ui-Tei K, Naito Y, Takahashi F, Haraguchi T, Ohki-Hamazaki H, Juni A, Ueda R, Saigo K (2004) Guidelines for the selection of highly effective siRNA sequences for mammalian and chick RNA interference. *Nucleic Acids Res* 32:936–948.

van der Krol AR, Mur LA, Beld M, Mol JN, Stuitje AR (1990) Flavonoid genes in petunia: addition of a limited number of gene copies may lead to a suppression of gene expression. *Plant Cell* 2:291–299.

Vickers TA, Koo S, Bennett CF, Crooke ST, Dean NM, Baker BF (2003) Efficient reduction of target RNAs by small interfering RNA and RNase H-dependent antisense agents. A comparative analysis. *J Biol Chem* 278:7108–7118.

Vickers TA, Wyatt JR, Freier SM (2000) Effects of RNA secondary structure on cellular antisense activity. *Nucleic Acids Res* 28:1340–1347.

Vollmer J, Jepsen JS, Uhlmann E, Schetter C, Jurk M, Wader T, Wullner M, Krieg AM (2004) Modulation of CpG oligodeoxynucleotide-mediated immune stimulation by locked nucleic acid (LNA). *Oligonucleotides* 14:23–31.

Wang YL, Liu W, Wada E, Murata M, Wada K, Kanazawa I (2005) Clinico-pathological rescue of a model mouse of Huntington's disease by siRNA. *Neurosci Res* 53:241–249.

Warby SC, Montpetit A, Hayden AR, Carroll JB, Butland SL, Visscher H, Collins JA, Semaka A, Hudson TJ, Hayden MR (2009) CAG expansion in the Huntington disease gene is associated with a specific and targetable predisposing haplogroup. *Am J Hum Genet* 84:351–366.

Wickstrom EL, Bacon TA, Gonzalez A, Freeman DL, Lyman GH, Wickstrom E (1988) Human promyelocytic leukemia HL-60 cell proliferation and c-myc protein expression are inhibited by an antisense pentadecadeoxynucleotide targeted against c-myc mRNA. *Proc Natl Acad Sci U S A* 85:1028–1032.

Wu H, Lima WF, Crooke ST (2001) Investigating the structure of human RNase H1 by site-directed mutagenesis. *J Biol Chem* 276:23547–23553.

Wu H, Lima WF, Zhang H, Fan A, Sun H, Crooke ST (2004) Determination of the role of the human RNase H1 in the pharmacology of DNA-like antisense drugs. *J Biol Chem* 279:17181–17189.

Xia H, Mao Q, Eliason SL, Harper SQ, Martins IH, Orr HT, Paulson HL, Yang L, Kotin RM, Davidson BL (2004) RNAi suppresses polyglutamine-induced neurodegeneration in a model of spinocerebellar ataxia. *Nat Med* 10:816–820.

Yakubov LA, Deeva EA, Zarytova VF, Ivanova EM, Ryte AS, Yurchenko LV, Vlassov VV (1989) Mechanism of oligonucleotide uptake by cells: involvement of specific receptors? *Proc Natl Acad Sci U S A* 86:6454–6458.

Yang SP, Song ST, Tang ZM, Song HF (2003) Optimization of antisense drug design against conservative local motif in simulant secondary structures of HER-2 mRNA and QSAR analysis. *Acta Pharmacol Sin* 24:897–902.

Yazaki T, Ahmad S, Chahlavi A, Zylber-Katz E, Dean NM, Rabkin SD, Martuza RL, Glazer RI (1996) Treatment of glioblastoma U-87 by systemic administration of an antisense protein kinase C-alpha phosphorothioate oligodeoxynucleotide. *Mol Pharmacol* 50:236–242.

Yoo H, Juliano RL (2000) Enhanced delivery of antisense oligonucleotides with fluorophore-conjugated PAMAM dendrimers. *Nucleic Acids Res* 28:4225–4231.

Yu JY, DeRuiter SL, Turner DL (2002) RNA interference by expression of short-interfering RNAs and hairpin RNAs in mammalian cells. *Proc Natl Acad Sci U S A* 99:6047–6052.

Zamecnik P, Aghajanian J, Zamecnik M, Goodchild J, Witman G (1994) Electron micrographic studies of transport of oligodeoxynucleotides across eukaryotic cell membranes. *Proc Natl Acad Sci U S A* 91:3156–3160.

Zamore PD, Tuschl T, Sharp PA, Bartel DP (2000) RNAi: double-stranded RNA directs the ATP-dependent cleavage of mRNA at 21 to 23 nucleotide intervals. *Cell* 101:25–33.

Zhang Y, Boado RJ, Pardridge WM (2003) In vivo knockdown of gene expression in brain cancer with intravenous RNAi in adult rats. *J Gene Med* 5:1039–1045.

Zhang Y, Leavitt BR, van Raamsdonk JM, Dragatsis I, Goldowitz D, MacDonald ME, Hayden MR, Friedlander RM (2006) Huntingtin inhibits caspase-3 activation. *EMBO J* 25:5896–5906.

Zhang Y, Zhang YF, Bryant J, Charles A, Boado RJ, Pardridge WM (2004) Intravenous RNA interference gene therapy targeting the human epidermal growth factor receptor prolongs survival in intracranial brain cancer. *Clin Cancer Res* 10:3667–3677.

10 Recombinant Intrabodies as Molecular Tools and Potential Therapeutics for Huntington's Disease

Ali Khoshnan, Amber L. Southwell, Charles W. Bugg, Jan C. Ko, and Paul H. Patterson

CONTENTS

INTRODUCTION

The therapeutic potential of intracellularly expressed, recombinant, or single-chain fragment variable (scFv) antibodies (intrabodies) is being explored for several diseases, including cancer, HIV, and neurodegenerative disorders. Intrabodies can bind and inactivate toxic intracellular proteins, prevent misfolding, promote degradation, and block aberrant protein–protein interactions with extreme molecular specificity. Neurodegenerative disorders are particularly attractive candidates for these reagents because many of these diseases involve protein misfolding, oligomerization, and aggregation [1]. In particular, intrabodies have shown efficacy in blocking the toxicity of the amyloidogenic protein fragment Aβ in cell culture and mouse models of Alzheimer's disease (AD), paving the way for clinical trials of these reagents in brain disorders [2]. In addition to their therapeutic potential, intrabodies are also useful molecular tools to identify the pathogenic epitopes in toxic proteins, which can be targets for other types of therapy. In this chapter, we will review the strategies that have been used to develop intrabodies specific for the huntingtin (htt) protein, and describe their testing in models of Huntington's disease (HD) and their development as potential therapeutic agents for clinical use in HD.

STRATEGIES FOR INTRABODY CONSTRUCTION

Intrabodies are recombinant antibody molecules usually derived from a monoclonal antibody of interest by cDNA cloning of the antigen binding domain; the variable heavy and light chains (V_H and V_L) from the monoclonal antibody are then joined together by a synthetic cDNA encoding a flexible polypeptide linker (Figure 10.1). Alternatively, naive intrabody libraries have been constructed and cloned in phage or displayed on yeast for selection and binding to specific antigens.

However, a major problem with intracellular expression of intrabodies is proper folding and low solubility in the reducing cytoplasmic environment [3]. This is because of the presence of disulfide bonds in both the V_H and V_L, which are required for efficient folding. Although some intrabodies are inherently stable in the cytoplasm, selection of stable intrabody frameworks, which fold efficiently in the absence of disulfide bonds, has also been achieved [4]. Additionally, a process known as *in vitro* maturation or re-engineering, where the disulfide bonds are removed, can be used to correct low solubility through several rounds of random mutagenesis and antigen binding selection [5].

SINGLE-DOMAIN INTRABODIES

Recently, functional single-domain (V_H or V_L but not both) intrabodies have also been developed and selected for specific targets. These single-domain intrabodies can block protein–protein interaction and are favored for their stability and better

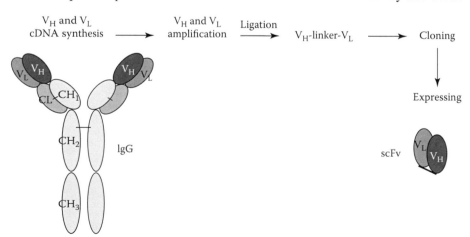

FIGURE 10.1 Schematic representation of cloning of scFvs.

folding [6]. Moreover, *in vitro* maturation of single-domain intrabodies can further enhance their folding, specificity, and solubility [5].

DEVELOPMENT OF EPITOPE-SPECIFIC INTRABODIES AGAINST HUNTINGTIN

EPITOPES IN MUTANT HUNTINGTIN FOR INTRABODY DEVELOPMENT

Intrabodies recognizing a variety of epitopes within mutant huntingtin (Htt) exon 1 (HDx1) have been isolated and tested for their ability to block toxicity and aggregation (Figure 10.1). Lecerf et al. [7] have isolated an intrabody recognizing the 17 N-terminal amino acids (AA) of Htt (C4) from a synthetic phage library and tested this intrabody for efficacy in cell culture. C4 was found to block aggregation and interfere with malonate-enhanced toxicity of mutant HDx1 [8]. Because of its modest efficacy, C4 was further matured and examined in a fly model of HD, where it was found to protect against the toxicity of HDx1 during the larval stage and to significantly increase lifespan [9].

Surprisingly, C4 increased the level of soluble mutant HDx1 in both fly and culture models [9]. This property of C4 raises the possibility that long-term exposure to this intrabody could lead to buildup of soluble HDx1 and could promote oligomerization and toxicity. Recent studies suggest that mutant HDx1 monomers can acquire a toxic conformation by switching from an α-helical to a β-sheet conformation [10]. Furthermore, blocking the 17 N-terminal AA of Htt may also have other undesirable consequences; for example, this motif is essential for vesicle localization and Htt cytoplasmic retention and turnover [11,12], and removal of the N-terminal domain results in nuclear localization of HDx1, which has been associated with enhanced toxicity [11,12]. Therefore, long-term expression studies in transgenic HD mice will be important in examining whether binding of C4 to the 17 N-terminal AA of Htt has any detrimental effects in a therapeutic setting.

Another intrabody that binds the N1–17 domain of Htt (V_L12.3) was isolated from a yeast surface display library as a single-domain light chain and matured *in vitro* through random mutagenesis and selection by yeast surface display [5]. V_L12.3, engineered for efficient intracellular expression and folding by removal of its disulfide bond, is a more potent inhibitor of mutant HDx1 aggregation and toxicity in cell culture than C4 [13]. However, like C4, V_L12.3 increases the level of soluble mutant HDx1; moreover, V_L12.3 promotes nuclear localization of mutant HDx1 [14]. This paradoxical inhibition of toxicity and aggregation together with enhancement of nuclear localization of Htt may eventually shed light on the role of nuclear Htt in toxicity. In fact, intrabodies such as V_L12.3 may have important research and clinical potential in blocking association of soluble mutant HDx1 with nuclear targets. Examination of V_L12.3 in animal models of HD is crucial for validating its protective effects and further understanding of the role of N-17 AA in mutant Htt toxicity.

THE POLYGLUTAMINE AND POLYPROLINE DOMAINS OF HTT

Intrabodies recognizing the polyglutamine (polyQ)- and proline-rich motifs of HDx1 have also been shown to influence toxicity. We generated a number of

V$_L$12.3   MW1&2   Happ1&3   EM48

N-MATLEKLMKAFESLKSFQQQQQQQQQ (n)PPPPPPPPPPPPQLPQPPPQAQPLLPQPQPPPPPPPPPPPGPAVAEEPLHRPK-//-C

      C4   MW7   MW7

FIGURE 10.2 Binding domains of different intrabodies that have been developed against the HDx1 peptide sequence.

monoclonal antibodies using either polyQ peptides or HDx1 recombinant proteins as antigens (Figure 10.2) [15]. The intrabodies cloned from these antibodies display striking, epitope-specific differences in their effects on mutant HDx1 toxicity. The MW7 intrabody, which recognizes the polyproline (polyP) motifs of mutant HDx1, protects against toxicity in several models of HD, including cell culture (Figure 10.3), acute brain slice culture, and *Drosophila* models [16,17, P. H. Reinhart et al., unpublished data]. This protection is correlated with reduced aggregation and increased turnover of mutant HDx1 [14,16].

In contrast, intrabodies that bind the expanded polyQ domain exacerbate the toxicity and aggregation of mutant HDx1 in cell culture [16]. One possible explanation for this effect is that the MW1 and MW2 intrabodies may bind and stabilize a novel confirmation in HDx1 with expanded polyQ. In fact, several anti-polyQ antibodies bind Htt in different cellular compartments, supporting the presence of distinct conformations of expanded polyQ [15]. On the other hand, anti-polyQ intrabody binding could aid in nucleation of monomeric mutant HDx1 and accelerate oligomerization. In a study of the crystal structure of MW1 bound to polyQ, the polyQ domain adopts an extended, coil-like structure with short sections of polyproline type II helix and β-strand. Consistent with the linear lattice model [18] for polyQ, linking MW1 intrabodies together in a multimeric form results in tighter binding to longer compared with shorter polyQ domains and, compared with monomeric Fv, binds expanded polyQ with higher apparent affinity [19]. Whether the affinity of

C   MW2   MW7

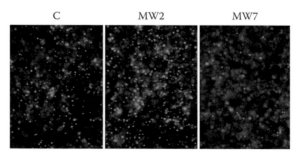

FIGURE 10.3 **(See color insert following page 172.)** MW7 prevents whereas MW2 promotes aggregation of mutant HDx1-enhanced green fluorescent protein (EGFP) in PC12 cells. MW7 and MW2 cDNAs were cloned into ecdysone-inducible vectors and transfected into PC12 cells that were engineered to express HDx1 in response to ecdysone [26]. Selected PC12 cell clones were then treated with ecdysone to induce simultaneous expression of HDx1 and the scFv. A luciferase construct was used as control (far left panel).

the monomeric versus multimeric form of MW1 influences the oligomerization of mutant Htt remains unknown.

Clearly, a unified view on the role of aggregates in HD pathology will be required to understand better how anti-polyQ intrabodies could be used to regulate mutant htt toxicity. The initial studies on the effects of MW1 and MW2 on HDx1 toxicity and aggregation were done in non-neuronal cells and with 103 polyQ HDx1, which may require a high concentration of intrabody to counteract its toxicity. Thus, re-evaluation of anti-polyQ intrabodies is worthy of investigation, possibly with shorter polyQ repeats or a multimeric form of MW1 [18]. Indeed, in light of recent findings that mutant HDx1 aggregation can be neuroprotective, anti-polyQ intrabodies will be ideal tools to dissect the role of aggregation and toxicity in neuronal models [20].

CONFORMATION-SPECIFIC INTRABODIES

Isolation of conformation-specific polyQ intrabodies may help in determining whether expanded polyQ can be a potential target for intrabody therapy. This approach has recently been reported for α-synuclein oligomers [21]. These oligomer-specific intrabodies inhibit both aggregation and toxicity of α-synuclein and have been useful tools for identifying the pathogenic epitopes. Our laboratory, in collaboration with Ron Wetzel's group, isolated a panel of monoclonal antibodies that specifically recognize oligomeric forms of polyQ proteins. Interestingly, some of these antibodies also react with fibrils formed by prion proteins and Aβ amyloid [22]. This cross-reactivity suggests the presence of common structural motifs in the fibrils of misfolded proteins that cause neurodegeneration. A similar antiserum that also reacts with amyloid fibrils of various misfolded proteins has been reported by Glabe's laboratory [23]. It will be interesting to see whether intrabodies derived from these antibodies can block oligomerization and the toxicity of these diverse proteins *in vivo*.

THE PROLINE-RICH DOMAIN OF HTT

Finally, we have recently isolated two V_L domain intrabodies from a human scFv phage display library [24] that specifically bind to the proline-rich epitope in HDx1 (which is between the two pure polyP domains discussed previously) [14]. These single-domain intrabodies (Happ1 and Happ3; see Figure 10.1) are efficient in reducing HDx1 toxicity and aggregation. A novel feature of these intrabodies, and of the anti-polyP intrabody MW7, is their reduction of soluble mutant HDx1 levels by increasing its turnover. It is intriguing that although the proline-rich epitope is identical in mutant and wild-type (WT) Htt, the Happ intrabodies have a greater effect on turnover of the mutant versus WT Htt [14]. In addition, the inhibitory effects of Happ1 and Happ3 suggest that the proline-rich domain of Htt also contributes to Htt toxicity and may be involved in the misfolding of mutant HDx1 or in its binding to partners critical for toxicity.

INTRABODIES AS RESEARCH TOOLS TO DISSECT MECHANISMS OF HTT DISEASE PATHOGENESIS

Although the therapeutic potential of intrabodies in mouse HD models remains to be explored, anti-Htt intrabodies are powerful molecular tools that can be used to identify and characterize the pathogenic epitopes in HDx1 that regulate oligomerization, toxicity, and interactions with other disease mechanisms and pathways. The findings that intrabodies directed against various epitopes of HDx1 can either block or enhance aggregation and toxicity underscore the importance of these domains.

For example, it is known that the first 17 AA of HDx1 regulate not only its nuclear targeting but also its endoplasmic reticulum and mitochondrial localization [11,12]. Thus, one hypothesis is that $V_L 12.3$ reduces toxicity by blocking the localization of mutant HDx1 to mitochondria, thereby reducing mitochondrial permeability. On the other hand, consistent with the role of the first 17 AA in cytoplasmic retention of HDx1, coexpression of $V_L 12.3$ with HDx1 also promotes HDx1 nuclear localization [14]. Thus, as noted previously, although studies with $V_L 12.3$ confirm the importance of 17 N-terminal AA in cellular distribution of Htt and the contribution of this motif to aggregation and toxicity [11,12], they also raise questions regarding whether and how nuclear localization contributes to toxicity. One possibility is that although $V_L 12.3$ promotes nuclear localization of HDx1, it may also prevent its association with the transcriptional apparatus.

Similarly, the ability of anti-polyQ intrabodies to promote aggregation and toxicity of mutant HDx1 [16] may be relevant for understanding the mechanism of *in vivo* oligomerization. One theory is that anti-polyQ intrabodies function as nucleating centers and recruit soluble HDx1, which then forms oligomers and eventually aggregates. Alternatively, binding of anti-polyQ intrabodies may induce or stabilize a conformation in the expanded polyQ domain that enhances oligomerization. If so, this raises the question of whether there are endogenous cellular modifiers that induce such conformation changes in this domain. Thus, understanding how MW1 and MW2 promote aggregation may shed light on this process *in vivo* and enable discovery of modifiers of polyQ oligomerization. In this context, it is intriguing that intracellular expression of a polyQ binding peptide blocks the toxicity of mutant HDx1 in tissue culture [25]; this peptide interferes with conversion of a nontoxic α-helical structure of polyQ to a toxic β-sheet conformation [10]. Although this conformation switch occurs *in vitro* with purified protein, the existence of an endogenous modifier of polyQ toxicity is an attractive area of investigation, and intrabodies will help with the identification of these potential regulators of toxicity.

The MW7 intrabody, on the other hand, was instrumental in identifying the HDx1 polyP domain as a pathogenic epitope [16,26]. Several important signaling proteins, including NF-κB essential modulator (NEMO)/IκB-kinase (IKK) γ, CREB-binding protein, WW domain proteins, dynamin, and FIP-2 (14.7K-interacting protein-2) require the HDx1 polyP domain for binding to HDx1 [26–29]. Therefore, the protective mechanism of the MW7 intrabody may work through its reducing the sequestration of important cellular proteins by mutant HDx1. In fact, we have shown that MW7 blocks binding of the IKK complex to the proline-rich domain of mutant HDx1 and subsequently reduces HDx1-induced nuclear factor κB activation [26]

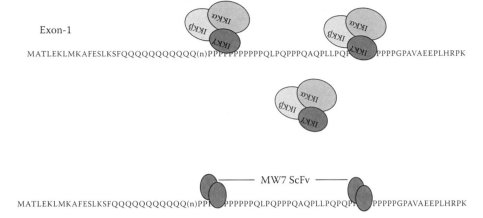

FIGURE 10.4   Schematic diagram showing the interaction of IKK complex and Htt. Binding of HDx1 to the IKK complex requires the polyP domain of Htt and the N terminus of IKKγ. Blocking the interaction of mutant HDx1 with the IKK complex reduces the toxicity in a brain slice culture model of HD. Binding of MW7 intrabody to the polyP domains of Htt also prevents IKK–HDx1 interaction and thereby reduces the toxicity of mutant HDx1. (Based on data from Khoshnan A., et al., *Proc Natl Acad Sci U S A* 99, 1002–1007, 2002.)

(Figure 10.4). Moreover, both MW7 and genetic inhibitors of the IKK complex have similar inhibitory effects on mutant HDx1 in cell and brain slice cultures [26]. These findings underscore the importance of intrabodies as molecular tools that can lead to the identification of novel pathogenic epitopes and therapeutic targets.

NOVEL TARGETS FOR INTRABODY THERAPY IN HD

To date, most of the intrabodies developed to perturb Htt function have been targeted to HDx1, which is generated by proteolytic processing of full-length Htt. However, a more upstream, primary therapeutic goal would be to prevent proteolytic processing of mutant Htt using specific intrabodies. Htt is cleaved by several proteases, including caspases-3 and -6, and the calpains [30,31]. Cleaved mutant Htt fragments are precursors to oligomers, and the species that accumulate in the nucleus likely contribute to transcriptional dysregulation [2]. Therefore, blocking the cleavage of full-length Htt by intrabodies may be an effective strategy to reduce the generation of fragments that misfold and induce toxicity.

Indeed, such inhibition of Htt cleavage by intrabody binding to cleavage sites may be preferred over small molecule inhibitors of the relevant proteases because of the target specificity of antibody binding, and because small molecule inhibitors can have systemic side effects. This technology has already been applied to reduce production of β-amyloid in AD models. Intracellular expression of an intrabody that binds an epitope in close proximity to the β-secretase cleavage site of amyloid precursor protein blocks production of amyloidogenic fragments and promotes cleavage with α-secretase, which generates nonamyloidogenic Aβ [32]. For HD, intrabodies specific to Htt cleavage sites can readily be isolated from phage display libraries and tested in tissue culture for their effects on Htt processing. This approach could be used to validate the role of these caspases on Htt processing and toxicity and, importantly, would generate potential therapeutics for HD.

DELIVERY OF INTRABODIES TO THE HD BRAIN

In principle, viral vector-based gene therapy is the ideal method for the delivery of therapeutic intrabodies to the brain. Optimal delivery of gene therapy vectors into the diseased brain remains an important research area and represents the best mode of delivery for long-term expression. Among these, adeno-associated viruses (AAVs) are the most promising vectors because they are largely nonpathogenic and because the virus is already widespread and nontoxic in human populations. AAV is capable of infecting both dividing and nondividing cells and generating long-term expression of transgenes. The existence of several serotypes offers varied tropism, allowing expression in a wide range of cell and tissue types. AAV vectors also appear to be safe and well tolerated, as no obvious side effects have been reported following a Phase I clinical trial of AAV-mediated delivery of glutamic acid decarboxylase to the brains of human Parkinson's disease patients [33]. Intracerebral delivery may also avoid systemic complications outside of the central nervous system (CNS).

In animal models of AD, several successful approaches have been reported for delivery of anti-Aβ intrabodies [2,34]. Intracranial delivery of AAV encoding anti-Aβ scFvs, which can be secreted and enter the circulation, has been effective in reducing amyloid plaque loads and neurotoxicity, as well as correcting behavioral abnormalities [2]. Viral injections in this model were performed at P0, which allowed widespread distribution and expression. Intrabody delivery to HD models may be more challenging because the toxic protein remains intracellular, in contrast to Aβ, which is secreted. Nonetheless, success has been obtained with systemic vaccination approaches in mouse models of Parkinson's disease, in which the targeted antigen, α-synuclein, is also thought to be intracellular [35]. Moreover, we find that anti-Htt antibodies display specific binding to the surfaces of live cells expressing mutant HDx1, suggesting that systemic and/or extracellular delivery of intrabodies may also be beneficial in combating Htt toxicity (Figure 10.5).

Direct viral delivery to the striatum has also proven to be effective. A single injection of an AAV vector encoding an RNA interference (RNAi) targeted against Htt results in extensive spread, reduced HDx1 oligomerization, enhanced DARPP-32 expression in striatal neurons, and amelioration of HD neuropathology [36,37]. Significant neuroprotection by AAV-mediated delivery of the neurotrophins glial cell line-derived neurotrophic factor and brain-derived neurotrophic factor to striatum has also been demonstrated in the quinolinic acid model of HD [38]. Thus, direct delivery of intrabody viral vectors to the striatum may be realistic, and it is expected that intrabodies will have fewer off-target effects than either RNAi or neurotrophins because of the high degree of specificity of antibodies. In fact, an intrabody constructed from the EM48 monoclonal antibody, which targets the C terminus of HDx1, provides significant protection against mutant HDx1 *in vivo* [39]. Injection of an adenovirus expressing EM48 intrabody in the striatum of N-171-82Q HD mice reduces the overall toxicity and decreases the aggregation of mutant Htt

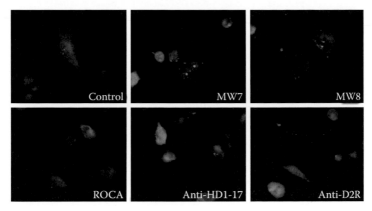

FIGURE 10.5   (See color insert following page 172.) The anti-huntingtin antibodies/intrabodies, anti-N1–17, MW7, and MW8, stain living striatal cells with a punctate pattern (red) similar to an anti-dopamine D2 receptor (D2R) antibody. The striatal ST-14 cell line was transduced with HDx1-enhanced green fluorescent protein (EGFP) (PQ103) lentivirus, and live cells were incubated with either control antibodies (mouse Ig2b) and a non-neuronal anti-CD9 (ROCA) or anti-Htt antibodies/intrabodies as indicated. A polyclonal antibody against D2R was used as positive control for cell surface staining. Alexa 568-conjugated secondary antibody was used to visualize staining (red); the green fluorescence is native HDx1-EGFP.

in the neuropil. Moreover, expression of EM48 in the striatum improves some of the behavioral deficits in the HD mice. However, EM48 does not extend the lifespan [39]. In a lentiviral model of mutant HDx1, which causes substantial degeneration in the striatum of injected mice, coinjection with an AAV expressing $V_L12.3$ or Happ1 reduces aggregation and ameliorates the loss of DARPP-32 expression in the adult. Both intrabodies also prevent the amphetamine-induced rotation bias seen with unilateral mutant Htt lentivirus injection [40]. Despite promising results in acute HD models, striatal delivery of AAV-$V_L12.3$ to transgenic models either has no effect or increases disease severity. This indicates that the negative effects of $V_L12.3$, that is, stabilizing soluble mHtt and increasing its nuclear localization [14], may have detrimental effects in long-term studies. Conversely, delivery of AAV-Happ1 to the striatum of transgenic HD models is highly beneficial to many aspects of the HD-like phenotype. Happ1 treatment restores normal motor performance in N171-82Q, YAC128, and BACHD mice in rotarod and beam-crossing tests and YAC128 and BACHD mice in climbing tests. Significant improvement is also seen in R6/2 mice in the rotarod and beam-crossing tests, as well as N171-82Q mice in the clasping test. Happ1 treatment also restores normal cognitive performance in YAC128 and BACHD mice in open-field tests and in YAC128 mice in novel object location and novel object preference tests. Although learning deficits are seen in BACHD mice by these tests, there is no effect of Happ1 treatment. Happ1 treatment dramatically reduces Htt aggregation in the R6/2 model, and normalizes ventricle size in R6/2, YAC128, and BACHD mice. While Happ1 has no effect on R6/2, YAC128, or BACHD body weight or R6/2 survival, it does significantly increase both body weight and life span of N171-82Q mice [40].

FUTURE DIRECTIONS FOR INTRABODIES IN HD THERAPY

Development of intrabodies for therapeutic purposes and as novel molecular tools to perturb protein function *in vivo* is an exciting emerging field. Some intrabodies have already reached clinical trials, and others have been used as novel diagnostic tools [41]. As optimization of delivery vehicles progresses, anti-Htt intrabodies will hold great promise for HD therapy in the future. However, many milestones, including the identification of the best targets, the most potent and effective intrabodies, and the most effective methods to ensure a widespread delivery to the CNS, must first be achieved. With rapid progress in proteomics, intrabodies can also serve as excellent tools for *in vivo* functional knockdown, for inactivating specific protein domains, and for inhibiting interactions between particular proteins. The HD field can also benefit from intrabody technology for inactivating other intracellular targets that enhance Htt toxicity, such as caspases, p53, and IKKs.

REFERENCES

1. Ross CA, Poirier MA (2005) What is the role of protein aggregation in neurodegeneration? *Nat Rev Mol Cell Biol* 6:891–8.
2. Levites Y, Jansen K, Smithson LA, Dakin R, Holloway VM, Das P, Golde TE (2006) Intracranial adeno-associated virus-mediated delivery of anti-pan amyloid beta, amyloid beta40, and amyloid beta42 single-chain variable fragments attenuates plaque pathology in amyloid precursor protein. *J Neurosci* 26:11923–8.
3. Biocca S, Ruberti F, Tafani M, Pierandrei-Amaldi P, Cattaneo A (1995) Redox state of single chain Fv fragments targeted to the endoplasmic reticulum, cytosol and mitochondria. *Biotechnology* 13:1110–5.
4. Tanaka T, Rabbitts TH (2003) Intrabodies based on intracellular capture frameworks that bind the RAS protein with high affinity and impair oncogenic transformation. *EMBO J* 22:1025–35.
5. Colby DW, Garg P, Holden T, Chao G, Webster JM, Messer A, Ingram VM, Wittrup KD (2004) Development of a human light chain variable domain (V(L)) intracellular antibody specific for the amino terminus of huntingtin via yeast surface display. *J Mol Biol* 342:901–12.
6. Tanaka T, Lobato MN, Rabbitts TH (2003) Single domain intracellular antibodies: a minimal fragment for direct in vivo selection of antigen-specific intrabodies. *J Mol Biol* 331:1109–20.
7. Lecerf JM, Shirley TL, Zhu Q, Kazantsev A, Amersdorfer P, Housman DE, Messer A, Huston JS (2001) Human single-chain Fv intrabodies counteract in situ huntingtin aggregation in cellular models of Huntington's disease. *Proc Natl Acad Sci U S A* 98:4764–9.
8. Murphy RC, Messer A (2004) A single-chain Fv intrabody provides functional protection against the effects of mutant protein in an organotypic slice culture model of Huntington's disease. *Mol Brain Res* 121:141–5.
9. Wolfgang WJ, Miller TW, Webster JM, Huston JS, Thompson LM, Marsh JL, Messer A (2005) Suppression of Huntington's disease pathology in Drosophila by human single-chain Fv antibodies. *Proc Natl Acad Sci U S A* 102:11563–8.
10. Nagai Y, Inui T, Popiel HA, Fujikake N, Hasegawa K, Urade Y, Goto Y, Naiki H, Toda T (2007) A toxic monomeric conformer of the polyglutamine protein. *Nat Struct Mol Biol* 14:332–40.

11. Rockabrand E, Slepko N, Pantalone A, Nukala VN, Kazantsev A, Marsh JL, Sullivan PG, Steffan JS, Sensi SL, Thompson LM (2007) The first 17 amino acids of Huntingtin modulate its sub-cellular localization, aggregation and effects on calcium homeostasis. *Hum Mol Genet* 16:61–77.

12. Atwal RS, Xia J, Pinchev D, Taylor J, Epand RM, Truant R (2007) Huntingtin has a membrane association signal that can modulate huntingtin aggregation, nuclear entry and toxicity. *Hum Mol Genet* 16:2600–15.

13. Colby DW, Chu Y, Cassady JP, Duennwald M, Zazulak H, Webster JM, Messer A, Lindquist S, Ingram VM, Wittrup KD (2004) Potent inhibition of huntingtin aggregation and cytotoxicity by a disulfide bond-free single-domain intracellular antibody. *Proc Natl Acad Sci U S A* 101:17616–21.

14. Southwell AL, Khoshnan A, Dunn D, Bugg C, Lo D, Patterson PH (2008) Novel intrabodies block aggregation and toxicity of mutant huntingtin by increasing its turnover. *J Neurosci* 28:9013–20.

15. Ko J, Ou S, Patterson PH (2001) New anti-huntingtin monoclonal antibodies: implications for huntingtin conformation and its binding proteins. *Brain Res Bull* 56:319–29.

16. Khoshnan A, Ko J, Patterson PH (2002) Effects of intracellular expression of anti-huntingtin antibodies of various specificities on mutant huntingtin aggregation and toxicity. *Proc Natl Acad Sci U S A* 99:1002–7.

17. Jackson GR, Sang TK, Ko J, Khoshnan A, Patterson PH (2004) Inhibition of mutant huntingtin-induced neurodegeneration in vivo by expression of a polyproline-binding single chain antibody. *Soc Neurosci* abstract 938–5.

18. Bennett MJ, Huey-Tubman KE, Herr AB, West AP Jr, Ross SA, Bjorkman PJ (2002) A linear lattice model for polyglutamine in CAG-expansion diseases. *Proc Natl Acad Sci U S A* 99:11634–9.

19. Li P, Huey-Tubman KE, Gao T, Li X, West AP Jr, Bennett MJ, Bjorkman PJ (2007) The structure of a polyQ-anti-polyQ complex reveals binding according to a linear lattice model. *Nat Struct Mol Biol* 14:381–7.

20. Arrasate M, Mitra S, Schweitzer ES, Segal MR, Finkbeiner S (2004) Inclusion body formation reduces levels of mutant huntingtin and the risk of neuronal death. *Nature* 431:805–10.

21. Emadi S, Barkhordarian H, Wang MS, Schulz P, Sierks MR (2007) Isolation of a human single chain antibody fragment against oligomeric alpha-synuclein that inhibits aggregation and prevents alpha-synuclein-induced toxicity. *J Mol Biol* 368:1132–44.

22. Geva M, O'Nuallain B, Dice L, Ko J, Ou S, Patterson PH, Wetzel R (2005) Monoclonal antibodies that bind different polyglutamine aggregate conformations. *Protein Science* 14: Supplement 1, Abstr. 404:199.

23. Kayed R, Head E, Thompson JL, McIntire TM, Milton SC, Cotman CW, Glabe CG (2003) Common structure of soluble amyloid oligomers implies common mechanism of pathogenesis. *Science* 300:486–9.

24. Griffiths AD, Williams SC, Hartley O, Tomlinson IM, Waterhouse P, Crosby WL, Kontermann RE, Jones PT, Low NM, Allison TJ, et al. (1994) Isolation of high affinity human antibodies directly from large synthetic repertoires. *EMBO J* 13:3245–60.

25. Popiel HA, Nagai Y, Fujikake N, Toda T (2007) Protein transduction domain-mediated delivery of QBP1 suppresses polyglutamine-induced neurodegeneration in vivo. *Mol Ther* 15:303–9.

26. Khoshnan A, Ko J, Watkin EE, Paige LA, Reinhart PH, Patterson PH (2004) Activation of the IkappaB kinase complex and nuclear factor-kappaB contributes to mutant huntingtin neurotoxicity. *J Neurosci* 24:7999–8008.

27. Qin ZH, Wang Y, Sapp E, Cuiffo B, Wanker E, Hayden MR, Kegel KB, Aronin N, DiFiglia M (2004) Huntingtin bodies sequester vesicle-associated proteins by a polyproline-dependent interaction. *J Neurosci* 24:269–81.

28. Steffan JS, Bodai L, Pallos J, Poelman M, McCampbell A, Apostol BL, Kazantsev A, Schmidt E, Zhu YZ, Greenwald M, et al. (2001) Histone deacetylase inhibitors arrest polyglutamine-dependent neurodegeneration in Drosophila. *Nature* 413:739–43.

29. Faber PW, Barnes GT, Srinidhi J, Chen J, Gusella JF, MacDonald ME (1998) Huntingtin interacts with a family of WW domain proteins. *Hum Mol Genet* 7:1463–74.

30. Gafni J, Hermel E, Young JE, Wellington CL, Hayden MR, Ellerby LM (2004) Inhibition of calpain cleavage of huntingtin reduces toxicity: accumulation of calpain/caspase fragments in the nucleus. *J Biol Chem* 279:20211–20.

31. Graham RK, Deng Y, Slow EJ, Haigh B, Bissada N, Lu G, Pearson J, Shehadeh J, Bertram L, Murphy Z, et al. (2006) Cleavage at the caspase-6 site is required for neuronal dysfunction and degeneration due to mutant huntingtin. *Cell* 125:1179–91.

32. Paganetti P, Calanca V, Galli C, Stefani M, Molinari M (2005) Beta-site specific intrabodies to decrease and prevent generation of Alzheimer's Abeta peptide. *J Cell Biol* 168:863–8.

33. Kaplitt MG, Feigin A, Tang C, Fitzsimons HL, Mattis P, Lawlor PA, Bland RJ, Young D, Strybing K, Eidelberg D, et al. (2007) Safety and tolerability of gene therapy with an adeno-associated virus (AAV) borne GAD gene for Parkinson's disease: an open label, phase I trial. *Lancet* 369(9579):2097–105.

34. Fukuchi K, Tahara K, Kim HD, Maxwell JA, Lewis TL, Accavitti-Loper MA, Kim H, Ponnazhagan S, Lalonde R (2006) Anti-Abeta single-chain antibody delivery via adeno-associated virus for treatment of Alzheimer's disease. *Neurobiol Dis* 23:502–11.

35. Masliah E, Rockenstein E, Adame A, Alford M, Crews L, Hashimoto M, Seubert P, Lee M, Goldstein J, Chilcote T, et al. (2005) Effects of alpha-synuclein immunization in a mouse model of Parkinson's disease. *Neuron* 46:857–68.

36. Harper SQ, Staber PD, He X, Eliason SL, Martins IH, Mao Q, Yang L, Kotin RM, Paulson HL, Davidson BL (2005) RNA interference improves motor and neuropathological abnormalities in a Huntington's disease mouse model. *Proc Natl Acad Sci U S A* 102:5820–5.

37. Machida Y, Okada T, Kurosawa M, Oyama F, Ozawa K, Nukina N (2006) rAAV-mediated shRNA ameliorated neuropathology in Huntington disease model mouse. *Biochem Biophys Res Commun* 343:190–7.

38. Kells AP, Fong DM, Dragunow M, During MJ, Young D, Connor B (2004) AAV-mediated gene delivery of BDNF or GDNF is neuroprotective in a model of Huntington disease. *Mol Ther* 9:682–8.

39. Wang CE, Zhou H, McGuire JR, Cerullo V, Lee B, Li SH, Li XJ (2008) Suppression of neuropil aggregates and neurological symptoms by an intracellular antibody implicates the cytoplasmic toxicity of mutant huntingtin. *J Cell Biol* 181:803–16.

40. Southwell AL, Ko J, Patterson PH (2009) Intrabody gene therapy ameliorates motor, cognitive, and neuropathological symptoms in multiple mouse models of Huntington's disease. *J Neurosci* 29(43):13589–90.

41. Holliger P, Hudson PJ (2005) Engineered antibody fragments and the rise of single domains. *Nat Biotechnol* 23:1126–36.

11 Biomarkers to Enable the Development of Neuroprotective Therapies for Huntington's Disease

Steven M. Hersch and H. Diana Rosas

CONTENTS

INTRODUCTION

The ultimate therapeutic goal for Huntington's disease (HD) is to develop disease-modifying therapies able to (1) delay or prevent clinical illness in those who are at genetic risk; and (2) slow the progression and permit some recovery in those who have manifest clinical illness. Rapidly advancing basic and translational research has identified numerous potential targets for neuroprotection. Some targets may be generically neuroprotective and relevant for a variety of neurological insults, whereas others may be more selective for HD. None yet stand out sufficiently to enable concentrating efforts on just a few of these, with the exception of the huntingtin protein itself, which does not have a conventional pharmacology with which to work.

Each potential target for HD is approached by multiple strategies, primarily small molecules but also by RNA interference, antisense, gene therapy, or cellular therapy.

These strategies start with families of compounds or biologicals. Medicinal chemistry, pharmacology, and biological assays winnow these families down by ordering them in terms of potency, favorable pharmacological properties, toxicity, teratogenicity, off-target effects, bioavailability, central nervous system (CNS) penetration, and so on. However, as helpful as *in silico, in vitro,* and *in vivo* models are, they provide only an incomplete understanding of target and treatment properties and disease modifying potential.

Indeed, for neurological disease, there is much more history with compounds working in cellular and in animal models and subsequently not working in human disease than there is of models successfully predicting effective therapies (e.g., in stroke or amyotrophic lateral sclerosis [ALS]). There may be more hope for HD because of its dominant genetic nature and the greater relevance of the models. In the end, target validation and prioritization, as well as discerning the potential risks and benefits of individual compounds, will have to come from clinical experiments in human subjects, particularly those with premanifest or manifest HD.

However, the capacity to conduct clinical trials is not close to keeping up with the numbers of compounds for which there are already some rationale and likelihood of safety and tolerability—and the gap is quickly widening. There are many reasons for this gap, and these mostly come down to limited resources of time, effort, money, investigators, and subjects. At-risk and affected individuals inexorably progressing toward clinical disease or through increasing disability provide an underlying urgency not only to do more testing of potential therapies but also to improve the process. In this context, the development and use of biomarkers in clinical trials for HD will have profound potential to increase the rate and accuracy with which treatments and by implication their targets, can be assessed.

Thus, there is a great need for the development of biomarkers for HD which are useful in early- and late-phase clinical trials. Although finding a dose range for a treatment in HD patients and testing for safety and tolerability are straightforward, it can be difficult to find signals that indicate that the desired pharmacological activity is occurring and is optimal, or that compare one agent with another. A further difficulty has been the lack of clinical or other outcome measures able to provide preliminary evidence of efficacy for neuroprotection that would help in the prioritization of compounds for large Phase III studies. Indeed, without such signals it is also very difficult to stop development of a compound short of its failure in a large-scale study.

Moreover, in premanifest HD, there may be no clinical measures to provide useful assessment of efficacy. In manifest HD, clinical symptoms progress slowly and are extremely variable, and their modulation does not intrinsically correspond to disease modification. Thus, although modulation of symptoms may point to a symptomatic benefit, improving symptoms does not necessarily predict slowing the disease process. For example, haloperidol can suppress chorea yet hasten death by worsening dysphagia.

Biomarkers that can indicate whether a potential disease-modifying therapy interacts with its target or affects disease processes (state) or progression in smaller, shorter early-phase trials are urgently needed to help decision making about therapeutic development, such that not every candidate has to be tested in large futility or Phase III studies. Biomarkers able to provide supportive or even primary evidence for efficacy would also facilitate late-phase trials. Indeed, the National Institutes of Health (NIH) Neuroscience Blueprint has made biomarker development for neurodegenerative diseases a high priority (http://neuroscienceblueprint.nih.gov).

However, although biomarker research is being embraced, there remains some confusion about biomarkers. This chapter will provide a framework for considering the development, assessment, and use of different types of biomarkers that could facilitate the development of neuroprotective therapies for HD.

Definitions

The field of biomarkers has been hampered by varying definitions of what a biomarker is and by the use of various adjectives to describe how thoroughly they have been assessed. A working group on biomarker definitions was convened by the NIH in 2000 to propose definitions for common use and to provide a framework for assessing their correspondence to clinical outcome measures.[1] Biomarkers were defined as a characteristic that is objectively measured and evaluated as an indicator of normal biological processes, pathogenic processes, or pharmacological responses to a therapeutic intervention. Biomarkers can also be considered to encompass technological improvements in looking at clinical signs, such as neuropsychological measures, eye movements, or quantitative movement measurements. Although these may also have value as outcome measures, for the purposes of this review these are better considered as refined clinical measures rather than biomarkers because they remain removed from the biology of the disease or of the treatments and may primarily measure symptoms, although with more sensitivity.

Surrogate endpoints were defined by the NIH Working Group on Biomarkers as a special subset of biomarkers intended to *substitute* for a clinical endpoint. A surrogate endpoint sufficiently correlated to a meaningful clinical outcome measure can be acceptable as a substitute for the clinical endpoint by the Food and Drug Administration (FDA) when considering whether an intervention is efficacious. Thus, surrogate endpoints can serve as a basis for regulatory approval, although this has been rare because establishing sufficient correspondence between a biomarker and a meaningful clinical endpoint is difficult and requires extensive validation in natural history and therapeutic studies. In other disease areas, established surrogate endpoints include blood pressure and cholesterol levels for cardiovascular drugs, blood glucose and glycohemoglobin for diabetes, viral RNA load and CD4 counts for HIV, and tumor size for antineoplastic agents. Although this standard is infrequently met, biomarkers need not become surrogate endpoints to be tremendously valuable. Likewise for HD, surrogate endpoints would be useful if they could improve the efficiency of efficacy studies. However, the main focus for biomarker research should be the development of useful purpose-driven biomarkers. These may ultimately include some able to serve also as surrogate endpoints.

Classification of Biomarkers

Biomarkers can serve many purposes in the development of treatments for HD. Thus, it is useful to classify biomarkers to help convey the different purposes for which they can serve. To a large degree, these purposes are intrinsic to what is being measured. The more global the biomarker, the more able it may be to capture a large portion of the disease process and thus be more predictive of disease progression. More specific biomarkers linked to particular facets of disease biochemistry may have great appeal and measurement precision but also have a danger of capturing only a fraction of the disease process, and it can be difficult to understand what that fraction is.

Global biomarkers, such as tumor size for oncology or the size of the brain or of brain structures in neurodegeneration, are removed from the pathophysiology but can capture the global direction and severity of disease. Accordingly, global biomarkers are most likely to be useful in staging the presence, progression, or severity of disease, or as outcome measures able to reflect the response to a treatment in studies seeking evidence of possible clinical efficacy.

Process biomarkers are more specific laboratory measures such as levels of proteins, gene expression, compounds, or metabolites that capture a molecular/biochemical aspect of the pathogenesis of the disease or of biological responses to the disease process. Process biomarkers can usefully demonstrate mechanisms of disease and pharmacological/pharmacodynamic responses but may only represent a component of the many contributors to disease. In a complex disorder like HD, in which a variety of pathogenic mechanisms have been implicated, a process biomarker may only capture a slice of the entire set of biochemical influences that make up the disease. Process biomarkers can be powerful for examining whether a compound appropriately affects its target (or other targets), for assessing dose, for comparing compounds for potency or toxicity, and for examining the biological bases of response variation between subjects (pharmacodynamic biomarkers).

However, the great specificity of a process biomarker often prevents it from reflecting the entire clinical disease process sufficiently for it to be informative about the impact a treatment might have on the clinical onset or progression of disease. For example, a process biomarker could show that a possible treatment successfully modulates its molecular target in an early-stage assessment, but the same biomarker may have an uncertain relationship to clinical efficacy, which might be better examined with other types of biomarkers. This discussion also illustrates the importance of matching biomarkers to the particular therapeutic development tasks at hand.

Another way of looking at this considers the known value of a biomarker rather than the biological modality from which it comes, in a more hierarchical manner. For HD, we can envision a hierarchical organization of biomarkers, as has been discussed for oncology.[2] A *diagnostic biomarker* can identify processes that distinguish HD from non-HD. For example, the genetic test for HD can diagnose the presence or absence of risk for developing HD; other tests useful for clinical trials might diagnose the presence of the HD prodrome in premanifest individuals or the conversion from premanifest to manifest HD. Diagnostic biomarkers can serve classification purposes but may not be able to measure clinical aspects of HD such as severity, magnitude, and progression.

Pharmacodynamic biomarkers additionally have a connection to likely pathogenic, neuroprotective, or pharmacological mechanisms and thus can provide indications of a response to a therapy. This might include levels of compounds in blood or the brain, signs of toxicity or off-target interactions, or evidence that a compound is on target for its desired effects. For HD, pharmacodynamic biomarkers are especially useful in early-phase clinical trials to assess whether a treatment modulates a desired target, thus revealing pharmacological efficacy. However, even if a treatment were to have pharmacological efficacy, it may not translate to a clinical benefit if the biomarker measures an inconsequential process.

Progression biomarkers additionally have predictive power for a desired clinical outcome that has been confirmed in observational or therapeutic trials sufficiently to carry weight in decisions about late-phase development or to serve as potential surrogate endpoints. For HD, such biomarkers would correspond to clinical or biological progression of disease and thus could serve as primary or secondary endpoints for assessing clinical efficacy. Clearly, these biomarker types require increasing rigor of validation yet also provide increasing confidence about their ability to reflect a significant portion of the disease phenotype or even of treatment efficacy.

BIOMARKERS AND CLINICAL STAGES OF HD

The focus of this review is on individuals who are either premanifest or in the earlier stages of being symptomatic because they constitute the HD populations with the greatest potential for benefiting from neuroprotective treatments. Individuals are born at risk for HD by virtue of possessing the causative CAG expansion; therefore, CAG length is a *diagnostic biomarker* that classifies individuals as someday developing manifest HD symptoms. The actual length of the CAG expansion carries further information about age of onset and can be used to model predictions of symptom onset for clinical research[3]; however, there is still considerable spread in actual age of onset even for people close to expected onset,[4] presumably as a result of other genetic and environmental influences.

Premanifest individuals are indistinguishable clinically from gene-negative individuals until they begin to approach the onset of unequivocal symptoms. There is a prodromal period lasting at least 10 years or more in which there may be cognitive, emotional, functional, and motor signs but little functional decline. Experimental neuropsychological, neuropathological, and neuroimaging studies are able to document changes in this period, and there is evidence that the brain is working harder as it compensates functionally for ongoing degenerative changes.[5–15] Thus, the HD prodrome may ultimately be the most opportune time for neuroprotective therapy because it marks a period when individuals have active brain disease but remain completely functional. Because the HD prodrome has an underlying biology, *diagnostic biomarkers* that mark its onset or presence, *pharmacodynamic biomarkers* that indicate treatment responses, and *progression biomarkers* that correspond to its evolution toward manifest HD can be expected to provide the most useful outcome measures.

Currently, a therapeutic trial seeking to delay the clinical onset of HD ("phenoconversion") would look for a differential rate of diagnoses of manifest HD between placebo and active treatment groups. To examine whether a treatment delays the onset of clinical symptoms in premanifest individuals with the HD genetic mutation, it has been estimated that 1,000–3,000 subjects and 3–6 years of follow-up evaluation are necessary to detect a large 30%–40% decrease in the frequency of symptom onset. Although the HD prodrome ends with the unequivocal presence of a movement disorder that permits a clinical diagnosis of manifest HD, the underlying biological processes that make up neurodegeneration are more continuous such that diagnosis likely marks some threshold being passed at which physiological compensation fails or observers gain sufficient sensitivity to detect symptomatic HD with confidence.

Thus, phenoconversion is essentially a binary variable of very low power but very high clinical relevance. It is of such low power that early-phase clinical trials in premanifest subjects would have little possibility of providing preliminary evidence for efficacy. In contrast, biomarkers could be used to detect disease and pharmacodynamic responses to treatment during the prodrome, without awaiting phenoconversion. Furthermore, biomarkers could provide continuous variables with slopes of change during the HD prodrome and early symptomatic period that could be monitored, and comparisons could be made between treatment groups. Biological responses to treatments or changes of slope in biomarker measures as a result of treatment could be detected much more sensitively than phenoconversion and provide evidence of an intervention being disease modifying in Phase II-sized studies. Such data would be essential to help justify the effort and expense that an efficacy study in premanifest HD would entail. It may also provide evidence that a treatment shown to slow clinical progression in manifest HD is behaving similarly in premanifest HD. Thus, clinical and biological markers able to detect the HD prodrome and measure its progression would enable the identification of subjects in this period and facilitate performing informative therapeutic trials.

An important issue in designing a clinical trial in the premanifest population is that the vast majority of individuals (>95%) at risk for carrying the HD genetic mutation have not desired genetic testing.[16] Some of the reasons for this include fear of genetic discrimination, the lack of effective treatment, and concern about the negative consequences of testing. Focus groups with at-risk individuals have revealed that many would be averse to taking part in clinical trials if informative genetic testing is required. In fact, many would willingly take an experimental medication and risk side effects in a clinical trial without genetic testing, understanding that there would be a 50% chance of not having the gene mutation. Performing a clinical trial only in subjects who have had genetic testing thereby raises a concern about creating an incentive for genetic testing along with its negative consequences in subjects wishing to participate. Currently, because relatively few individuals have pursued genetic testing, it is necessary to include individuals who do not know their genetic status in clinical research. Because a biomarker can be a surrogate (albeit a less sensitive one) for the HD genetic test, it will be important to treat biomarker data with the same concerns about confidentiality and blinding.

Once diagnosed clinically, manifest HD has a highly variable phenotype. Some of this variation is inherent in the disease. For example, different affected individuals can have predominant motor, predominant cognitive, or predominant psychiatric presentations and different rates of progression. Moreover, the severity of symptoms can be modulated by many temporary factors such as mood, nutrition, medications, and sleep disturbances. Despite great day-to-day variability in symptoms, progression is slow when assessed by an integrated measure of functional capacity (e.g., the Total Functional Capacity [TFC] scale).[17] For manifest HD, significant reduction of functional decline over time is the *sine qua non* of slowing progression and has been acceptable to the FDA as meeting their legal mandate that a potential treatment must have clinical significance.

Currently, neuroprotection efficacy studies in symptomatic patients with HD rest on a primary clinical outcome measure (TFC scale) requiring, for example, about 600 subjects and 5 years of follow-up evaluation to detect a 20% slowing of functional decline (1:1 randomized placebo controlled trial). Thus, the expense, time, and great magnitude of effort needed to test efficacy means that few interventions can be tested. More sensitive clinical measures have been explored and may help, but none has demonstrated clear promise for increasing clinical trial power dramatically, and none has been validated as improving on the TFC scale. Furthermore, having large numbers of subjects on placebo treatment for years in these trials is an unfortunate necessity.

In this context, biomarkers corresponding to disease progression could help assess efficacy, could supplement the TFC scale and other clinical endpoints in symptomatic subjects, and may ultimately serve as surrogate endpoints enabling the testing of disease modification in fewer subjects more quickly. Biomarkers of disease biological activity can be used in these studies to help answer whether treatments have the desired pharmacodynamic effects, whether dosing is optimal, and whether there are potential biological explanations for response heterogeneity. Finally, pharmacodynamic and other biomarker responses in early-phase studies suggestive of disease modification could help greatly in deciding whether to proceed to large efficacy studies and in enabling refinement of study design.

BIOMARKER DEVELOPMENT AND CHARACTERIZATION

The attributes of an ideal biomarker include pathophysiological relevance, sensitivity and specificity for HD and for treatment effects, reliability (accuracy, precision, robustness, reproducibility), practicality (must be minimally invasive, affordable), and simplicity (does not require unusual skills or equipment, scalable to multicenter trials). For HD, regional brain atrophy has very high pathophysiological relevance as it likely reflects neurodegeneration directly. However, the relevance of process biomarkers, such as levels of small molecules in blood or cerebrospinal fluid (CSF), enzyme activities, or gene expression may be uncertain because the exact manner in which mutant huntingtin causes neurodegeneration has not yet been established. Nevertheless, connections can be made to processes known to play roles in pathogenesis, such as energy compromise, oxidative stress, and transcriptional disruption, and such levels can serve as diagnostic, progression, and pharmacodynamic biomarkers. Reliability, practicality, sensitivity, and specificity of biomarkers must be assessed experimentally or statistically. There are no hard criteria for these, but their positive and negative attributes must be balanced against the applications for which they are needed and any available alternatives.

Criteria for biomarker assessment and validation are defined by the nature of the questions that the biomarker is intended to address, the degree of certainty required, and assumptions about relationships to clinical endpoints. Validation is a continuous process that evolves as new information is accumulated from preclinical, early-phase clinical, and late-phase clinical studies. Brooks et al.[18] presented criteria for the assessment of potential biomarkers in studies of progression in Parkinson's disease (PD), which can be adapted to the characterization of biomarkers for HD as follows:

1. The biomarker should correlate with the likelihood or progression of premanifest disease activity, with the development of clinical HD in premanifest individuals, or with clinical deterioration in symptomatic individuals.
2. The biomarker should be objective (amenable to blinded or centralized assessment).
3. The biomarker should be reproducible (reliable).
4. Biomarker changes should be specific to changes in disease status (e.g., they should not be distorted by pharmacological treatments or coexisting medical conditions).
5. Biomarker assessment itself should be safe and well tolerated.
6. Ideally, the biomarker should be inexpensive and easy to use.
7. The biomarker should be able to be assessed repeatedly in the same patient so as to provide measures of change and progression.

Discovery and validation of biomarkers proceed in phases.[19] First, promising directions are identified and prioritized in preclinical and retrospective or cross-sectional clinical material; these could be nonbiased screening ("omic") approaches or candidate-based approaches. Examples include metabolomic and gene expression screening analyses, brain-wide regional morphometry, and testing of mechanistic assays such as oxidative stress markers or indole levels. Evaluation of sensitivity, specificity, and dynamic range helps prioritize candidates, and replication in additional sample or subject sets is important to validate the potential biomarkers and their prioritization. Validation at this level can be sufficient to identify diagnostic and pharmacodynamic biomarkers.

Next, identified biomarkers are tested in relevant prospective clinical cohorts for their assessment as continuous variables and to examine their correspondence to desired clinical indicators, such as time, specific clinical features, clinical progression, response to therapy, and response to covariates (e.g., gender, age, smoking, medications, and concomitant illnesses). These studies enable comparing markers, developing algorithms for combining them, and determining slopes and variability, which help define useful sample intervals and sizes.

Finally, prospective biomarkers are tested in definitive large-scale observational or therapeutic studies to assess their feasibility and power to predict the onset of disease, disease progression, or the response to a disease-modifying therapy. These progressive analyses enable true estimates of the power of the biomarkers to make diagnostic and therapeutic predictions and lead to an accumulation of information about their feasibility, generalizability, and cost.

It is well known that population-level associations between biomarkers and clinical outcomes, even when strong and highly statistically significant, do not necessarily translate into precise predictions at an individual level. For example, the mean age at HD onset for a population of at-risk subjects can be estimated precisely as a function of CAG repeat length, but the variability around this prediction for any single individual remains high. Predictive values of tests for disease status based on a continuous biomarker can be described by receiver operating characteristic (ROC) curves,[20] which plot sensitivity against specificity. The area under an ROC curve represents the probability that an HD subject will have a higher (or lower) value of the biomarker than a healthy subject and is a useful summary measure of predictive ability that can be used for rank ordering biomarker effectiveness. Using clinically meaningful criteria, the predictive ability of biomarkers can be calculated for distinguishing controls from premanifest or manifest subjects, premanifest subjects with detectable disease activity from those without, premanifest subjects who develop manifest disease from those who do not, and "fast progressors" from "slow progressors" among premanifest and manifest subjects (intrinsically or in response to treatment). Importantly, ROC curves do not depend on the absolute scale of raw data measurements, which enables comparisons among different types of biomarkers. They also display true- and false-positive rates, which are particularly relevant.

Biomarker Modalities—Neuroimaging

Modern neuroimaging, including structural (mainly computed tomography and magnetic resonance imaging [MRI]), biochemical (magnetic resonance spectroscopy), and functional neuroimaging (mainly positron emission tomography [PET], single-photon emission computed tomography, and functional MRI [fMRI]), may enable the visualization of early brain changes *in vivo*. These methods are largely noninvasive, are increasingly accessible in clinical practice, and, together with the clear relevance to the neuropathology, make neuroimaging an appealing potential biomarker for HD. Until recently, most efforts to use morphometric neuroimaging in HD have been focused on the striatum because it is so severely affected and because it is composed of structures that can be easily delineated. However, if HD is truly a "polytypic process," any imaging analysis limited to the study of a small number of structures may fail to fully characterize the distribution and temporal course of pathology or to explain the clinical symptoms of HD accurately enough to be of value in clinical trials. Thus, while the relative contributions of the cortex and striatum to clinical symptoms of HD remain to be elucidated, treatments aimed solely at preserving striatal function or striatal anatomy may fail to ameliorate clinical symptoms (Figure 11.1).

Development of neuroimaging biomarkers for HD has been strongly supported by recent technological advancements in structural neuroimaging methods, which have led to improvements in the acquisition of images with better spatial resolution and gray/white contrast, as well as in the availability of automated methods for detailed morphometric analyses, with improved sensitivity and reliability compared with traditional manual methods and which enable evaluation of the entire brain. Diffusion tensor imaging is another emerging technology that enables examination of the brain at a microstructural level. MRI measures of cortical, white matter and subcortical neurodegeneration are detectable years before symptom onset[15,21–23] and provide a sensitive measure of progression in premanifest and symptomatic individuals. These

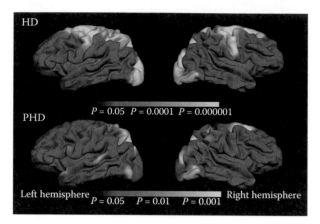

FIGURE 11.1   (See color insert following page 172.) Cortical changes occur early and are extensive in HD. Top, maps of statistically significant thinning in 33 patients with early HD compared with age- and sex-matched subjects. Bottom, maps of statistically significant thinning during the premanifest period, a total of 35 subjects. The most yellow areas correspond to loss of the cortical gray matter more than 20% thinner than controls. The distribution of cortical change and their correspondence with clinical symptoms suggest that cortical atrophy may play an important role in the clinical manifestations of HD.

changes correlate with disabling symptoms and functional decline as measured using the TFC scale[24] and may be responsive to neuroprotective treatments providing potentially far greater experimental power than clinical outcome measures. These readily available, noninvasive methods allow for multiple repeated evaluations over the course of longitudinal studies and may lend themselves well as biomarkers in clinical drug trials.

Several studies in both premanifest and early symptomatic studies using fMRI (which uses changes in regional blood flow during neuronal activity to identify regions of brain activated during performance of a cognitive task) have shown alterations in patterns of activation in HD.[25–27] However, although potentially appealing as a biomarker of early neuronal dysfunction, there are several limitations to its use in clinical trials: fMRI activation patterns can vary significantly across subjects[28]; fMRI is highly susceptible to trivial differences in conditions and study design; there are inherent limitations to its use in longitudinal studies; and, finally, it is unclear how changes in activation may actually reflect neuropathology directly.[29,30]

PET imaging has the potential to provide information about neurochemical, hemodynamic, or metabolic processes that may be disrupted in HD. Several studies have shown abnormalities in glucose utilization, striatal raclopride binding, and microglial activation.[31–34] Several PET tracers have been used in PD to study pathophysiological mechanisms; however, no existing tracer has yet been identified that could serve as a surrogate endpoint in clinical trials in PD.[35] Finally, PET imaging has been proposed as a potential biomarker for disease-modifying trials in Alzheimer's disease (AD)[36,37]; however, the generalizability of these methods for multicenter studies and for HD is uncertain.

BIOLOGICAL FLUIDS AND CANDIDATE MOLECULAR BIOMARKERS

CSF is in contact with the brain and thus is likely to pick up biochemical signals from brain disease. It is relatively acellular, so its potential lies primarily in the measurement of small molecules and proteins. CSF biomarkers have been developed for AD that are diagnostic, may correspond to progression, and are potentially pharmacodynamic. There are also possible biomarkers in CSF for ALS and PD. CSF has some limitations in that its collection is a significantly inconvenient and time-consuming procedure with some morbidity and expense, and so may be best suited for studies without many repeated measurements. Some changes have been noted in the CSF of HD patients, including monoamine metabolites[38]; tryptophan pathway metabolites[39]; F2-isoprostanes, a marker of lipid peroxidation and oxidative stress[40]; and measures of transglutaminase activity,[41] which is involved in protein cross-linking and antioxidant responses. However, none has yet been studied extensively enough to determine whether it has potential as a clinically useful biomarker.

Blood and urine are peripheral fluids with the obvious advantage of being easily obtained with a minimum of effort and risk. However, it may be counterintuitive that useful biomarkers of HD could be found in blood or urine because we are most interested in detecting and treating neurodegeneration. For a peripheral marker to represent the "state" of HD, it would either have to leak from the brain or represent pathology occurring in the periphery. Leakage can be considered broadly to include molecules escaping the brain, the effects of escaped molecules on peripheral biochemistry, and disturbances of the means by which the brain customarily acts on the periphery (e.g., neuroendocrine, metabolic, autonomic, and behavior). Any of these paths have the potential to indicate the presence and progression of HD and its response to treatment.

Effects of mutant huntingtin in the periphery independent from the brain could also readily indicate disease activity and responses to treatments. Progressive processes could also occur in the periphery, although they may or may not mirror progression in the brain sufficiently to serve as biomarkers corresponding to neurodegeneration. In the end, the correspondence of a biomarker to disease features is determined experimentally by correlation with clinical endpoints; therefore, fully understanding their source is not a requirement. Nevertheless, the closer such a biomarker is to a relevant disease mechanism, the more likely it is to be useful.

Blood consists of cellular (red blood cells, leukocytes, or platelets) and liquid components (serum or plasma) that are typically separated for analyses. Most studies have either examined small molecules or proteins free in plasma or contained in leukocytes (buffy coat preparation separated from red blood cells by centrifugation) or platelets. Because leukocytes are cellular, they also provide an opportunity to examine markers associated with organelles, cell membranes, and cellular and nuclear processes (e.g., proteolysis, gene expression). Platelets can also be isolated and express an interesting subset of proteins and receptors that are in common with neurons. Lymphocytes survive from weeks to lifelong (as is the case for memory lymphocytes); thus, biochemical and molecular changes may accumulate as the disease progresses over time. Moreover, in HD, accumulating oxidative damage to nuclear and mitochondrial DNA and accumulating transcriptional alterations in longer-lived erythrocytes and erythropoietic progenitor cells may further contribute to progressive biochemical and gene expression changes in blood cells. Thus, the turnover of blood cells may not hamper their potential for providing biomarkers of disease progression.

Blood liquids can pick up molecules from the entire body, as well as the brain. Although the blood–brain barrier (BBB) may limit CNS-to-blood transfer, other tissues and organs in the body have their own diverse leakiness into the blood stream. Because the mutant huntingtin protein is expressed ubiquitously throughout the body and because it is so promiscuously interactive,[42] it would be surprising if it does not affect biochemistry and gene expression much more widely than just in the brain, even if these effects are mostly clinically silent. In fact, effects of HD on peripheral cells such as lymphocytes and peripheral tissues, such as muscle, have been detected.[43–45] Although studies in blood in HD are in their infancy, a number of promising candidate biomarkers have already been identified in survey studies using various "omic" platforms and candidate-based studies measuring specific molecules. Not surprisingly, these potential HD biomarker candidates mostly cluster around pathways involved in the pathogenesis or expression of HD as understood in neurons. These include oxidative stress, energy compromise, transcriptional alterations, neurotrophin alterations, cell death pathways, inflammation, and proteolysis.

Oxidative Stress and Inflammation

Evidence for energy compromise and oxidative stress being significant contributors to the pathogenesis of HD, and not just fallout from neurodegeneration, has progressively accumulated. Peripheral markers of these processes have included measurement of oxidized molecules (DNA, lipids) in blood and urine,[46–50] endogenous antioxidants,[48,51] branch chain amino acids as products of metabolism,[52] and levels of the hormones ghrelin and leptin, which are involved in energy balance and weight maintainance.[53] Plasma levels of 8-hydroxy-2-deoxy guanosine, a marker of oxidized DNA, have been used as a pharmacodynamic biomarker in a study of creatine monohydrate as a potential neuroprotective therapy (Figure 11.2).[46] Several studies have identified proinflammatory molecules as being elevated in HD, although it is uncertain whether there is a primary connection between inflammatory responses in the brain or the periphery to pathogenesis, or whether there may be inflammatory responses to oxidative or other stresses. These molecules include C-reactive protein in plasma,[47] and a proteomic analysis[54] identified interrelated cytokine and complement pathway activations (clusterin, α2-macroglobulin, interleukin 6, C7, C9). These latter molecules have had some correlation with HD stage and so are candidate progression biomarkers awaiting further validation.

Transcriptional Markers

Many studies have demonstrated direct and indirect effects on gene transcription by mutant huntingtin, and there are growing connections between these effects and the pathogenesis of HD. The greatest potential for transcriptional biomarkers lies in measuring gene transcription in cellular blood, particularly leukocytes, or in measuring gene products more generally in blood cells and fluids. Skeletal muscle

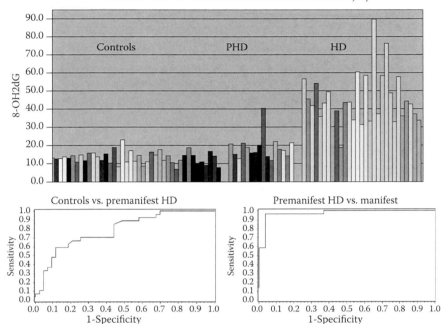

FIGURE 11.2 (See color insert following page 172.) Individual 8-hydroxy-2-deoxyguanosine (8-OHDG) levels in plasma from age- and gender-matched controls (mean = 13.5), premanifest HD (PHD) subjects (mean = 18.1), and early manifest HD subjects (mean = 45.3). Levels are higher in PHD subjects than in controls ($P = .003$) and almost fourfold higher in manifest HD subjects ($P = 5.013E-13$). Partial discrimination of controls from premanifest HD subjects (area under the receiver operating characteristic curve = 0.784) suggests that 8-OHDG elevates during the HD prodrome. Complete discrimination of premanifest HD from manifest HD subjects (area under the receiver operating characteristic curve = 0.973) suggests that 8-OHDG elevations accelerate as individuals become symptomatic. 8-OHDG has potential as a biomarker of a disease activity (DNA damage), as a diagnostic biomarker of phenoconversion, and as a pharmacodynamic biomarker because it can be suppressed by antioxidant and energy-buffering treatments.

obtained by muscle biopsy also has potential,[45] although variability and changes in nutrition, weight, and muscle use in the course of HD are potential complicating factors. Survey and targeted approaches to transcription have used microarrays for discovery and quantitative polymerase chain reaction for confirmation.[45,55,56] Borovecki et al.[55] developed a 12-gene marker set that differentiated control and HD subjects, correlated with progression, and responded pharmacodynamically to treatment with an experimental histone deacetylase inhibitor. However, this same marker set was not confirmed by Runne et al.,[56] who failed to observe any strong signals in blood.

Such studies suggest great potential in transcriptomic markers but also caution about their sensitivity. The genes in these studies arose mostly from microarray screens. Other genes already known to be affected in the brain in HD have been examined as possible biomarkers by assessing their gene products in blood. Brain-derived neurotrophic factor (BDNF) is a neurotrophin that could play a role in neurodegeneration; it is down-regulated in the HD brain[57] and can cross the BBB. Ciammola et al.[58] have studied serum levels of BDNF in HD subjects and demonstrated that it is reduced in HD and that this reduction could be related to CAG length and to disease severity. Another gene that is profoundly repressed in the brain in HD is the adenosine A2a receptor. Maglioni et al.[59] and Varani et al.[60] studied A2a receptor function in platelets from HD subjects and found an increase in receptor binding and a potential correspondence with proximity to clinical onset of HD.

Small Molecules

Biological tissues and fluids contain thousands of small molecules. Metabolomic profiling enables the comparison of small molecules between different groups or conditions, as well as determination of which specific molecules might differentiate among them. Metabolomic profiling has been applied to serum[61] and plasma from HD subjects, and it is clear that there are rich changes in small molecule populations that can differentiate between HD and controls and could potentially serve as biomarkers. The promise in metabolomics is to identify candidate small molecules or metabolic pathways for further assessment as task-specific biomarkers in clinical studies. One such promising pathway is kynurenine metabolism,[47] in which multiple alterations have been identified in HD blood.

Huntingtin

Finally, worth special mention as a potential biomarker is the mutant huntingtin protein itself, although this has been technically difficult to measure so far because of its insolubility. Measurement of huntingtin would be extremely important as a variety of potential therapies are targeted against either huntingtin itself or against some of its functional properties. Measuring huntingtin levels, its physical state, its proteolytic products, or its interactions with other molecules would all have distinct values as biomarkers, especially as pharmacodynamic indicators.[62] Thus, developing reliable and sensitive methods for measuring huntingtin is a high priority in the HD biomarker field.

REFERENCES

1. De Gruttola, V.G., P. Clax, D.L. DeMets, et al., Considerations in the evaluation of surrogate endpoints in clinical trials: summary of a National Institutes of Health workshop. *Controlled Clinical Trials,* 2001. 22: p. 485–502.
2. Floyd, E. and T.M. McShane, Development and use of biomarkers in oncology drug development. *Toxicol Pathol,* 2004. 32 Suppl 1: p. 106–15.
3. Langbehn, D.R., R.R. Brinkman, D. Falush, et al., A new model for prediction of the age of onset and penetrance for Huntington's disease based on CAG length. *Clin Genet,* 2004. 65: p. 267–77.
4. Andrew, S., J. Theilmann, E. Almqvist, et al., DNA analysis of distinct populations suggests multiple origins for the mutation causing Huntington disease. *Clin Genet,* 1993. 43: p. 286–94.
5. Aylward, E.H., A.M. Codori, A. Rosenblatt, et al., Rate of caudate atrophy in presymptomatic and symptomatic stages of Huntington's disease. *Mov Disord,* 2000. 15: p. 552–60.

6. Blekher, T., S.A. Johnson, J. Marshall, et al., Saccades in presymptomatic and early stages of Huntington disease. *Neurology*, 2006. 67: p. 394–99.
7. Blekher, T.M., R.D. Yee, S.C. Kirkwood, et al., Oculomotor control in asymptomatic and recently diagnosed individuals with the genetic marker for Huntington's disease. *Vision Res*, 2004. 44: p. 2729–36.
8. Brandt, J., B. Shpritz, A.M. Codori, et al., Neuropsychological manifestations of the genetic mutation for Huntington's disease in presymptomatic individuals. *J Int Neuropsychol Soc*, 2002. 8: p. 918–24.
9. Feigin, A., M.F. Ghilardi, C. Huang, et al., Preclinical Huntington's disease: compensatory brain responses during learning. *Ann Neurol*, 2006. 59: p. 53–59.
10. Gutekunst, C.A., S.H. Li, H. Yi, et al., Nuclear and neuropil aggregates in Huntington's disease: relationship to neuropathology. *J Neurosci*, 1999. 19: p. 2522–34.
11. Ho, A.K., B.J. Sahakian, T.W. Robbins, et al., Random number generation in patients with symptomatic and presymptomatic Huntington's disease. *Cogn Behav Neurol*, 2004. 17: p. 208–12.
12. O'Donnell, B.F., M.A. Wilt, A.M. Hake, et al., Visual function in Huntington's disease patients and presymptomatic gene carriers. *Mov Disord*, 2003. 18: p. 1027–34.
13. Paulsen, J.S., J.L. Zimbelman, S.C. Hinton, et al., fMRI biomarker of early neuronal dysfunction in presymptomatic Huntington's disease. *AJNR Am J Neuroradiol*, 2004. 25: p. 1715–21.
14. Reading, S.A., A.C. Dziorny, L.A. Peroutka, et al., Functional brain changes in presymptomatic Huntington's disease. *Ann Neurol*, 2004. 55: p. 879–83.
15. Rosas, H.D., D.S. Tuch, N.D. Hevelone, et al., Diffusion tensor imaging in presymptomatic and early Huntington's disease: selective white matter pathology and its relationship to clinical measures. *Mov Disord*, 2006. 21: p. 1317–25.
16. Tibben, A., M.F. Niermeijer, R.A. Roos, et al., Understanding the low uptake of presymptomatic DNA testing for Huntington's disease. *Lancet*, 1992. 340: p. 1416.
17. Marder, K., H. Zhao, R.H. Myers, et al., Rate of functional decline in Huntington's disease. Huntington Study Group. *Neurology*, 2000. 54: p. 452–58.
18. Brooks, D.J., K.A. Frey, K.L. Marek, et al., Assessment of neuroimaging techniques as biomarkers of the progression of Parkinson's disease. *Exp Neurol*, 2003. 184 Suppl 1: p. S68–79.
19. Pepe, M.S., R. Etzioni, Z. Feng, J.D., et al., Phases of biomarker development for early detection of cancer. *J Natl Cancer Inst*, 2001. 93: p. 1054–61.
20. Pepe, M., *The Statistical Evaluation of Medical Tests for Classification and Prediction.* 2003, Oxford: Oxford University Press.
21. Douaud, G., V. Gaura, M.J. Ribeiro, et al., Distribution of grey matter atrophy in Huntington's disease patients: a combined ROI-based and voxel-based morphometric study. *Neuroimage*, 2006. 32: p. 1562–75.
22. Reading, S.A., M.A. Yassa, A. Bakker, et al., Regional white matter change in pre-symptomatic Huntington's disease: a diffusion tensor imaging study. *Psychiatry Res*, 2005. 140: p. 55–62.
23. Rosas, H.D., N.D. Hevelone, A.K. Zaleta, et al., Regional cortical thinning in preclinical Huntington disease and its relationship to cognition. *Neurology*, 2005. 65: p. 745–47.
24. Rosas, H.D., D.H. Salat, S.Y. Lee, et al., Cerebral cortex and the clinical expression of Huntington's disease: complexity and heterogeneity. *Brain*, 2008. 131: p. 1057–68.
25. Georgiou-Karistianis, N., A. Sritharan, M. Farrow, et al., Increased cortical recruitment in Huntington's disease using a Simon task. *Neuropsychologia*, 2007. 45: p. 1791–800.
26. Thiruvady, D.R., N. Georgiou-Karistianis, G.F. Egan, et al., Functional connectivity of the prefrontal cortex in Huntington's disease. *J Neurol Neurosurg Psychiatry*, 2007. 78: p. 127–33.

27. Zimbelman J.L., J.S. Paulsen, A. Mikos, et al., fMRI detection of early neural dysfunction in preclinical Huntington's disease. *J Int Neuropsychol Soc,* 2007. 13: p. 758–69.

28. Kim J.S., S.A. Reading, T. Brashers-Krug, et al., Functional MRI study of a serial reaction time task in Huntington's disease. *Psychiatry Res,* 2004. 131: p. 23–30.

29. Gavazzi, C., R.D. Nave, R. Petralli, et al., Combining functional and structural brain magnetic resonance imaging in Huntington disease. *J Comput Assist Tomogr,* 2007. 31: p. 574–80.

30. Wolf R.C., N. Vasic, C. Schonfeldt-Lecuona, et al., Dorsolateral prefrontal cortex dysfunction in presymptomatic Huntington's disease: evidence from event-related fMRI. *Brain,* 2007. 130: p. 2845–57.

31. Feigin, A., C. Tang, Y. Ma, et al., Thalamic metabolism and symptom onset in preclinical Huntington's disease. *Brain,* 2007. 130: p. 2858–67.

32. Pavese, N., T.C. Andrews, D.J. Brooks, et al., Progressive striatal and cortical dopamine receptor dysfunction in Huntington's disease: a PET study. *Brain,* 2003. 126: p. 1127–35.

33. van Oostrom, J.C., R.P. Maguire, C.C. Verschuuren-Bemelmans, et al., Striatal dopamine D2 receptors, metabolism, and volume in preclinical Huntington disease. *Neurology,* 2005. 65: p. 941–3.

34. Tai, Y.F., N. Pavese, A. Gerhard, et al., Microglial activation in presymptomatic Huntington's disease gene carriers. *Brain,* 2007. 130: p. 1759–66.

35. Ravina, B., D. Eidelberg, J.E. Ahlskog, et al., The role of radiotracer imaging in Parkinson disease. *Neurology,* 2005. 64: p. 208–15.

36. Cummings J.L., R. Doody, and C. Clark, Disease-modifying therapies for Alzheimer disease: challenges to early intervention. *Neurology,* 2007. 69: p. 1622–34.

37. Pike, K.E., G. Savage, V.L. Villemagne, et al., Beta-amyloid imaging and memory in non-demented individuals: evidence for preclinical Alzheimer's disease. *Brain,* 2007. 130: p. 2837–44.

38. Kurlan, R., E. Caine, A. Rubin, et al., Cerebrospinal fluid correlates of depression in Huntington's disease. *Arch Neurol,* 1988. 45: p. 881–83.

39. Schwarcz R., C.A. Tamminga, R. Kurlan, et al., Cerebrospinal fluid levels of quinolinic acid in Huntington's disease and schizophrenia. *Ann Neurol,* 1988. 24: p. 580–82.

40. Montine, T.J., M.F. Beal, D. Robertson, et al., Cerebrospinal fluid F2-isoprostanes are elevated in Huntington's disease. *Neurology,* 1999. 52: p. 1104–05.

41. Jeitner, T.M., M.B. Bogdanov, W.R. Matson, et al., N(epsilon)-(gamma-L-glutamyl)-L-lysine (GGEL) is increased in cerebrospinal fluid of patients with Huntington's disease. *J Neurochem,* 2001. 79: p. 1109–12.

42. Kaltenbach, L.S., E. Romero, R.R. Becklin, et al., Huntingtin interacting proteins are genetic modifiers of neurodegeneration. *PLoS Genet,* 2007. 3: p. e82.

43. Gizatullina, Z.Z., K.S. Lindenberg, P. Harjes, et al., Low stability of Huntington muscle mitochondria against Ca2+ in R6/2 mice. *Ann Neurol,* 2006. 59: p. 407–11.

44. Saft, C., J. Zange, J. Andrich, et al., Mitochondrial impairment in patients and asymptomatic mutation carriers of Huntington's disease. *Mov Disord,* 2005. 20: p. 674–79.

45. Strand, A.D., A.K. Aragaki, D. Shaw, et al., Gene expression in Huntington's disease skeletal muscle: a potential biomarker. *Hum Mol Genet,* 2005. 14: p. 1863–76.

46. Hersch, S.M., S. Gevorkian, K. Marder, et al., Creatine in Huntington disease is safe, tolerable, bioavailable in brain and reduces serum 8OH2'dG. *Neurology,* 2006. 66: p. 250–52.

47. Stoy, N., G.M. Mackay, C.M. Forrest, et al., Tryptophan metabolism and oxidative stress in patients with Huntington's disease. *J Neurochem,* 2005. 93: p. 611–23.

48. Chen, C.M., Y.R. Wu, M.L. Cheng, et al., Increased oxidative damage and mitochondrial abnormalities in the peripheral blood of Huntington's disease patients. *Biochem Biophys Res Commun,* 2007. 359: p. 335–40.

49. Christofides, J., M. Bridel, M. Egerton, et al., Blood 5-hydroxytryptamine, 5-hydroxyindoleacetic acid and melatonin levels in patients with either Huntington's disease or chronic brain injury. *J Neurochem,* 2006. 97: p. 1078–88.

50. Liu, C.S., W.L. Cheng, S.J. Kuo, et al., Depletion of mitochondrial DNA in leukocytes of patients with poly-Q diseases. *J Neurol Sci,* 2008. 264: p. 18–21.

51. Klepac, N., M. Relja, R. Klepac, et al., Oxidative stress parameters in plasma of Huntington's disease patients, asymptomatic Huntington's disease gene carriers and healthy subjects: a cross-sectional study. *J Neurol,* 2007. 254: p. 1676–83.

52. Mochel, F., P. Charles, F. Seguin, et al., Early energy deficit in Huntington disease: identification of a plasma biomarker traceable during disease progression. *PLoS ONE,* 2007. 2: p. e647.

53. Popovic, V., M. Svetel, M. Djurovic, et al., Circulating and cerebrospinal fluid ghrelin and leptin: potential role in altered body weight in Huntington's disease. *Eur J Endocrinol,* 2004. 151: p. 451–55.

54. Dalrymple, A., E.J. Wild, R. Joubert, et al., Proteomic profiling of plasma in Huntington's disease reveals neuroinflammatory activation and biomarker candidates. *J Proteome Res,* 2007. 6: p. 2833–40.

55. Borovecki, F., L. Lovrecic, J. Zhou, et al., Genome-wide expression profiling of human blood reveals biomarkers for Huntington's disease. *Proc Natl Acad Sci U S A,* 2005. 102: p. 11023–28.

56. Runne, H., A. Kuhn, E.J. Wild, et al., Analysis of potential transcriptomic biomarkers for Huntington's disease in peripheral blood. *Proc Natl Acad Sci U S A,* 2007. 104: p. 14424–29.

57. Zuccato, C., D. Liber, C. Ramos, et al., Progressive loss of BDNF in a mouse model of Huntington's disease and rescue by BDNF delivery. *Pharm Res,* 2005. 52: p. 133–39.

58. Ciammola, A., J. Sassone, M. Cannella, et al., Low brain-derived neurotrophic factor (BDNF) levels in serum of Huntington's disease patients. *Am J Med Genet B Neuropsychiatr Genet,* 2007. 144: p. 574–77.

59. Maglione V., P. Giallonardo, M. Cannella, et al., Adenosine A2A receptor dysfunction correlates with age at onset anticipation in blood platelets of subjects with Huntington's disease. *Am J Med Genet B Neuropsychiatr Genet,* 2005. 139: p. 101–5.

60. Varani, K., A.C. Bachoud-Levi, C. Mariotti, et al., Biological abnormalities of peripheral A(2A) receptors in a large representation of polyglutamine disorders and Huntington's disease stages. *Neurobiol Dis,* 2007. 27: p. 36–43.

61. Underwood, B.R., D. Broadhurst, W.B. Dunn, et al., Huntington disease patients and transgenic mice have similar pro-catabolic serum metabolite profiles. *Brain,* 2006. 129: p. 877–86.

62. Weiss, A., D. Abramowski, M. Bibel, et al., Single-step detection of mutant huntingtin in animal and human tissues: A bioassay for Huntington's disease. *Anal Biochem,* 2009. 395: p. 8–15.

12 Huntington's Disease
Clinical Experimental Therapeutics

E. Ray Dorsey and Ira Shoulson

CONTENTS

INTRODUCTION

Successful treatment of a neurodegenerative disorder depends on three interdependent factors: (1) the "right" treatment candidates; (2) the ability to detect the therapeutic effects in an appropriate clinical population; and (3) clinically meaningful treatment effects. Each contribution is necessary but insufficient by itself to realize therapeutic gains and ultimately clinical benefits.

These issues are illustrated for hepatolenticular degeneration, or Wilson's disease, which in 1911 was first recognized by Samuel Alexander Kinnier Wilson as a disorder that was familial, progressive, fatal, and associated with softening of the lenticular nucleus (pallidum and putamen) and cirrhosis.[1] Hepatolenticular degeneration was linked in 1913 to an accumulation of hepatic copper[2] and hypothesized in 1921 to be inherited in an autosomal recessive fashion.[3] However, chelating therapy was not developed and reasoned to be an effective experimental treatment until 1948–1958. Coincidentally, the copper-binding protein ceruloplasmin was found to be deficient in individuals who inherited two copies of the gene responsible for Wilson's disease. Experimental decoppering treatments would eventually be found to slow and even reverse neurological and hepatic deterioration in clinically affected individuals.[4–6]

In the case of Wilson's disease, the right treatment emerged from an incremental knowledge base and the rational understanding that copper accumulated in vital organs could be removed or prevented by effective and reasonably safe treatments. The therapeutic effects of decoppering therapy (2,3-dimercaptopropanol or British anti-Lewisite and later penicillamine) were of such large magnitude and relatively small variance that benefits could be detected in just a few patients without placebo controls.[4] The clinical meaningfulness of penicillamine therapy became clear by 1968 when this intervention was demonstrated to prevent the onset of clinical features in premanifest individuals who had deficient ceruloplasmin, increased hepatic copper, and were presumed to carry the homozygotic genetic defect accounting for Wilson's disease.[6] Again, the treatment effects were sufficiently strong and uniform as to be convincing without placebo controls. As may occur with disease-modifying therapies, treatment may result in initial clinical worsening (as can be the case with penicillamine[7]) before clinical improvement. Improvements in therapy would later ensue in the form of trientine and zinc treatments.[8,9] Although success in the experimental therapeutics of Wilson's disease seemed in retrospect to come in quantum scientific leaps, therapeutic benefits accrued slowly but steadily through incremental gains in clinical research. Interestingly, a molecular understanding of the genetic defects underlying Wilson's disease only followed the development of successful treatment.[10]

The clinical and hereditary aspects of Huntington's disease (HD) were first described by George Huntington in 1872. It took nearly a century to unravel the relatively selective pattern of neuronal loss and gliosis and the associated neurochemical abnormalities that characterized the neurodegeneration of HD. In the 1980s and 1990s, studies of large and multiple families and the development and application of molecular genetic techniques led eventually to identification of the gene responsible for HD.[11] In the past two decades, a remarkable and collaborative scientific inquiry has elucidated the key relationship of genetic dosage (CAG repeat length) to clinical features (age at onset), identified the mutant huntingtin protein, provided insights into the mechanisms underlying neuronal degeneration, and enabled the development of genetic animal models. The ability to detect premanifest HD in individuals who have inherited the mutant gene and its expanded CAG repeat has provided the opportunity to develop preventive therapies aimed at postponing or preventing the onset of illness.

CANDIDATES FOR EXPERIMENTAL THERAPEUTICS

Despite the major scientific advances in understanding HD and the efforts of many, the treatments available for HD are limited.[12] The development of treatments is costly, time intensive, and risky.[13] Given the considerable resources required to evaluate experimental therapeutics, identifying and prioritizing candidates are important for strategic research and development.

Broadly speaking, experimental therapeutic agents emerge from two principal routes (see Table 12.1). The first route is treatments for which therapeutic evidence is largely derived from empirical observation. In many cases, the benefits of a treatment are discovered by empirical observation or what some have called planned serendipity. In contrast to empirical observation, rational treatment design has grown in influence during the past two decades, especially in situations where the underlying (frequently genetic) abnormality responsible for a particular disease has become elucidated. With pathogenetic knowledge in hand, targets can be identified, and "druggable" treatments can be specifically designed to interact with that target to ameliorate the resulting disease process. This approach of "rational treatment design" has largely supplanted efforts at mass screening of compounds against an array of

TABLE 12.1
Sources of Experimental Therapeutics

	Empirical Observation	Rational Drug Design
Examples outside of HD	Diazepam (Valium)	Levodopa
	Lithium	Imatinib (Gleevec)
	Sildenafil	
Examples within HD		
Prospects	Near term	Long term

possible targets (see Chapter 8, this volume). Treatments derived from empirical observation and rational drug design both have a role in the development of therapies for HD, just as they did for the successful treatments for Wilson's disease.

Systematic descriptions of empirical reasoning date back to at least Aristotle, who, in contrast to Plato, emphasized observations that could be perceived by the senses. Empiricism has evolved since Aristotle's time, and many modern medical advances have their roots in empirical observations. For example, benzodiazepines were developed in a "purely empirical manner" as scientists, including Dr. Leo Sternbach, at Hoffmann-La Roche Inc. sought to develop a new class of tranquilizers to supplant barbiturates.[14] Because knowledge of brain processes was limited, the investigators could not think of "an intelligent working hypothesis." With expertise in chemical synthesis, the investigators sought to search for a new class of tranquilizers through molecular modification of existing compounds. Among the criteria the researchers used in identifying possible compounds were that the new class be relatively unexplored, be readily accessible, and "look" as if it could lead to biologically active products. The compounds that were first developed had their origins as dyes. The compound that led to the eventual development of chlordiazepoxide (Librium) was found during a "cleanup operation" where a precursor compound was sent for pharmacological testing with little hope of efficacy. The compound was found to cause muscle relaxation and sedation in animals. This serendipitous discovery led to refinement of the compound, the development of a whole class of compounds, and, like Wilson's disease, the advancement of basic neuroscience by eventually leading to the identification of the benzodiazepine receptor.[14]

The development of lithium for bipolar disorder also relied principally on empirical observation. Lithium was initially used for the treatment of gout and uric acid disorders. A psychiatrist in Australia, John Cade,[15] observed that guinea pigs given lithium carbonate became lethargic. Dr. Cade was originally working on the hypothesis that mania was related to intoxication by normal body products, such as urea, and that the lithium urate salt was very soluble. Dr. Cade subsequently injected lithium carbonate into fully conscious animals and found that they became extremely lethargic. Although the mechanism of action of lithium and its target were not known and not what Dr. Cade had suspected, he turned this discovery into an effective treatment for mania. Subsequent randomized, placebo-controlled trials performed two decades later confirmed the efficacy of lithium in mood disorders.[16] Dr. Cade later commented, "It [seems] a long way from lethargy in guinea pigs to the control of manic excitement."[17] Although Dr. Cade's initial observation was published in 1949, such serendipity continues. Sildenafil (Viagra) was being examined as an antianginal compound, and the effect on erectile dysfunction was discovered only by chance.[18]

In HD, tetrabenazine, a monoamine-depleting agent,[19] was developed initially as an antipsychotic drug by Hoffmann-La Roche Inc. in the late 1950s. Its benefit for hyperkinetic movement disorders was only discovered a decade later, and 50 years later, in December 2007, the Peripheral and Central Nervous System Advisory Committee to the U.S. Food and Drug Administration (FDA) unanimously recommended its approval for the treatment of chorea associated with HD. Although the FDA has (as of June 2008) yet to act formally on the recommendation, tetrabenazine is poised to be the first treatment approved specifically for HD.

Empirical observation by astute clinicians will continue to play a role in the development of experimental therapeutics for HD. The limit of empirical observation is that it relies largely on chance and a prepared mind.[20] Designing experimental therapeutics solely on the basis of empirical observation is neither efficient nor sufficient.

In contrast to empirical observation, strategies based on rational design hold promise for developing effective, targeted treatments for HD. Levodopa treatment of Parkinson's disease (PD) is perhaps the most notable example of a "rationally" designed therapeutic drug in neurology. Based on evidence that postmortem brain samples of individuals with PD were deficient in dopamine, Arvid Carlsson, Oleh Hornykiewicz, George Cotzias,[21,22] and others showed that levodopa, a precursor to dopamine, was effective in relieving parkinsonian symptoms. However, this success came after several false starts and the eventual recognition of the need to achieve adequate dosage and duration of levodopa treatment.

Rational design grew with the increased understanding of genetics, especially in oncology. The discovery of the Philadelphia chromosome,[23] a translocation of chromosomes 9 and 22, as the genetic abnormality underlying chronic myelogenous leukemia set the stage for rational drug design. The translocation and the resulting overexpression of tyrosine kinase and subsequent increased cell division provided investigators a therapeutic target. Research, led by Dr. Brian Drucker, demonstrated that the compound STI-571 inhibited the proliferation of hematopoietic cells expressing the genetic abnormality found in chronic myelogenous leukemia. This compound inhibited tyrosine kinase and was subsequently marketed by Novartis Pharmaceuticals as imatinib mesylate (Gleevec), a remarkably effective treatment for chronic myelogenous leukemia.[24]

Nascent clinical investigations of rationally designed experimental therapeutics are underway in HD. Based on evidence of bioenergetic mitochondrial defects in HD,[25,26] coenzyme Q_{10} at a dosage of 600 mg daily was examined in placebo-controlled clinical trials of 347 ambulatory patients with HD who were treated for up to 30 months.[27] This study marked the first time that a therapeutic signal was observed, consisting of about a 15% slowing in functional decline that was accompanied by a slowing in cognitive decline. A smaller, placebo-controlled trial of coenzyme Q_{10} in PD showed a slowing of disease-related disability, but this was achieved only at a dosage of 1200 mg daily.[28] These observations have prompted the design of a larger placebo-controlled study of coenzyme Q_{10}, 2400 mg daily, in early HD patients who will be followed for up to 5 years.[29] Other rationally designed and promising therapeutics for HD, using innovative technologies such as gene therapy and RNA interference, have yet to proceed into clinical trials.[30,31]

As our knowledge of HD has expanded, more promising targets for therapeutic action have been identified, as described in earlier chapters. To assess the current state of the therapeutic pipeline for HD, we identified all drugs with a primary indication for HD from a proprietary drug database (Pharmaprojects).[32] The results in Figure 12.1 and Table 12.2 show a growing pipeline. However, most of the therapeutic candidates are in the nonclinical phase of development. Rationally designed therapies for HD may still be years away from clinical trials and it may be even longer until successful treatments become more widely available. Although the nonclinical and clinical components of the pipeline are expected to grow considerably in the coming years, time remains the chief limitation in the development of therapeutics for HD.

DETECTION OF THERAPEUTIC EFFECTS

The ability to detect the therapeutic effects in an appropriate clinical population is an essential part of development. Unlike Wilson's disease, the magnitude of clinical effects detected to date in HD has been small in magnitude and somewhat variable. Clinical trials aimed at disease modification require years of follow-up evaluation and estimates of relatively large sample sizes in the hundreds and even thousands of research participants. With identification of better candidates, the magnitude of the treatment effects is expected to increase. With improved clinical research methodology, the variability of outcome measures is expected to narrow.

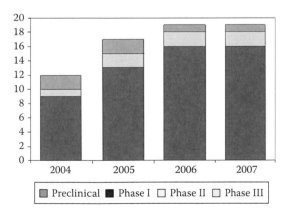

FIGURE 12.1 **(See color insert following page 172.)** Number of drugs in development by phase for Huntington's disease, 2004–2007.

TABLE 12.2
Drugs in Development for Huntington's Disease, 2007

Drug Name	Company	Development Phase	Proposed Mechanism of Action
AC-0523	Accera	Preclinical	Not available
Unnamed experimental therapeutic	Alnylam	Preclinical	RNA interference
Unnamed experimental therapeutic	CombinatoRx	Preclinical	Not available
CoQ₁₀ analogues	Edison	Preclinical	Antioxidant in mitochondrial chain
Dekafin-2	Enkam Pharmaceuticals	Preclinical	Target the fibroblast growth factor receptor
EVPK-0003547	EnVivo Pharmaceuticals	Preclinical	Histone deacetylase inhibitor; neuroprotectant
Unnamed experimental therapeutic	FoldRx	Preclinical	Not available
NeurotrophinCell (NtCell)	Living Cell Technologies	Preclinical	Live cell transplant providing neuroprotecting hormones
AAV-siRNA	Merck & Co	Preclinical	RNA interference
Huntington gene therapy	Neurologix	Preclinical	Gene therapy
XIAP gene therapy	Neurologix	Preclinical	Gene therapy
ReN008	ReNeuron	Preclinical	Stem cell therapy
NTx-889	Stem Cell Therapeutics	Preclinical	Stem cell therapy
TopoTarget-2	TopoTarget	Preclinical	Histone deacetylase inhibitor; anti-inflammatory
TRO-51646	Trophos	Preclinical	Neurotrophic factors that combat cell death mechanisms triggered by the mutant *huntingtin* gene
Creatine	Avicena	Phase II	Neuroprotectant, reduces oxidative damages
ACR-16	NeuroSearch	Phase II	Dopaminergic stabilizer
Miraxion	Amarin	Phase III	Stabilizes mitochondrial membrane and reduces enzymes associated with apoptosis
Tetrabenazine	Prestwick Pharmaceuticals	Preregistration	Inhibition of vesicular monoamine transporter 2, a brain synaptic vesicular monoamine transporter

Source: Pharmaprojects; company websites.

There are several approaches to narrowing variability of outcome measures. One is to lessen the variability between observers and ensure more consistency of ratings over time, which can be facilitated by ongoing training of clinical raters. The statistical power of the study can also be enhanced by blocking the randomization of experimental treatments according to the investigators, the largest source of variability. However, there are limits to reducing the variability in outcome measures, which after all reflect the clinical heterogeneity of the HD phenotype.

A major strategy to improve experimental therapeutics came about in the past decade as observational studies were undertaken to define the clinical precursors of premanifest HD. This is being accomplished by longitudinal examination of large research populations of adults, including more than 1,000 (1,021 as of June 2008) participants at 32 sites in North America and Australia who had become aware of their HD gene status through predictive DNA testing (PREDICT-HD)[33] and 1,001 participants at 40 North American sites who are at risk to develop HD but have chosen not to learn their gene status (PHAROS).[34] These innovative studies are expected to provide clearer definition of the clinical onset of HD and the clinical precursors that antedate onset. In turn, the profile and pace of clinical onset observed in PREDICT-HD and PHAROS can be used to design controlled trials aimed at delaying the earliest manifestations of illness.

An important way of detecting therapeutic effects is to develop and identify biological markers that parallel the pathogenesis of HD. So-called state biomarkers can be used in early-phase clinical trials to better identify promising treatment candidates and optimal dosages. Efforts are underway to identify "wet" biomarkers in body fluids and tissues of individuals with premanifest and manifest HD that could be measures of disease state.[35] Similarly, neuroimaging approaches with MRI technology and PET ligands will further enhance the state and pace of disease.[36] Such biomarker development is taking place in the context of many of the observational and clinical trials listed in Table 12.2 and through concerted approaches such as the COHORT observational study, a prospective study collecting phenotypic and genetic data from people with HD or individuals in an HD family,[37] and TRACK-HD, which is designed to determine the combination of outcome measures that is the most sensitive at detecting changes as HD state progresses.[38] To identify premanifest disease and to determine gene dosage in the form of the expanded CAG repeats that predicts age at clinical onset are key advantages in research and the development of biomarkers for HD. Promising biomarkers will require validation in the context of clinical trials, where learning[39] how these biomarkers change over time and respond to interventions can take place. Through the test of successful clinical trials, validated biomarkers may eventually take on the role of surrogate endpoints that will further facilitate therapeutic development. Some authorities have even advocated that well-developed biomarkers might serve as the initial basis for approval, pending verification of safety, effectiveness, and clinically meaningful outcomes.[40]

CLINICALLY MEANINGFUL OUTCOMES

This review has focused on so-called disease-modifying treatments that are aimed at slowing neurodegeneration and that will likely take years to demonstrate effectiveness

in slowing progression of manifest illness or delaying onset of clinical manifestations. In any event, the clinical outcomes will need to be relevant in slowing or
preventing the onset of functional decline or progressive disability. For example,
despite tetrabenazine's substantial reduction in chorea,[19] the largely insensitive (at
least over the short term) functional measures used in the study did not demonstrate
any associated improvement, an issue that was raised by the FDA in its Advisory
Committee review of the drug.

Efforts are currently underway to develop and catalogue outcome measures and
rating scales that are clinically meaningful and have been validated in clinical trials.
The HD Toolkit project[41] is a cooperative approach focused on providing investigators with outcome measures appropriate to their clinical trial objectives. Researchers
of the HD Toolkit project seek to identify tests that are sensitive to the subtle changes
of prediagnostic and early HD based on validity, reliability, feasibility, and evidence
of linear change with disease progression.[42]

SPECIAL CHALLENGES AND OPPORTUNITIES
IN HD CLINICAL TRIALS

HD as an inherited and fully penetrant, adult-onset neurodegenerative disorder has
unique characteristics that pose special challenges and create valuable opportunities
in the design and conduct of clinical trials. Because of its insidious course, demonstrating the efficacy of potentially disease-modifying therapies may take many years.
For example, an earlier study of coenzyme Q_{10} showed a modest benefit on "total
functional capacity" of individuals after 30 months,[27] which is now being investigated in a 5-year study of coenzyme Q_{10}.[29] In addition to the long duration of many
HD studies, the genetics of HD can influence recruitment, especially in observational studies of individuals at risk for HD. The fear of health insurance loss or other
negative ramifications (e.g., job loss) of participation in studies can be very real for
research participants. The recently passed Genetic Information Nondiscrimination
Act protects individuals and family members participating in research that includes
genetic testing and may help alleviate some of that fear.[43] Finally, because of the high
risk of suicide and other neuropsychiatric problems that are associated with HD,[44–46]
monitoring the safety and mental health of research participants during clinical trials
is especially important. With the increasing promise of biologicals and their application to HD, safety issues will likely take on heightened importance.[47]

In the future, trials in HD will have special opportunities. Among them will be
the incorporation of promising biomarkers in early-stage trials, which has already
begun.[35] Although these biomarkers will not substitute for clinical outcomes, they
will provide additional insight into the disease and evidence of whether an experimental therapeutic is worthy of future investigation in lengthy (and expensive) clinical trials. In addition to biomarkers, the genetics of the disease will allow HD to
serve as a paradigm for investigating therapies among individuals with "premanifest"
(known carriers of the genetic mutation responsible for HD but do not yet have clear
symptoms of the disease) HD. One upcoming study, PREQUEL, aims to investigate
the safety and tolerability of coenzyme Q_{10} in this population. Another opportunity
for future HD clinical trials is the use of ongoing observational studies, such as

COHORT and its European counterpart REGISTRY,[48] to identify research participants in those studies who may be eligible for participation in interventional studies. Thus, these large observational studies can serve as a valuable pool and introduction to clinical research for those affected by HD. Finally, involving individuals with and affected by HD more directly in the conduct of clinical trials (e.g., through participation on steering committees) will provide a unique perspective to clinical research and likely generate other secondary benefits. Research participants in HD trials have recently expressed satisfaction with learning the results of trials in a timely and accurate manner.[49] Continuing to meet their needs and encouraging their participation in clinical research in new ways is highly desirable for all stakeholders.

Involving research participants more directly in the conduct of research is another example of the collaborative nature of the HD community. Those affected by HD have fueled investments in research into the etiology and pathogenesis of HD. This scientific inquiry continues to advance at a rapid pace, catalyzed in large part by research efforts of the High Q Foundation (New York, NY) and its nonprofit discovery group CHDI, Inc. (Los Angeles, CA), the Hereditary Disease Foundation (Santa Monica, CA), the Huntington's Disease Society of America's Coalition for the Cure (New York, NY), the Huntington Society of Canada (Kitchener, ON), and governmental support through the National Institutes of Health, the FDA Orphan Products Division, and the European Union. These research undertakings are expected to provide more and better-defined targets for therapeutic intervention. The scientific developments have been attended by heightened interest and initiative among individuals and families affected by HD. In record numbers, patients with HD and individuals at risk for or with known premanifest HD are volunteering to participate in many large-scale interventional and observational studies (see Table 12.3). By and large, clinical trials in HD have enrolled research participants on schedule or in advance of projected timetables.

This growth of clinical research has been enabled by several novel collaborations aimed at developing effective treatments for HD. The Huntington Study Group[50] established in 1993 and the European Huntington's Disease Network[51] established in 2004 are not-for-profit consortia of academic investigators who are working cooperatively to improve methodology and trial design and to efficiently conduct controlled trials of experimental therapeutics for HD. The High Q Foundation, Inc.[52] is a private philanthropic foundation established in 2002 aimed at bringing industry, academia, government agencies, and funding organizations together to facilitate the development of treatments for HD. CHDI, Inc. is a not-for-profit drug discovery arm of the High Q Foundation that identifies rational targets for experimental intervention and "druggable" treatments that show promise of becoming safe and effective treatments for HD.

A major advantage of HD research is the collegial approach that has been forged worldwide in recent years by scientists, clinical investigators, and, importantly, by the community of patients and families affected by HD. This cooperative effort will further enhance the prospects for ensuring that promising therapeutic candidates will be identified, their safety and efficacy will be detected in a reliable fashion, and that meaningful clinical outcomes will ensue. Someday, expectedly sooner rather than later, HD, like Wilson's disease, will be taken off the rolls of untreatable

TABLE 12.3

Multicenter, Placebo-Controlled Clinical Trials for Huntington's Disease

Name of Study	Number of Participants	Experimental Drug	Mechanism of Action	Status	Results	Conducted by
PHEND-HD	58	Phenylbutyrate	Histone deacetylase inhibitor	Phase II complete	Safety and tolerability confirmed	HSG
TREND-HD	316	Ethyl eicosapentaenoate	Stabilizes mitochondrial membrane	Phase III complete	No significant benefit in double-blind study	HSG and EHDN in parallel studies
DIMOND-A	9	Dimebolin hydrochloride	Antihistamine with NMDA inhibitory properties	Phase I–IIA complete	Safety and tolerability confirmed	HSG
TETRA-HD	84	Tetrabenazine	Depletes central monoamines by binding reversibly to the type 2 vesicular monoamine transporter	Phase III complete	Significantly reduced chorea and improved CGI	HSG
EHDI	537	Riluzole	Inhibits glutamate release	Phase III complete	No significant difference in TFC, CGI, or motor scores	EHDN
DOMINO	114	Minocycline	Antiapoptotic and antioxidant	Phase II active	Study ongoing	HSG
DIMOND-B	91	Dimebolin hydrochloride	Antihistamine with NMDA inhibitory properties	Phase II active	Study ongoing	HSG
2-CARE	Upcoming	Coenzyme Q_{10}	Facilitates mitochondrial bioenergetics and antioxidative	Phase III upcoming	Enrollment in progress	HSG
CREST-E	Upcoming	Creatine	Facilitates cellular bioenergetics and antioxidative	Phase III upcoming	Upcoming study	HSG
PREQUEL	Upcoming	Coenzyme Q_{10} in premanifest HD	Facilitates mitochondrial bioenergetics and antioxidative	Upcoming	Upcoming study	HSG
ACR-16	Upcoming	ACR-16	Stabilizes dopaminergic neurotransmission	Phase III upcoming	Upcoming study	HSG and EHDN in parallel studies

CGI, Clinical Global Improvement; EHDN, European Huntington's Disease Network; HSG, Huntington Study Group; NMDA, *N*-methyl-D-aspartic acid; TFC, total functional capacity.

Source: Huntington Study Group (http://www.huntington-study-group.org); company websites.

neurodegenerative disorders. The ultimate success of these undertakings remains to be realized, but at a minimum, these efforts have added energy and support to the efforts of many in developing effective treatments for HD and other conditions.

ACKNOWLEDGMENTS

We thank Joel Thompson and Lisa Deuel for assistance in analyzing and preparing the data in the tables and figures.

REFERENCES

1. Wilson SAK. Progressive lenticular degeneration: a familial nervous disease associated with cirrhosis of the liver. *Brain* 34, 295–507. 1912.
2. Scheinberg IH and Sternlieb I. *Wilson's Disease*. Philadelphia: W.B. Saunders Company. 1984.
3. Anonymous. Review of *La Degenerescence Hepato-lenticulaire. Maladie de Wilson-Pseudosclerose. Archives of Neurology* 7, 549–550. 1922.
4. Walshe JM. Penicillamine. A new oral therapy for Wilson's disease. *American Journal of Medicine* 21, 487–495. 1956.
5. Walshe JM. Prophylactic use of penicillamine. *New England Journal of Medicine* 278, 795–796. 1968.
6. Sternlieb I and Scheinberg I. Prevention of Wilson's disease in asymptomatic patients. *New England Journal of Medicine* 278, 352–359. 1968.
7. Brewer GJ, Terry CA, Aisen AM, et al. Worsening of neurologic syndrome in patients with Wilson's disease with initial penicillamine therapy. *Archives of Neurology* 44, 490–493. 1987.
8. Brewer GJ, Dick RD, Johnson VD, et al. The treatment of Wilson's disease with zinc: XV: long-term follow-up studies. *The Journal of Laboratory and Clinical Medicine* 132, 264–278. 1998.
9. Walshe JM. Treatment of Wilson's disease with trientine (triethylene tetramine) dihydrochloride. *Lancet* 319, 643–647. 1982.
10. de Bie P, Muller PA, Wijmenga C, et al. Molecular pathogenesis of Wilson and Menkes disease: correlation of mutations with molecular defects and disease phenotypes. *Journal of Medical Genetics* 44, 673–688. 2007.
11. Macdonald ME, Ambrose CM, Duyao MP, et al. A novel gene containing a trinucleotide repeat that is expanded and unstable on Huntington's disease chromosomes. *Cell* 72, 971–983. 1993.
12. Bonelli RM and Wenning GK. Pharmacological management of Huntington's disease: an evidence-based review. *Current Pharmaceutical Design* 12, 2701–2720. 2006.
13. DiMasi JA, Hansen RW, and Grabowski HG. The price of innovation: new estimates of drug development costs. *Journal of Health Economics* 22, 151–185. 2003.
14. Sternbach LH. Benzodiazepine story. *Journal of Medicinal Chemistry* 22, 1–7. 1979.
15. Cade JFJ. Lithium salts in the treatment of psychotic excitement. *Medical Journal of Australia* 36, 349–352. 1949.
16. Coppen A. 50 years of lithium. *Bipolar Disorders* 1, 3–4. 1999.
17. Walter G. John Cade and lithium. *Psychiatric Services* 50, 969. 1999.
18. Ban TA. The role of serendipity in drug discovery. *Dialogues in Clinical Neuroscience* 8, 335–344. 2006.
19. Marshall FJ, Walker F, Frank S, et al. Tetrabenazine as antichorea therapy in Huntington disease—a randomized controlled trial. *Neurology* 66, 366–372. 2006.

20. Pasteur, L. Inaugural lecture. University of Lille. Douai, France, 7 December 1854.
21. Cotzias GC, Van Woert MH, and Schiffer LM. Aromatic amino acids and modification of parkinsonism. *New England Journal of Medicine* 276, 374–379. 1967.
22. Cotzias GC, Papavasiliou PS, and Gellene R. Modification of parkinsonism—chronic treatment with L-dopa. *New England Journal of Medicine* 280, 337–345. 1969.
23. Nowell P and Hungerford D. A minute chromosome in chronic granulocytic leukemia. *Science* 132, 1497. 1960.
24. Kurzrock R, Kantarjian HM, Druker BJ, et al. Philadelphia chromosome-positive leukemias: from basic mechanisms to molecular therapeutics. *Annals of Internal Medicine* 138, 819–830. 2003.
25. Beal MF, Henshaw DR, Jenkins BG, et al. Coenzyme Q(10) and nicotinamide block striatal lesions produced by the mitochondrial toxin malonate. *Annals of Neurology* 36, 882–888. 1994.
26. Koroshetz WJ, Jenkins BG, Rosen BR, et al. Energy metabolism defects in Huntington's disease and effects of coenzyme Q(10). *Annals of Neurology* 41, 160–165. 1997.
27. Kieburtz K, Koroshetz W, McDermott M, et al. A randomized, placebo-controlled trial of coenzyme Q(10) and remacemide in Huntington's disease. *Neurology* 57, 397–404. 2001.
28. Shults CW, Oakes D, Kieburtz K, et al. Effects of coenzyme Q(10) in early Parkinson disease—evidence of slowing of the functional decline. *Archives of Neurology* 59, 1541–1550. 2002.
29. Huntington Study Group. Coenzyme Q_{10} in Huntington's Disease (2CARE). Available at http://www.huntington-study-group.org/NEW%20CLINICAL%20TRIAL%20 INITIATIVES.html. Accessed on September 20, 2007.
30. Martin JB. Gene therapy and pharmacological treatment of inherited neurological disorders. *Trends in Biotechnology* 13, 28–35. 1995.
31. MacMillan JC, Snell RG, and Harper PS. Clinical considerations in gene therapy of Huntington's disease. *Gene Therapy* 1, Suppl 1, S88. 1994.
32. Pharmaprojects—tracking global pharmaceutical R&D since 1980. Available at http://www.pjbpubs.com/pharmaprojects/web.htm. Accessed on March 21, 2007.
33. Paulsen JS, Hayden M, Stout JC, et al. Preparing for preventive clinical trials—the Predict-HD study. *Archives of Neurology* 63, 883–890. 2006.
34. Shoulson I, Kieburtz K, Oakes D, et al. At risk for Huntington disease—the PHAROS (Prospective Huntington At Risk Observational Study) cohort enrolled. *Archives of Neurology* 63, 991–998. 2006.
35. Hersch SM, Gevorkian S, Marder K, et al. Creatine in Huntington disease is safe, tolerable, bioavailable in brain and reduces serum 80H2'dG. *Neurology* 66, 250–252. 2006.
36. Rosas HD, Feigin AS, and Hersch SM. Using advances in neuroimaging to detect, understand, and monitor disease progression in Huntington's disease. *NeuroRx* 1, 263–272. 2004.
37. Huntington Study Group. At-risk and observational research studies. Available at http://www.huntington-study-group.org/CLINICAL%20TRIALS%20IN%20PROGRESS.html. Accessed on October 1, 2007.
38. University College London. TRACKHD. Available at http://www.track-hd.net. Accessed on October 1, 2007.
39. Sheiner LB. Learning versus confirming in clinical drug development. *Clinical Pharmacology & Therapeutics* 61, 275–291. 1997.
40. Wood AJJ. A proposal for radical changes in the drug-approval process. *New England Journal of Medicine* 355, 618–623. 2006.
41. Stout JC, Tomusk A, Queller S, et al. Evidence-based selection of outcome measures for clinical trials: the Huntington's disease toolkit project. *Journal of Neuropsychiatry and Clinical Neuroscience* 19, 209–212. 2007.

42. Julie Stout's Laboratory, Clinical and Cognitive Neuroscience Research. Huntington's Disease toolkit. Available at http://www.indiana.edu/~ccns/research_hdtk.html. Accessed on June 11, 2008.
43. Hudson KL, Holohan MK, and Collins FS. Keeping pace with the times—The Genetic Information Nondiscrimination Act of 2008. *New England Journal of Medicine* 358, 2661–2663. 2008.
44. Paulsen JS, Hoth KF, Nehl C, et al. Critical periods of suicide risk in Huntington's disease. *American Journal of Psychiatry* 162, 725–731. 2005.
45. Paulsen JS, Nehl C, Hoth KF, et al. Depression and stages of Huntington's disease. *Journal of Neuropsychiatry & Clinical Neurosciences* 17, 496–502. 2005.
46. van Duijn E, Kingma EM, and van der Mast RC. Psychopathology in verified Huntington's disease gene carriers. *Journal of Neuropsychiatry & Clinical Neurosciences* 19, 441–448. 2007.
47. Pearson H. Tragic drug trial spotlights potent molecule. Available at http://www.nature.com/news/2006/060313/full/news060313-17.html. Accessed on July 28, 2008.
48. European Huntington Disease Network. European HD registry. Available at http://www.euro-hd.net/html/registry. Accessed on June 30, 2008.
49. Dorsey ER, Beck CA, Adams M, et al. Communicating clinical trial results to research participants. *Archives of Neurology* 65, 1590–1595. 2008.
50. Huntington Study Group. About us. Available at http://www.huntington-study-group.org. Accessed on August 31, 2007.
51. European Huntington's Disease Network. Network. Available at http://www.euro-hd.net. Accessed on August 31, 2007.
52. High Q Foundation. About CHDI. Available at http://www.highqfoundation.org. Accessed on August 31, 2007.

Index

Page numbers followed by f indicate figures; those followed by t indicate tables.

A

A2A receptors. *See* Adenosine 2A (A2A) receptors
AAV. *See* Adeno-associated virus (AAV)
Acetylcholine (ACh) receptor, 242
HACF1. *See* Bromodomain adjacent to zinc finger domain 1A (BAZ1A)
ACh receptor. *See* Acetylcholine (ACh) receptor
Acivicin, 135
Acute promyelocytic leukemia (APL), 188
AD. *See* Alzheimer's disease (AD)
Adeno-associated virus (AAV), 110, 239, 240, 241, 262, 263
Adenosine 2A (A2A) receptors, 14, 105, 208, 216, 280
Adenosine receptor A2B, 208
Adenovirus, 240, 262
Adult-onset HD, 7, 166, 180
  intervention model for, 20f
Aggregation-based assays, 130–131
Ago2 protein. *See* Argonaute 2 (Ago2) protein
Allelic discrimination, in HD, 237–238, 239f
α2-Macroglobulin, 278
α-Lipoic acid, 187
α-Synuclein, 259, 262
ALS. *See* Amyotrophic lateral sclerosis (ALS)
Alzforum, 111
Alzheimer's disease (AD), 7, 15, 17, 29, 40, 43, 44, 67, 68, 122, 131, 155, 255, 261, 262, 276, 277
Amantadine, 199
Amino acid glutamine, 3
Amphora Discovery Corporation, 210
Amyotrophic lateral sclerosis (ALS), 40, 43, 44–45, 68, 130, 229–230, 231, 268, 277
Anacardic acid, 75
Androgen receptor (AR), 42, 132, 202
Anti-Aβ intrabodies, 262
Anti-polyQ intrabodies, 258, 259, 260
Antisense *in vitro* studies, for HD, 228–229
Antisense oligodeoxynucleotides (ASO), 211, 212
Antisense oligonucleotide gene silencing, 225
  anticipated side effects of, 232
  antisense *in vitro* studies, for HD, 228–229
  brain, injecting into, 231–232

cells, entering, 230–231
chemical modifications of, 230–231
delivery issues, 230
design, 227–228
immune stimulation by, 232
in vivo studies, for antisense therapy, 229–230
meaning of, 226–227
mechanical techniques, 231
opinions and speculations, 245–246
vs. RNAi, 244–245
and target mRNA, 226, 228
Antisense technologies, HD therapeutics based on, 225, 226–232
Anxiety, 8, 17, 171, 180
APL. *See* Acute promyelocytic leukemia (APL)
AR. *See* Androgen receptor (AR)
Argonaute 2 (Ago2) protein, 234
Aristotle, 287
ASO. *See* Antisense oligodeoxynucleotides (ASO)
Aspiny interneurons, 14
Aspiration pneumonia, 7
Assay Guidance Manual Version 4.1, 205
Astrocytes, in HD, 43, 44, 46
Atomic force microscopy, 31
Autophagy, 32–33, 71, 155
Axonal trafficking dysfunction, 41
Axonal transport
  vs. cargo, 40
  and cytoskeletal defects, 39–42
Axons, 39–40

B

BAC. *See* Bacterial artificial chromosome (BAC)
BAC103 models. *See* Bacterial artificial chromosome 103 (BAC103) models
BACHD mice, 185
  behavioral phenotypes, 180
  genetic construct, 179–180
  neuropathology, 180
  survival and body weight, 181
Bacterial artificial chromosome (BAC), 103, 178, 179, 201
  models, 103, 104, 106, 107, 108
Bates group, 33

299